MICROEMULSION SYSTEMS

SURFACTANT SCIENCE SERIES

CONSULTING EDITORS

MARTIN J. SCHICK
Consultant
New York, New York

FREDERICK M. FOWKES
Department of Chemistry
Lehigh University
Bethlehem, Pennsylvania

MICROEMULSION SYSTEMS

Edited by

HENRI L. ROSANO
City College of the City University of New York
New York, New York

MARC CLAUSSE
Université de Technologie de Compiègne
Compiègne, France

MARCEL DEKKER, INC.   New York and Basel

Library of Congress Cataloging-in-Publication Data

Microemulsion systems.

   Proceedings of the 59th Colloid and Surface Science
Symposium and the 5th International Conference on
Surface and Colloid Science, both held at Potsdam, N.Y.,
in June, 1985.
   1. Emulsions--Congresses. I. Rosano, Henri L.,
[date]. II. Clausse, Marc. III. Colloid and
Surface Science Symposium (59th : 1985 : Potsdam, N.Y.)
IV. International Conference on Surface and Colloid
Science (5th : 1985 : Potsdam, N.Y.)
TP156.E6M49   1987   660.2'94514   87-544
ISBN 0-8247-7439-6

MARCEL DEKKER, INC.
270 Madison Avenue, New York, New York 10016

Current printing (last digit):
10 9 8 7 6 5 4 3 2 1

PRINTED IN THE UNITED STATES OF AMERICA

D

660.29454 14

MIC

HJ

Preface

The 59th Colloid and Surface Science Symposium and the 5th International Conference on Surface and Colloid Science were held on the campus of the Clarkson College of Technology in Potsdam, New York, June 24–28, 1985. The meeting was organized by the Clarkson's Institute of Colloid and Surface Science. Among the eighteen special topics discussed was one entitled Microemulsion Systems; this monograph reports most of the papers presented at that symposium, where the discussions ranged from fundamental research to development efforts. The latest developments in the fields of the mechanisms of formation and stability of microemulsions, characterization, relation between phase behavior and structure, dynamic processes, interfacial properties, theoretical modeling, gas solubility, waterless microemulsions, diffusion phenomena between phases, and many more studies, are reported in this collection of twenty-five chapters on microemulsion systems.

We would like to express our appreciation to all those who participated in preparing this volume, especially the contributors. We would also like to thank Evelyn Cropper for her patience while we were gathering and editing the twenty-five chapters by scientists from the four corners of the world; without her collaboration these proceedings could not have been published. We feel that this universal interest in microemulsion systems, initially in the field of tertiary oil recovery, is starting to find more and more industrial applications, such as in pharmaceuticals, chemical manufacturing, detergents, personal care products, pesticides, herbicides, food products, and in many other important fields.

Henri L. Rosano
Marc Clausse

Contents

Contributors

C. A. ANGELL   Purdue University, West Lafayette, Indiana

LOÏC AUVRAY*   Laboratoire de Physique de la Matière Condensée, Collège de France, Paris, France

R. C. BAKER†   Jealotts Hill Research Station, Bracknell, Berkshire, United Kingdom

ALAIN BERTHOD   Laboratoire de Chimie Analytique 3, Université Claude Bernard, Villeurbanne, Lyon, France

CARLOS BORZI‡   Cornell University, Ithaca, New York

AMY L. R. BUG   Exxon Research and Engineering, Corporate Research Science Laboratories, Annandale, New Jersey

MARTINE BUZIER   Université de Nancy, Nancy, France

D. CANET   Université de Nancy, Nancy, France

E. CAPONETTI**   University of Tennessee, Knoxville, Tennessee

Present affiliations:
*Laboratoire Léon Brillouin CEA-CEN, Saclay, Gif sur Yvette, France
†Coopers Animal Health Ltd., Berkhampsted, Herts, United Kingdom
‡Instituto de Física de Líquicos y Sistemas Biológicos, La Plata, Argentina
**Postdoctoral Fellow at the University of Tennessee, 1984. Permanent affiliation: Instituto di Chimica Fisica, University of Palermo, Palermo, Italy

J. CARNALI   Lund University, Lund, Sweden

JOHN L. CAVALLO   City College of The City University of New York, New York

A. M. CAZABAT   Laboratoire de Physique, ENS, Paris, France

D. Y. CHAO   Industrial Technology Research Institute, Hsinchu, Taiwan, Republic of China

D. CHATENAY   Laboratoire de Physique, ENS, Paris, France

M. L. CHEN   Industrial Technology Research Institute, Hsinchu, Taiwan, Republic of China

MARC CLAUSSE   Université de Technologie de Compiègne, France

J. P. COTTON   Laboratoire Léon Brillouin, CEA-CEN, Saclay, Gif sur Yvette, France

P. DELORD   University of Montpellier 2, Montpellier, France

JEAN-MARC DI MEGLIO   Laboratoire de Physique de la Matière Condensée, Collège de France, Paris, France

M. DUPEYRAT   Université Pierre et Marie Curie, Paris, France

GASTON DUPONT   Université de Nancy, Nancy, France

J. L. DUSSOSSOY*   University of Montpellier 2, Montpellier, France

MAYA DVOLAITZKY   Laboratoire de Physique de la Matière Condensée, Collège de France, Paris, France

STIG E. FRIBERG   University of Missouri at Rolla, Rolla, Missouri

GARY S. GREST   Exxon Research and Engineering, Corporate Research Science Laboratories, Annandale, New Jersey

P. GUERING   Laboratoire de Physique, ENS, Paris, France

J. HEIL[†]   Université de Technologie de Compiègne, France

Present affiliation:
*Centre d'Etudes Nucléaires de Cadarache, Saint Paul-Les-Durance, France
†PUM S.A., Bonneuil sur Marne, France

U. HENRIKSSON   Royal Institute of Technology, Stockholm, Sweden

E. A. HILDEBRAND   Monash University, Clayton, Victoria, Australia

HU MEI-LONG   Xinjiang Institute of Chemistry, Academy of Science of China, Urumuqui, Xinjiang, People's Republic of China

M. KAHLWEIT   Max-Planck-Institut fuer biophysikalische Chemie, Geottingen, Federal Republic of Germany

M. KAMIONER   Université Pierre et Marie Curie, Paris, France

D. LANGEVIN   Laboratoire de Physique, ENS, Paris, France

F. C. LARCHE   University of Montpellier 2, Montpellier, France

A. LATTES   Université Paul Sabatier, Toulouse, France

LI ZHI-PING   Xinjiang Institute of Chemistry, Academy of Science of China, Urumuqui, Xinjiang, People's Republic of China

YUH-CHIRN LIANG   University of Missouri at Rolla, Rolla, Missouri

BJÖRN LINDMAN   Chemical Center, Lund University, Lund, Sweden

REINHARD LIPOWSKY*   Cornell University, Ithaca, New York

GEORGE B. LYONS   City College of The City University of New York, New York

D. R. MACFARLANE   Monash University, Clayton, Victoria, Australia

L. J. MAGID   University of Tennessee, Knoxville, Tennessee

I. R. MCKINNON   Monash University, Clayton, Victoria, Australia

J. MEUNIER   Laboratoire de Physique ENS, Paris, France

H. NERY   Université de Nancy, Nancy, France

Present affiliations:
*Sektion Physik der Universität München, FRG

L. NICOLAS-MORGANTINI* Université de Technologie de Compiègne, France, and Université de Pau et des Pays de L'Adour, Pau, France

RAYMOND OBER Laboratoire de Physique de la Matière Condensée, Collège de France, Paris, France

JEAN-CLAUDE RAVEY Université de Nancy, Nancy, France

I. RICO Université Paul Sabatier, Toulouse, France

HENRI L. ROSANO City College of The City University of New York, New York

J. ROUVIERE Université des Sciences et Techniques du Languedoc, Montpellier, France

S. A. SAFRAN Exxon Research and Engineering, Corporate Research Science Laboratories, Annandale, New Jersey

Y. G. SHEU Industrial Technology Research Institute, Hsinchu, Taiwan, Republic of China

DUANE H. SMITH[†] Phillips Petroleum Company, Bartlesville, Oklahoma

O. SÖDERMAN Lund University, Lund, Sweden

K. SOHOUNHLOUE Université des Sciences et Techniques du Languedoc, Montpellier, France

PETER STILBS Uppsala University, Uppsala, Sweden

R. STREY Max-Planck-Institute fuer biophysikalische Chemie, Goettingen, Federal Republic of Germany

TH. F. TADROS Jealotts Hill Research Station, Bracknell, Berkshire, United Kingdom

CHRISTIANE TAUPIN Laboratoire de Physique de la Matière Condensée, College de France, Paris, France

---

Present affiliations
*L'Oreal, Aulnay-sous-Bois, France
[†]Morgantown Energy Technology Center, Morgantown, West Virginia

D. TOURAUD   Université de Technologie de Compiègne, France

W. URBACH   Laboratoire de Physique, ENS, Paris, France

H. WALDERHAUG   Lund University, Lund, Sweden

T. WÄRNHEIM   Royal Institute of Technology, Stockholm, Sweden

ITZHAK WEBMAN   Rutgers University, New Brunswick, New Jersey

BENJAMIN WIDOM   Cornell University, Ithaca, New York

J. S. ZHOU   Université Pierre et Marie Curie, Paris, France

A. ZRADBA†   Université de Technologie de Compiègne, France, and Université de Pau et des Pays de l'Adour, Pau, France

---

Present affiliation:
†Ecole Normale Supérieure de Casablanca, Casablanca, Morocco.

Introduction

In reporting the proceedings concerned with "Microemulsion Systems" at the 59th Colloid and Surface Science Symposium and the 5th International Conference on Surface and Colloid Science (Potsdam, New York, June 24–28, 1985), this book also reflects the difficulty experienced by the participants in reaching a common ground on the definition of the term "microemulsions." Over 20 definitions were proposed, but no clear-cut definition emerged. The only common definition described these clear (or sometimes lactescent) systems as containing one polar liquid (generally water or saline), a hydrophobic liquid (the "oil phase"), and a selected (one or several) surface active agent.

The word "microemulsion" was originally proposed by Jack H. Schulman and co-workers in 1959, although the first paper on the topic dates from 1943. They prepared a quaternary solution of water, benzene, hexanol, and K oleate, which was stable, homogeneous, and slightly opalescent. These systems became clear as soon as a short chain alcohol was added. In the years between 1943 and 1965, Schulman and co-workers described how to prepare these transparent systems. Basically, a coarse (or macro) emulsion was prepared, and the system was then titrated to clarity by adding a cosurfactant (second surface active substance). When the combination of the four components was right, the system cleared *spontaneously*. Most of the work reported by Schulman dealt with four-component systems: hydrocarbons (aliphatic or aromatic); ionic surfactants; the cosurfactant, generally 4- to 8-carbon chain aliphatic alcohol; and an aqueous phase. Schulman had previously published extensively in the field of monolayers, and applied what he had learned in that field to explain the formation of a microemulsion. He proposed that the surfactant and the cosurfactant, when properly selected, form a mixed film at the oil/water interface, resulting in an interfacial pressure exceeding the initial positive interfacial tension. The concept of a *negative interfacial tension* was suggested by Schulman. Rosano and co-workers showed that transitory zero (or very low) interfacial tensions obtained during the redistribution of the cosurfactant often play a major role in the spontaneous formation of these systems. This conclusion was based on the differences in the observed results when the order of mixing the components was changed.

To summarize, the basic observation made by Schulman and co-workers was this: when a cosurfactant is titrated into a coarse emulsion (composed of a mixture of water/surfactant in sufficient quantity to obtain microdroplets/oil), the result may be a system which is low in viscosity, transparent, isotropic, and very stable. The transition from opaque emulsion to transparent solution is spontaneous and well defined. It was found that these systems are made of spherical microdroplets with a diameter between 60 to 800 angstroms. It was only in 1959 that Schulman proposed to call these systems *microemulsions*: previously, he used terms such as *transparent water and oil dispersion, oleopathic hydromicelles,* or *hydropathic oleomicelles.*

In 1943, T. P. Hoar and Jack H. Schulman describing an oil/water system wrote: "The disperse phase consists of submicroscopic micelles having a core of soap-in-water solution and a surface monolayer of soap ionic-pairs, interspersed with nonionized amphipathic molecules, with the hydrocarbon portions orientated outwards. The high soap/water ratio ensures that the soap is undissociated. The presence of the water/oil interface containing orientated nonionized amphipatic molecules allows undissociated soap ion-pairs also to orientate there, as an alternative to association in curd filers. The nonionized amphiphatic molecules separate the soap ion-pairs sufficiently to prevent the repulsion between them that would otherwise occur, and indeed convert the repulsion in the monolayers into an attraction by forming complexes with the soap."

In order to explain the "spontaneous" formation of these microdroplets and their stability, the following explanation was advanced by Schulman: recall that π-surface pressures can be obtained spontaneously by the monolayer penetration of an alkyl alcohol or cholesterol monolayer with ionic surface active agents such as salts of alkylsulfate and alkylamines and substituted amines injected into the underlying solution (at constant area of the insoluble monolayer). The value of π-surface pressures at the air/water interface can reach values of more than 60 dynes/cm (although the collapse π-pressure for single component monolayers on their own are below 35 dynes/cm). If the surface pressure is held constant, or below the 60 dynes/cm but above the 35 dynes/cm, immediate expansion of the interfacial area takes place as the molecules of ionic surfactant penetrate the monolayer at the air/water interface. The analogous penetration of the mixed film by oil molecules at the oil/water interface increasing the surface area is the basis for the formation of microemulsions. However, the penetrating pressure in this case must be greater than the oil/water surface tension γ o/w, which is always less than γ a/w tension of 72 dynes/cm, which is approximately 50 dynes/cm for hydrocarbon and 35 dynes/cm for aromatic hydrocarbon compounds.

Therefore, it is progressively easier to obtain negative interfacial tensions for these systems, provided that the oil molecules can penetrate the interfacial film. Rosano et al. measured the change in the water-oil interfacial tension while alcohol was injected into one of the phases. It was found that the interfacial tension may be temporarily lowered to zero while the alcohol diffused through the interface and redistributed itself between the water and oil phases. It would therefore be possible for a dispersion to occur spontaneously (while $\gamma_i = 0$). It is then possible for the interfacial tension of a system to drop to zero for a certain period of time due to redistribution of amphiphatic molecules while the equilibrium γ_i remains positive. A sufficiently low positive value of γ_i is always better for emulsion (macro- or microemulsion) formation; nevertheless, below a certain positive small equilibrium value of γ_i phase separation, sol or gel formation will be produced but not emulsification.

Finally, stability, in turn, is not dependent on the value of the interfacial tension, but on the structure of the interfacial film surrounding the individual droplet. Therefore, low interfacial tension appears to be required for stability to occur at any degree. This finding is similar to observations of Shinoda and Saito, who first suggested forming their emulsification process using nonionic surfactants at the phase inversion temperature (PIT) where minimal interfacial tension is reached and then lowering it for further stability. At the PIT, γ_i is minimum favor to o/w interfacial area formation. In Rosano's et al. work on microemulsification, the same concept seems to apply: low interfacial tension is initially produced to lower the initial work requirement but through transfer and redistribution at the interface. Eventually the interfacial tension increases while the o/w interface curls and droplets are formed, allowing a barrier to form at the interface and preventing coalescence. This observation was also described mathematically by Defay and Sanfeld.

In order to obtain spontaneous microemulsion formation, the selection of the primary surfactant and cosurfactant and the right procedure capable of favoring redistribution between phases is important. Rosano et al. have stressed that the spontaneous formation of these dispersed swollen micelles is not dependent on simple thermodynamic stability, but rather—at least in part—on the occurrence of kinetic conditions favorable to the dispersion of the dispersed phase into the o/w system.

Essentially, the pioneering work of Schulman and later, his
students, was to correlate chemical structure of the various components in the microemulsion system to their spontaneous formation. As was mentioned above, the use of the word "microemulsion"
has been disputed and other labels have been preferred: transparent emulsion micellar emulsion, micellar solution, swollen micellar
solution. Those who prefer to use micellar solution or swollen micellar solution consider the fact that systems of the same kind are
made of one surfactant with double chain (sodium dioctylsulfosuccin-
ate-Aerosol OT). It is possible to obtain a microemulsion-type system
by either incorporating a hydrocarbon into the aqueous micellar
solution or putting water into an organic solution containing the
surfactant (*inverse micelles*). For these authors, microemulsion
formation must be considered a micellization or micellar solubilization
process of a macroemulsion when a selected cosurfactant is added.
Winsor has described and studied monophasic media composed of
water/organic salt/additives/hydrocarbon. He has described these
systems as "solubilized." On the other hand, Friberg et al. and
several others have, from the analysis of phase diagrams, under-
scored the fact that systems studied by Schulman (water/surfactant/
cosurfactant/oil) evolve from ternary monophasic solutions (water/
ionic surfactant/alcohol).

In the final analysis, there is no real opposition to the concept
of microemulsion versus swollen micellar solution. It seems that the
word "microemulsion" is used more often to describe any multicomponent fluids made of water (or a saline solution), a hydrophobic liquid (oil),
and one or several surfactants, these systems being stable, with a
low viscosity, transparent, and isotropic. On the one hand, from
an application point of view, "spontaneous" formation of these systems (high energy drinks, O_2-carrier, imaging solutions, inks,
household and personal care products, and so on) remains an important characteristic. On the other hand, from a purely scientific
point of view a so-called microemulsion fluid may exist as one phase
of mutual solubilization of all the components. It may also constitute
one phase of a polyphasic system. The other phases can either be
water or organic phases in excess (Winsor systems), or mesophases
(lytropic liquid crystals). With certain polyphasic systems, two
such phases may be of the microemulsion type. It should be noted
that the word "microemulsion" has been used to describe media with
no strong amphiphile (waterless microemulsions) and also media
where formamide or glycerol were used instead of water. The lack
of discrimination in the use of a precise definition of the word
"microemulsion" may be unacceptable to some but it has not prevent-
ed people from investigating these systems. We must be grateful
to the late Jack H. Schulman for offering such a catchy word to his
colleagues. Maybe the ambiguity of this word, which is easy to

remember, has helped to awaken the scientific and industrial world to consider the fundamental and practical applications of these systems. To be sure, the word has helped to some degree to stimulate research on these systems, and it is hoped that this book will encourage more.

On June 20, 1967, the announcement of the tragic death of Jack H, Schulman, Professor at Columbia University, came as a great shock to his many friends and colleagues. We in the field of colloid and interface science hold him in high esteem for his impressive record of important achievements. It is fitting for us to dedicate this book to his memory.

Henri L. Rosano

MICROEMULSION SYSTEMS

1

The Phase Behavior of H_2O-Oil-Nonionic Amphiphile Ternary Systems

M. KAHLWEIT and R. STREY   *Max-Planck-Institut fuer biophysikalische Chemie, Goettingen, FRG*

I. INTRODUCTION

In this chapter, the qualitative features of the phase behavior of H_2O—oil—nonionic amphiphile ternary systems are briefly summarized as they depend on the carbon number of the oil and the amphiphilicity of the surfactant within homologous series. The results yield an explanation for the correlation between the maximum of the solubilization parameter and the minimum of the interfacial tension between H_2O and oil. This suggests a procedure for systematically searching for the most efficient amphiphile, assuming that temperature, oil, and brine concentration are given.

The problem is (1) to increase the mutual solubility between a polar and a nonpolar liquid—for example, H_2O and "oil"—by adding an appropriate amphiphile (or a mixture of amphiphiles) in order to prepare a homogeneous fluid mixture with as little amphiphile as possible; and (2) to achieve—with as little amphiphile as possible—a separation of the mixture into three fluid phases with low interfacial tensions and similar densities in order to prepare emulsions with a low coarsening rate.

As we have shown in a series of papers [1], these two problems are closely related to each other: A system that separates into three phases within a certain temperature interval also shows the highest mutual solubility between H_2O and oil and, accordingly, the lowest interfacial tensions in that interval.

In practice, such systems often include inorganic electrolytes. However, from the Gibbs—Duhem equation, it follows that the effect of an added electrolyte on the chemical potential of the other components is equivalent to that of changing either temperature or pressure. What can be achieved by adding an electrolyte can thus, in general, also be achieved by changing T or p. The problem thus reduces to that of studying the phase behavior of such systems in order to predict the positions and extensions of the three-phase bodies as they depend on the chemical nature of the oil, the amphiphile, and the salt. Since the pressure dependence of the phase behavior is weak, we shall restrict our considerations to atmospheric pressure.   .

In order to study the phase behavior of H_2O (A)−oil (B)−non-ionic amphiphile (C), it is convenient to change the oils as well as the amphiphiles within homologous series. The oils (B_k) are characterized by their carbon number k. As nonionic amphiphiles (C) we have chosen n-alkyl polyglycolethers C_iE_j. In this case one may change two parameters, either the carbon number i of the hydrophobic part of the molecule at constant j, or the number j of oxyethylene groups of the hydrophilic part at constant i. Accordingly, we have studied the dependence of the phase behavior of the ternary system on the parameters i, j, and k.

II. GENERAL FEATURES OF THE PHASE
BEHAVIOR OF TERNARY SYSTEMS

The phase behavior of the ternary system $H_2O-B_k-C_iE_j$ is determined by the interplay of the lower miscibility gap of the binary system $B_k-C_iE_j$ and of the upper loop of the binary system $H_2O-C_iE_j$ with the "central gap" between H_2O and B_k. Figure 1 shows these three binary phase diagrams schematically, representing the sides of the (unfolded) phase prism of the ternary system with the Gibbs triangle as base and temperature T as ordinate. The UCST (upper critical solution temperature) T_α of the $B_k-C_iE_j$ gap lies, in general, close to the melting point of the mixture, whereas the LCST (lower critical solution temperature) T_β of the $H_2O-C_iE_j$ loop lies between the melting and the boiling point of the mixture.

For a given oil, the positions of both T_α and T_β depend on the amphiphilicity of the C_iE_j molecules—that is, on i and j as well as on the ratio between these two numbers. The larger i at constant j, the lower both critical points, and vice versa. As a consequence, T_α and T_β cannot be changed independently by varying i or j.

Since T_β is higher than T_α, the amphiphile is, at low temperatures, mainly dissolved in the H_2O phase. With rising temperature, the influence of the $H_2O-C_iE_j$ loop on the phase behavior of the ternary system eventually overcomes that of the $B_k-C_iE_j$ gap. As a result, the plait point of the central gap moves from the oil-rich to the H_2O-rich side. Consequently, the phase volume ratio between the two phases shows a sigmoidal curve when plotted versus temperature.

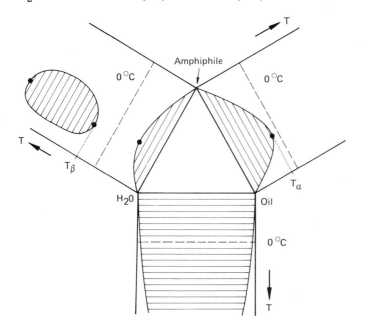

FIG. 1 Unfolded phase prism of a ternary system H_2O-oil-nonionic amphiphile with the phase diagrams of the three corresponding binary systems (schematic).

The inflection point of this curve indicates that temperature at which the influence of the B_k-C_iE_j gap and that the H_2O-C_iE_j loop compensate each other. This temperature depends, evidently, on all three parameters i, j, and k. For given C_iE_j, the inflection point rises with increasing k; whereas for given B_k, the temperature drops with increasing i (at constant j) or decreasing j (at constant i). This was demonstrated in Fig. 4 of Ref. 2, which shows the phase volume ratios for the rather hydrophilic aromats as oil and C_4E_3 as amphiphile: The inflection point rises with increasing hydrophobicity of the oil, the tangent at the inflection point becoming increasingly more horizontal.

From this it follows that, for the less hydrophobic oils, the inflection point may lie below the melting point of the mixture; whereas for the more hydrophilic C_iE_j, it may lie above the boiling point. The position of the system on the temperature scale with respect to the inflection point can be readily determined by adding an appropriate amount of the amphiphile to a mixture of equal volumes of

H_2O and oil. If the lower aqueous phase becomes larger than the upper oil phase, the system lies below its inflection point, and vice versa.

The critical line, that connects the plait points of the central gap with the phase prism, can be looked at as an elastic spring. If one increases its "bending tension" by either increasing k (thus raising T_α), or by decreasing j at constant i (thus lowering T_β), the critical line may break at what is called a tricritical point (tcp) [3]. This is shown schematically in Fig. 2, which shows the projection of the critical line from the H_2O-oil-T plane onto some parallel plane in the rear. Here C_iE_j is kept constant while T_α is raised by increasing the carbon number of the oil within a homologous series. With rising T_α (T_β remains constant), the inflection point of the "connected" critical line rises and the tangent at the inflection point becomes increasingly more horizontal, until the bending tension becomes too strong and the critical line breaks at T_{tcp}. With further rising T_α, the two end points of the now broken critical line move increasingly farther apart.

This breaking of the critical line gives rise to the formation of a three-phase body within the phase prism [1]. The body appears by separation of the aqueous phase at T_1 at the (lower) end point of the critical line cp_β into a H_2O-rich and an amphiphile-rich phase. With rising temperature, the amphiphile-rich (middle) phase moves clockwise around the surface of the body of heterogeneous phases until it merges with the oil-rich phase at T_u at the (upper) end point of critical line cp_α. Systems with a "broken" critical line thus show a Winsor type I phase diagram below T_1, a Winsor type III diagram between T_1 and T_u, and a Winsor type II diagram above T_u.

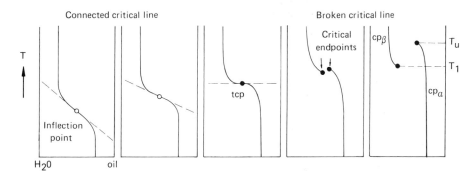

FIG. 2 Transition from a connected to a broken critical line. Projection of the critical line from the H_2O-oil-T plane of the phase prism onto some parallel plane in the rear (schematic).

The three-phase body is characterized by its position on the temperature scale as well as by its extension within the phase prism:

1. Its vertical extension, namely, the width of the three-phase temperature interval $\Delta T \equiv T_u - T_l$
2. The position of this interval on the temperature scale, namely, its mean temperature $\overline{T} \equiv (T_l + T_u)/2$.
3. Its horizontal extension, namely, the height of the three-phase triangle in its symmetric shape at some temperature between T_l and T_u.

Both the first and the third properties of the three-phase body grow with increasing distance from the tcp: Close to the tcp, the three-phase body is small, being located close to the surface of the body of heterogeneous phases. With increasing distance from the tcp, the three-phase body grows into the body of heterogeneous phases and the relative area covered by the three-phase triangles increases, as does ΔT.

The most convenient procedure for measuring these properties is to cut a vertical section through the phase prism erected on the center line of the base, that is, on the line H_2O/oil = 1/1 in the Gibbs triangle. If looked at from the oil edge of the prism, one can see the profile of the body of heterogeneous phases with a section through the three-phase body floating within that body like a "fish" (Fig. 3). This section yields both T_l and T_u, as well as the composition of the "head" and "tail" of the fish, the latter being the minimum mass of the amphiphile needed to prepare a homogeneous solution of equal masses of H_2O and oil.

The systems in Fig. 4 in Ref. 2 show connected critical lines only. In order to break the critical line between the melting point and the boiling point, one thus has to choose a series of oils that are a little more hydrophobic than the aromats, that is, raise T_α, and an amphiphile that is a little less hydrophylic than C_4E_3, that is, lower T_β. As such we have chosen the homologous series of the phenylalkanes $(C_6H_5)(CH_2)_kH$ as oils and C_4E_2 as amphiphile. Figure 7 in Ref. 4 shows the fishes of these systems. With low carbon numbers k, the systems show a connected critical line. As one increases k, the critical line breaks between k = 6 and k = 7. For k = 7, one accordingly finds a very thin fish, which grows with further increasing carbon number both in its horizontal and vertical extensions.

Figure 4 shows ΔT plotted versus the carbon number k of the oils. The upper curve represents the dependence of T_u on k, the lower one that of T_l. Both drop with decreasing hydrophobicity of the oil. T_u, however, drops a little steeper than T_l. As a consequence, ΔT shapes a cusp that terminates at the tcp. For this particular series of systems, one thus finds a tcp at about 52°C

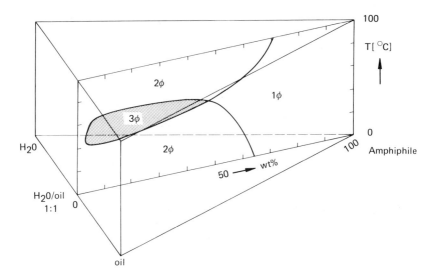

FIG. 3 Vertical section through the phase prism of a ternary system H_2O-oil-nonionic amphiphile at H_2O/oil = 1/1 with profile of the body of heterogeneous phases and section through the three-phase body (schematic).

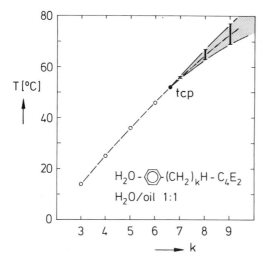

FIG. 4 Three-phase temperature intervals ΔT versus carbon number k of the system $H_2O-(C_6H_5)(CH_2)_kH-C_4E_2$ terminating in a tricritical point at about 52°C. Empty points: apexes of the grooves for the systems with connected critical lines.

with a phenylalkane having a carbon number between $k = 6$ and $k = 7$.

The groove of the "tail" of the fishes is a consequence of Schreinemakers' rule, which requires for thermodynamic reasons that the boundaries of each homogeneous phase must intersect at a corner of the three-phase triangle such that both their extensions either pass into the triangle or into the adjacent two-phase regions [5]. As a consequence, the amphiphile-rich phase cuts a groove into the body of heterogeneous phases on its way from the H_2O- to the oil-rich side. Its depth increases with increasing amphiphilicity of C_iE_j, that is, with increasing i and j. Below and above that groove, the mutual solubility between H_2O and oil is much lower and does not vary much from amphiphile to amphiphile.

But, as can be seen from Figure 7 of Ref. 4, even the connected critical line (for $k \leqslant 6$) cuts a groove into the body of heterogeneous phases as it changes from the oil-rich to the H_2O-rich side. The apexes of these grooves on the temperature scale are represented by the empty points in Fig. 4. The dashed line, connecting these points, continues smoothly into the three-phase cusp as a line connecting the mean temperatures $\overline{T}$ of the corresponding intervals. This raises the question of how to extend the thermodynamic reasoning for the existence of a groove in the presence of a three-phase body to systems with connected critical lines, that is, systems without three-phase bodies. The fact that this groove exists not only close to the H_2O/oil = 1/1 ratio but all along the critical line is demonstrated in Fig. 5, which shows the profiles of the body of heterogeneous phases of the ternary system H_2O-1-phenylbutane-C_4E_2 at different sections through the phase prism. As one can see, the groove ascends temperaturewise from the oil-rich to the H_2O-rich side. From this is follows that in systems with a three-phase body, the highest mutual solubility between H_2O and oil is found in the three-phase temperature interval, or, in systems with a connected critical line, in that temperature range in which the critical line changes from the oil-rich to the H_2O-rich side of the phase prism.

This readily explains the temperature dependence of the so-called solubilization parameter σ^*. This parameter is defined as the volume of H_2O (or oil) per unit volume of amphiphile at equal volumes of H_2O and oil in the amphiphile-rich (middle) phase [6] and is thus approximately proportional to the inverse of the minimum mass fraction of amphiphile needed to prepare a homogeneous solution of equal masses of H_2O and oil. If plotted at a given temperature and for a given amphiphile versus the carbon number of a homologous series of oils, one finds a maximum of σ^* at a certain carbon number. If, for nonionic amphiphiles, one raises the temperature, this maximum is shifted toward higher carbon numbers,

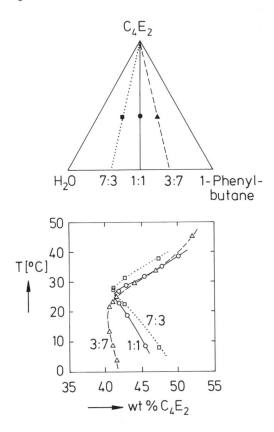

FIG. 5 Profiles of the two-phase body of the system $H_2O-(C_6H_5)$ $(CH_2)_4H-C_4E_2$ at different vertical sections through the phase prism with the groove ascending from the oil-rich to the H_2O-rich side.

and vice versa. Since the maximum mutual solubility between H_2O and oil is found at the mean temperature $\overline{T}$ of the three-phase temperature interval, the maximum of σ^* is, accordingly, found for that oil, $\overline{T}$ of which lies closest to the particular experimental temperature. $\overline{T}$, on the other hand, rises temperaturewise with increasing carbon number (see Fig. 4 in Ref. 7). The relative value of σ^* can thus be taken as a measure for the distance (temperaturewise) between $\overline{T}$ of each oil and the experimental temperature: The higher the σ^*, the closer the experimental temperature is to $\overline{T}$ of the corresponding oil for the given amphiphile.

The interfacial tensions between the coexisting phases, on the other hand, show a minimum in the three-phase temperature interval.

In such systems with a broken critical line, this can be readily explained by the fact that the aqueous (a) and the amphiphile-rich phase (c) emerge from the (lower) end point of the critical line cp_β (see Fig. 5 in Ref. 2). Consequently, the interfacial tension a/c between these two phases rises from zero at T_1 according to the scaling law for near critical phases. Phase (c) and the oil phase (b), on the other hand, merge at the (upper) end point of the critical line cp_α. Consequently, the interfacial tension c/b between these two phases vanishes at T_u. The closer to a tcp—that is, the smaller $(T_u - T_1)$—the lower the interfacial tensions between the phases of the three-phase triangle.

The same, however, holds for systems with a connected critical line, that is, without a three-phase body. Again one finds the lowest interfacial tension between the aqueous and the oil phase at the temperature at which the critical line changes from the oil-rich to the H_2O-rich side. This is demonstrated in Fig. 6 for the system H_2O-1-phenylbutane-C_4E_4. On the left we have redrawn the groove at H_2O/oil = 1/1 (see Fig. 5), on the right we have plotted the temperature dependence of the interfacial tension between H_2O and the oil phase, measured at 30 wt% C_4E_2, that is, sufficiently apart from the groove (42 wt%). As one can see, the interfacial tension passes through a minimum at that temperature at which the mutual solubility between H_2O and oil reaches its maximum, that is, where the critical line passes from the oil-rich to the H_2O-rich side of th phase prism.

This has an important consequence for the correlation between the solubilization parameter and the interfacial tension between the aqueous and the oil phases. Hitherto, the correlation has been established by measuring the interfacial tension between these two phases in the three-phase body. If measured at a given temperature and for a given amphiphile and then plotted versus the carbon number of the oil, one finds a minimum of the interfacial tension at the maximum of the solubilization parameter (see, e.g., Fig. 5 in Ref. 8). The interfacial tension between the aqueous and the oil phases in the three-phase body, however, is difficult to measure, since a slight variation in temperature may give rise to the formation of a very thin layer of the middle phase between these two phases. The above results, on the other hand, suggests measuring the interfacial tension between the aqueous and the oil phases at some distance from the three-phase body, namely, in the two-phase region between the three-phase body and the H_2O-oil-T plane of the phase prism. In this region one can be sure that the results will not be obstructed by a thin—possibly invisible—film of the middle phase. Figure 7 shows the results of such an experiment. On top we have redrawn the three-phase cusps of the system H_2O-n-alkane-C_4E_1 (taken from Fig. 4 in Ref. 7). On the bottom we have plotted the interfacial tension between the aqueous and the oil

phases, measured at 25°C at 10 wt% C_4E_1, versus the carbon number k of the alkanes. At this mass fraction of this particular amphiphile, one is for all these oils sufficiently far from the three-phase
body (see, e.g., the upper fish in Fig. 3 of Ref. 7). As expected,
the interfacial tension shows a minimum for those oils, the three-
phase intervals of which lie closest to the experimental temperature
of 25°C.

The same should be true if one keeps the oil constant and varies
the hydrophilicity of the amphiphile, or if one keeps both constant
and adds a lyotropic salt, such as NaCl, as a fourth component.
The addition of a lyotropic salt is equivalent to decreasing the effective hydrophilicity of the amphiphile. As a consequence, the three-
phase interval is lowered (and widened) temperaturewise (see, e.g.,
Fig. 29 of Ref. 1). Accordingly, if one measures the solubilization
parameter and the interfacial tension between the aqueous and the

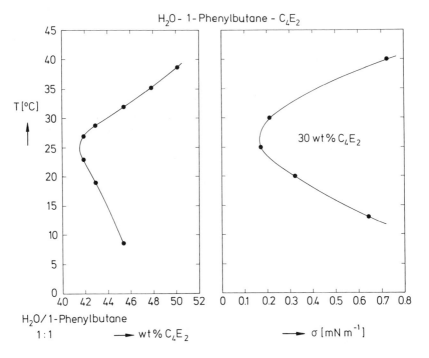

FIG. 6 Left: Profile of the two-phase body of the system H_2O-
$(C_6H_5)(CH_2)_4H$-C_4E_2 at H_2O/oil = 1/1. Right: Interfacial tension
between the aqueous and the oil phase versus T, measured at
30 wt% C_4E_2, with a minimum in a system without a three-phase
body.

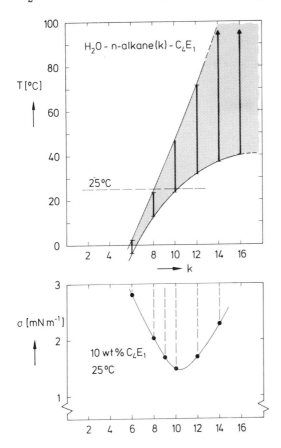

FIG. 7 Top: Three-phase temperature intervals ΔT of the systems H_2O-n-alkane(k)-C_4E_1 versus carbon number k of the alkanes. Bottom: Interfacial tension between the aqueous and the oil phase versus k, measured at 10 wt% C_4E_1, that is, between the three-phase bodies and the H_2O-oil-T plane of the phase prism.

oil phases, and plots them versus salinity, one finds a maximum of σ^* and a minimum of σ at that salt concentration at which the three-phase cusp passes through the experimental temperature (see, e.g., Fig. 25 of Ref. 6).

III. CONSEQUENCES WITH RESPECT TO APPLICATION

In application, the temperature, the oil (mixture), and the concentration of the salt (mixture) is given. The problem is to find that

amphiphile (mixture) that yields the highest mutual solubility between
H_2O and oil combined with the lowest interfacial tension between
the two. The above results suggest a procedure for systematically
searching for that amphiphile. One starts by studying the phase
behavior of the ternary system H_2O-oil-amphiphile with a "simple,"
well-studied amphiphile such as C_4E_1. By comparing the properties
of the three-phase body with those of "pure" oils (see, e.g., Fig. 4
of Ref. 7), one can find the effective carbon number of the oil mixture. If one then measures the temperature dependence of the three-
phase interval on NaCl concentration and compares the result with
the phase behavior in the presence of the given "brine," one will
find the effective salinity of the latter. The final step is then to
adjust the amphiphilicity of the surfactant by changing either i or j
in C_iE_j until the three-phase body lies in the desired temperature
range (Fig. 5 of Ref. 7) [10]. The efficiency of the amphiphile with
respect to the solubilization parameter σ^* and interfacial tension σ
may then be increased further by increasing both i and j, as was
demonstrated in Fig. 3 of Ref. 7, or by changing to another class
of nonionic compounds (e.g., alkylphosphine oxides, see Ref. 9)
and/or by adding an appropriate ionic surfactant.

  As a simple but instructive example, we shall consider an experiment performed by Benson et al. [11] in order to study the
efficiency of $C_{12}E_j$ for cold-water cleaning of various textiles. On

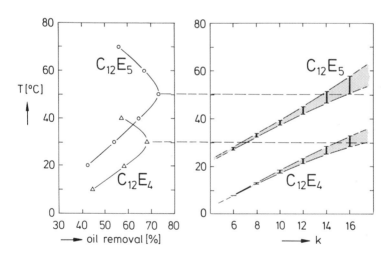

FIG. 8 Left: Oil removal (in percent) of a mineral oil blend from
65/35 polyester/cotton by $C_{12}E_4$ and $C_{12}E_5$ versus temperature [11].
Right: Three-phase temperature interval ΔT of the systems H_2O-
n-alkanes(k)-$C_{12}E_4$ and $C_{12}E_5$, respectively, versus carbon number
k of the alkanes.

the left of Fig. 8 we have redrawn one of their results, namely,
the removal (in percent) of a mineral oil blend from 65/35 polyester/
cotton versus temperature for $C_{12}E_4$ and $C_{12}E_5$ (Fig. 2 of Ref. 11).
On the right is shown the three-phase temperature intervals of the
two amphiphiles versus the carbon number of n-alkanes, taken from
Ref. 7. By comparing the two figures one finds that the maxima of
the oil removal both lie in the three-phase interval of n-hexadecane,
from which we conclude that the mineral oil had an effective carbon
number of $k \sim 16$. While the authors tried to correlate their results to the cloud points of the corresponding binary systems H_2O-
C_iE_j, it follows from our interpretation that the highest efficiency of
these nonionic amphiphiles is achieved when the ternary system H_2O-
oil-C_iE_j shows a three-phase body within the phase prism with high
mutual solubility between H_2O and oil and low interfacial tensions between the two phases.

IV. REFERENCES

1. For a review, see Kahlweit, M., and Strey, R., *Angew. Chem.*
   (Engl. Ed.) *24*, 654 (1985).
2. Kahlweit, M., Strey, R., and Haase, D., *J. Phys. Chem. 89*,
   163 (1985).
3. Griffiths, R. B., and Widom, B., *Phys. Rev. A8*, 2173 (1973).
4. Kahlweit, M., Strey, R., Firman, P., and Haase, D., *Langmuir*
   1, 281 (1985).
5. Wheeler, J. C., *J. Chem. Phys. 61*, 4474 (1974).
6. Reed, R. L., and Healy, R. N. in *Improved Oil Recovery by*
   Surfactant and Polymer Flooding, Shah, D. O., and Schechter,
   R. S. (eds.), Academic Press, New York, 1977, p. 383.
7. Kahlweit, M., Strey, R., and Firman, P., *J. Phys. Chem.*,
   90, 671 (1986).
8. Wade, W. H., Morgan, J. C., Schechter, R. S., Jacobson,
   J. K., and Salager, J. L., *SPE-AIME*, 1978, p. 242.
9. Pospischil, K. H., *Langmuir*, *2*, 170 (1986).
10. See also Kunieda, H., and Shinoda, K., *J. Colloid Interface*
    Sci. 107, 107 (1985).
11. Benson, H. L., Cox, K. R., and Zweig, J. E., *Happi*, March
    1985, p. 50.

2

Water/Ionic Surfactant/Alkanol/Hydrocarbon Systems: Influence of Certain Constitution and Composition Parameters upon the Realms-of-Existence and Transport Properties of Microemulsion-Type Media

M. CLAUSSE, L. NICOLAS-MORGANTINI,* A. ZRADBA,†
and D. TOURAUD   U.A. CNRS No. 858, Université de
Technologie de Compiègne, France

I. INTRODUCTION

The realms-of-existence of macroscopically homogeneous, stable, fluid, optically transparent, and isotropic media (so-called microemulsions) have been delineated, at T = 25°C, for a great number of systems incorporating water, sodium dodecylsulfate, various straight or branched alkanols, and various hydrocarbons.

From the results obtained, it is possible to analyze how certain constitution parameters, such as straight alkanol number of carbon atoms, alkanol isomery, normal aliphatic hydrocarbon chain length, and hydrocarbon aromaticity, influence the configuration of the microemulsion pseudo-ternary domain. Correlations can also be established between microemulsion domain configuration and microemulsion electroconductive and viscous behavior.

By adequately mixing water, a hydrophobic hydrocarbon, ("oil"), and suitable amphiphilic compounds, macroscopically homogeneous fluids may be formed that currently are called "microemulsions" [1–16]. These media are multicomponent liquids that exhibit long-term stability, have a low apparent viscosity (10 cP or so), and generally are optically transparent and isotropic.

Present affiliations:
*L'Oreal, Aulnay-sous-Bois, France
†Ecole Normale Supérieure de Casablanca, Casablanca, Morocco

In fact, the word "microemulsion" is rather confusing, since it is used indiscriminately to label a wide category of apparently similar media that may in fact differ greatly from one another by their local structures. The classical structural model for microemulsions is that of a monodisperse population of dynamic microglobules in interaction. Microemulsion optical transparency indicates that the microglobule diameter cannot be larger than 0.1 μm or so. Since hydrocarbon or water is the major constituent, the microglobules are either of the "direct" or of the "inverse" type. A "direct" microglobule consists of a spherical organic core surrounded by a monomolecular shell of amphiphilic molecules whose polar heads are in contact with the continuous aqueous medium. Symmetrically, an "inverse" microglobule consists of a spherical aqueous core surrounded by a monomolecular shell of amphiphilic molecules whose hydrophobic tails are oriented toward the continuous organic medium. Recent studies [17–19] have confirmed that this model, originally proposed by Schulman and co-workers [1], may satisfactorily describe the structure of certain hydrocarbon-rich or water-rich microemulsions. In the case of microemulsions containing comparable amounts of water and hydrocarbon, the model that best depicts the local organization of the medium is that of "dynamic bicontinuous structures," as suggested by Scriven [20–21]. This model, which has been further analyzed and refined by other authors [22–26], indeed appears in the light of quite recent data [27–28], to be applicable at least to microemulsions belonging to Winsor III systems [29]. The spherical microglobule model seems to be also inappropriate in depicting the structure of microemulsion-type media incorporating ionic surfactants and in which the water content is low [30–32]. This aspect of microemulsion structure will be discussed later [33]. Finally, it even has been suggested that certain microemulsion-type media that are very rich in amphiphilic compounds could possess a local lamellar organization [34].

It is quite obvious that, for a given kind of water/amphiphilic compound(s)/hydrocarbon system, the larger the microemulsion domain (i.e., the range of compositions corresponding to the existence of microemulsion-type media), the higher the diversity of microemulsion structure. This fact is important to the formulator interested in the design of specialty products or of multicomponent fluids to be used in industrial processes. For instance, the efficiency of a microemulsion-type fluid as a vehicle of some specific chemical or biochemical compound may be greatly optimized by formulating it in such a way that its local structure furthers the solubilization of the said compound. The optimum formulation may be obtained by adequately acting on certain key constitution and/or composition parameters of the system. This implies that some general rules must be known beforehand with regard to the influence of these parameters on the microemulsion domain configuration and also on some physicochemical properties that, in one way or another, reflect microemulsion local structure. With a view toward setting up such general rules, Zradba [35] and Nicolas-Morgantini [36] have carried out a thorough study of a great number of systems incorporating water, sodium dodecylsulfate as ionic surfactant, various straight or branched alkanols (cosurfactant), and various aliphatic or aromatic hydrocarbons. The most significant among the many results collected by these authors are reported and commented upon later in this chapter. It will be proved that two distinct types of water/ionic surfactant/cosurfactant/hydrocarbon systems can be defined, since the microemulsion domain in the phase tetrahedron consists of two disjoined volumes V_1 and V_2 or forms an all-in-one-block volume V that continuously spans the major portion of the phase tetrahedron.

II. EXPERIMENTAL

The full list of the systems studied is given in Table 1. Each system, containing water and sodium dodecylsulfate, is characterized by a certain hydrocarbon, a certain alkanol, and a certain sodium dodecylsulfate-to-alkanol mass ratio k_m (or molar ratio k_x).

TABLE 1   Systems Studied[a]

Aliphatic hydrocarbons

n-Octane

1-Pentanol (k_m = 1/2)

n-Dodecane

1-Butanol (k_m = 0; 1/9; 1/4; 1/2; 1/1.68; 1/1; 1/0.85; 1/0.82)

1-Pentanol (k_m = 0; 1/9; 1/4; 1/2.38; 1/2; 1/1.72; 1/1.50; 1/1; 1/0.43)

1-Hexanol (k_m = 0; 1/9; 1/4; 1/3.03; 1/2.32; 1/2; 1/1.76; 1/1.50; 1/1)

1-Heptanol (k_m = 0; 1/9; 1/4; 1/2.63; 1/2; 1/1)

n-Hexadecane

1-Pentanol (k_m = 1/2)

TABLE 1   (continued)

Aromatic hydrocarbons

Benzene

Straight alkanols

1-Butanol (k_m = 1/2.47; 1/2; 1/1.68)

1-Pentanol (k_m = 1/2.93; 1/2)

1-Hexanol (k_m = 1/3.40; 1/2.32; 1/2)

1-Heptanol (k_m = 1/3.87; 1/2.64; 1/2)

Branched alkanols

2-Propanol-2-methyl (k_m = 1/2; 1/1.68)

2-Butanol (k_m = 1/2; 1/1.68)

1-Propanol-2-methyl (k_m = 1/2; 1/1.68)

2-Butanol-2-methyl (k_m = 1/2.93; 1/2)

2-Butanol-3-methyl (k_m = 1/2)

3-Pentanol (k_m = 1/2)

2-Pentanol (k_m = 1/4; 1/2; 1/1)

1-Propanol-2,2-dimethyl (k_m = 1/2)

1-Butanol-2-methyl (k_m = 1/2)

1-Butanol-3-methyl (k_m = 1/2.93; 1/2)

3-Pentanol-3-methyl (k_m = 1/2.32)

2-Hexanol (k_m = 1/2.32)

Toluene

1-Butanol (k_m = 1/1.68)

1-Pentanol (k_m = 1/2)

1-Hexanol (k_m = 1/2.32)

[a]In all systems, the aqueous component was distilled water and the ionic surfactant was sodium dodecylsulfate. Each group of systems corresponds to a certain hydrocarbon. In a given group, each system is characterized by a certain alkanol and a certain value of k_m, the sodium dodecylsulfate/alkanol mass ratio.

A. Compounds

The water used was freshly bi-distilled water, (electrical conductivity σ around 10^{-4} Sm^{-1}).

The surfactant was sodium dodecylsulfate (99% purity), purchased from Touzart et Matignon, France.

All the alkanols used (number of carbon atoms n_a ranging from 2 to 7 were highly pure materials (purity above 98%), purchased from Fluka AG, Switzerland.

The aliphatic and aromatic hydrocarbons used (n-octane, n-dodecane, n-hexadecane, toluene, and benzene) were also highly pure materials (purity above 98%), purchased from Fluka AG, Switzerland.

B. Methods

The boundaries of the microemulsion domains in isothermal WSH-type triangular diagrams (see Figs. 1 and 2) were determined, for different values of the surfactant/alkanol mass ratio k_m, by progressive

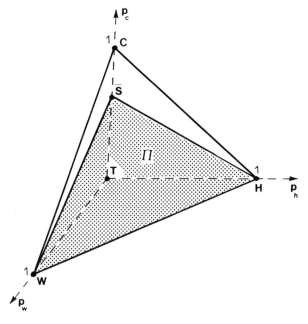

FIG. 1 Phase tetrahedron of a water/ionic surfactant/cosurfactant/hydrocarbon system. p_w, p_h, p_c: mass fraction of water, hydrocarbon, and cosurfactant, respectively. The triangle π is a section of the phase tetrahedron corresponding to a certain surfactant-to-surfactant mass ratio k_m, as defined by the point $\bar{S}$.

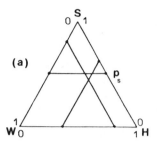 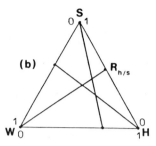

FIG. 2 Composition paths inside pseudo-ternary phase diagrams.
W: 100% water; H: 100% hydrocarbon; S: 100% surfactant/cosur-
factant combination characterized by a certain value of k_m. (a)
p-type paths (the mass fraction of one of the components is constant); (b) R-type paths (the mass ratio of two of the components
is constant).

titration of four-component mixtures. Usually, the experimental
paths followed in the triangular diagrams were of the $R_{h/s}$ type
(see Fig. 2), water being the titration component. The ratio k_m
being chosen, a mixture of surfactant, alkanol, and hydrocarbon,
characterized by a certain value of $R_{h/s}$,* was progressively enriched in water (added drop by drop) while being submitted to
gentle agitation by means of a magnetic stirrer. The concentrations
of water at which turbidity-to-transparency and transparency-to-
turbidity transitions occurred were derived from precise weight measurements. By repeating this experimental routine for other values of
$R_{h/s}$, the boundaries of the microemulsion domain corresponding to the
chosen value of k_m were determined. The entire process was further
repeated for other values of k_m. All the titration experiments were
performed in an airconditioned room in which the temperature was kept
at T = 25 ± 1°C. In some difficult cases, it was necessary to carry
out "static" titration experiments. Series of samples, whose compositions were close to the transition composition determined approximately
through "dynamic" titration experiments of the kind previously described, were prepared and stored at T = 25 ± 0.1°C in a temperature-
controlled chamber. The phase behavior of these samples was checked
at regular intervals until no evolution was observed. Samples that still
appeared monophasic at the end of a storage period of 3 months or so

---

*$R_{h/s}$ is the mass ratio of hydrocarbon to the surfactant/alkanol
combination, the latter itself being characterized by k_m, the sur-
factant/alkanol mass ratio.

were considered as being stable monophasic microemulsion-type media.
It was thus possible to pinpoint transition compositions that could not
be properly determined through "dynamic" titration experiments. It
may be estimated that, on an average, the relative uncertainty on the
components' mass fractions corresponding to transition points is less
than 2%.

Microemulsion specific mass ρ was measured, at T = 25°C, by
means of an automated densimeter (Anton Paar DMA 45), the relative uncertainty on the values of ρ being less than 0.1%.

Microemulsion kinematic viscosity ω was determined, at T = 25°C,
by means of calibrated capillary tubes connected to a semiautomated
data analyzer (Lauda S/1 viscotimer). Using the values of ω and
the corresponding values of ρ, the values of the dynamic viscosity
η were computed, with a relative uncertainty of about 1%.

Microemulsion low-frequency electrical conductivity σ was measured, at T = 25°C, by means of Mullard-type cells (Philips PR 9512/
00 or PR 9512/01), connected to a semiautomated conductance bridge
working at 1.592 kHz (Wayne-Kerr, B 331). The relative uncertainty
on the values of σ was less than 1%, on an average.

III. MICROEMULSION DOMAINS: INFLUENCE OF ALKANOL
AND HYDROCARBON MOLECULAR STRUCTURE

A. Basic Observations

The isothermal pseudo-ternary phase diagrams reported in Figs. 3
through 5 clearly exemplify the differences that exist between simple micellization or molecular solubilization, on the one hand, and
"microemulsification," on the other hand. From Figs. 3 and 4, it is
evident that neither the addition of an ionic surfactant such as sodium dodecylsulfate alone, nor the addition of an alkanol alone, can
efficiently promote the "mutual solubilization" of water and n-dodecane. Figure 5, in contrast, shows that, upon adequately combining
sodium dodecylsulfate with an alkanol ("cosurfactant"), it is possible
to enhance considerably this "mutual solubilization" and consequently
induce the formation and secure the existence of stable single-phased
fluid media ("monophasic" microemulsions), which may simultaneously
incorporate large amounts of water and n-dodecane.

It appears from Fig. 5 that, all other things being equal (i.e.,
surfactant/cosurfactant molar ratio k_m, hydrocarbon, and, of course,
temperature and pressure), the alkanol molecular structure has a
strong influence on microemulsion solubilization capacity as defined
by the shape and extent of the microemulsion pseudo-ternary domain.

With the longest n-alkanol (1-heptanol, Fig. 5a), the microemulsion domain consists of two strikingly dissimilar regions, N_1 and N_2,
which are separated from each other by a range of compositions

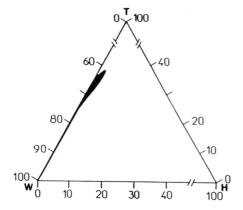

FIG. 3 Water/sodium dodecylsulfate/n-dodecane system (T = 25°C).
Realm-of-existence of monophasic direct micellar solutions (black
area).

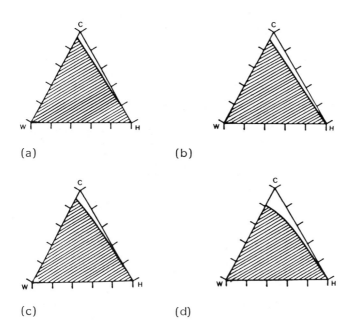

(a)   (b)

(c)   (d)

FIG. 4 Water/n-alkanol/n-dodecane systems (T = 25°C). Realms-
of-existence of monophasic areas (blank areas). (a) 1-heptanol;
(b) 1-hexanol; (c) 1-pentanol; (d) 1-butanol.

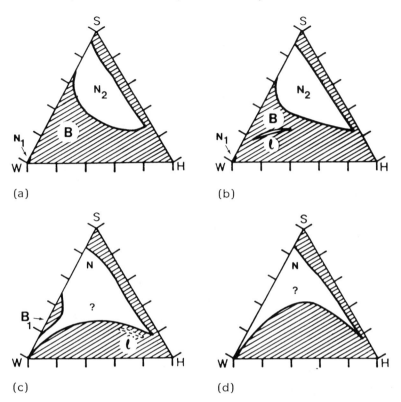

FIG. 5 Water/sodium dodecylsulfate/n-alkanol/n-dodecane systems, (k_x = 1/6.54; T = 25°C). Influence of n_a, the n-alkanol number of carbon atoms, on the configuration of the microemulsion pseudo-ternary domain (blank areas). (a) 1-heptanol; (b) 1-hexanol; (c) 1-pentanol; (d) 1-butanol.

corresponding to the existence of mesophases (region B). Considering the locations of N_1 and N_2, it can reasonably be asserted that N_1 is the realm-of-existence of microemulsions of the "direct" kind and N_2 is the realm-of-existence of microemulsions of the "inverse" kind. This situation has been labeled "type S" by Clausse et al. [37]. With the shortest alkanol (1-butanol, Fig. 5d), the microemulsion pseudo-ternary domain is an all-in-one-block triangular region N that spans a great part of the pseudo-ternary diagram, continuously from its WS edge to near its HS edge. In this case, contrary to what is observed with 1-heptanol, no "solubilization gap," corresponding to the existence of mesomorphous media, separates the region of water-rich microemulsions from the region of water-poor microemulsions.

This evidently raises the question of the microstructure of the microemulsion-type media, which, by their compositions, belong to the central part of the microemulsion domain. This situation has been labeled "type U" by Clausse et al. [37].

With both 1-hexanol and 1-pentanol, hybrid situations are observed. The microemulsion pseudo-ternary domain of the 1-hexanol system (Fig. 5b) consists of three disjoined regions, N_1, l, and N_2. As in the case of 1-heptanol, it may be considered a priori that N_1 and N_2 correspond respectively to direct and inverse microemulsion-type media. Taking into account the location of the small lenticular region l, it can reasonably be assumed that l corresponds to micro-emulsions of the direct kind. This will be fully demonstrated later in this section. The pseudo-ternary microemulsion domain of the 1-pentanol system (Fig. 5c) consists of a large triangular region N, which by shape and extent is roughly similar to the region N de-limited in the case of 1-butanol, and of a tiny lenticular region l that lies close to the main region N but is completely inserted in the part of the phase diagram corresponding to hydrocarbon-rich polyphasic media. It will be proved [38] that in spite of its strange location in the phase diagram, this lenticular region l defines a range of compositions that probably correspond to water-external microemulsion-type media.

The influence of alkanol nature on microemulsion pseudo-ternary domain configuration becomes clearer, especially with regard to the cases of the 1-hexanol and 1-pentanol systems, if one considers the ternary diagrams shown in Fig. 6. These diagrams show that two types of water/sodium dodecylsulfate/alkanol systems may be defined. With the longer alkanols (1-heptanol and 1-hexanol), the realm-of-existence of the monophasic ternary solutions consists of two dis-joined regions; L_1 is the region corresponding to alkanol solubiliza-tion in water ("direct" solubilization), and L_2 is the region corres-ponding to water solubilization in alkanol ("inverse" solubilization). Between the L_1 and L_2 regions, there is a broad range of composi-tions that correspond to the existence of monophasic mesomorphous media or of polyphasic media of different kinds. With 1-butanol (the shortest of the alkanols considered), the L_1 and L_2 regions merge so as to form a wide solubilization area, L, which extends in the ter-nary diagram continuously from the W apex (100% water) to the C apex (100% alkanol). In the case of 1-pentanol, the situation is rather similar, but for the fact that the L_1 and L_2 regions are con-nected via a very narrow channel C_L. Since it is possible to obtain quaternary microemulsion-type media by adding hydrocarbon to water/ ionic surfactant/alkanol monophasic solutions, it can reasonably be assumed that the configuration of the microemulsion domain in the phase tetrahedron is determined primarily by the configuration of the water/ionic surfactant/alkanol solubilization area. Since this

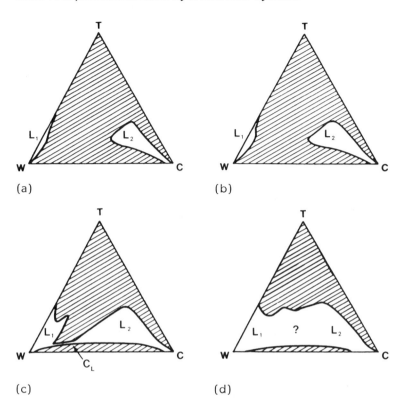

FIG. 6 Water/sodium dodecylsulfate/n-alkanol systems (T = 25°C).
Realms-of-existence of monophasic ternary solutions (blank areas).
(a) 1-heptanol; (b) 1-hexanol; (c) 1-pentanol; (d) 1-butanol.

solubilization area consists of two disjoined regions, L_1 and L_2, or
forms a vast region L, the microemulsion domain consists of two disjoined volumes, V_1 and V_2, that spring respectively from the L_1
or L_2 region (type S), or forms an all-in-one-block volume V whose
basis is the L region (type U). The validity of this conjecture has
been established by Zradba [35,39], who showed that the configuration of the water/ionic surfactant/alkanol mutual solubilization area
is indeed a good criterion for determining whether a water/ionic sur-
factant/alkanol/hydrocarbon system belongs to the type U category
or to the type S one. This is illustrated conclusively by Figs. 7-2
and 8-2, which show, for 1-hexanol (Fig. 7-2) and 1-butanol (Fig.
8-2), the evolution of the microemulsion pseudo-ternary domain as
the surfactant-to-cosurfactant ratio increases. The values of k_m,

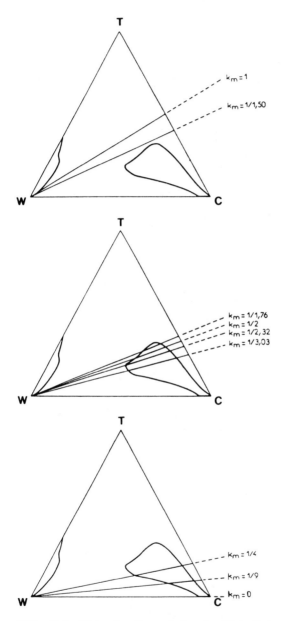

FIG. 7-1 Water/sodium dodecylsulfate/1-hexanol system (T = 25°C).
Sections of the phase diagram corresponding to different values of
the surfactant-cosurfactant mass ratio k_m.

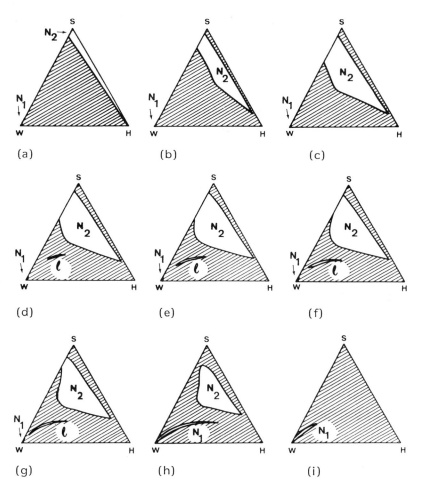

FIG. 7-2 Water/sodium dodecylsulfate/1-hexanol-n-dodecane system (T = 25°C). Influence of the surfactant-to-cosurfactant ratio on the configuration of the microemulsion pseudo-ternary domain (blank areas). (a) $k_m = 0$, $k_x = 0$; (b) $k_m = 1/9$, $k_x = 1/25.40$; (c) $k_m = 1/4$, $k_x = 1/11.29$; (d) $k_m = 1/3.03$, $k_x = 1/8.55$; (e) $k_m = 1/2.32$, $k_x = 1/6.54$; (f) $k_m = 1/2$, $k_x = 1/5.64$; (g) $k_m = 1/1.76$; $k_x = 4.97$; (h) $k_m = 1/1.50$, $k_x = 1/4.23$; (i) $k_m = 1/1$, $k_x = 1/2.82$.

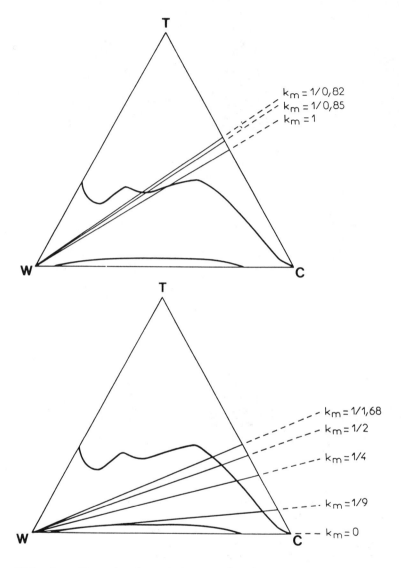

FIG. 8-1 Water/sodium dodecylsulfate/1-butanol system (T = 25°C). Sections of the phase diagram corresponding to different values of the surfactant-to-cosurfactant mass ratio k_m.

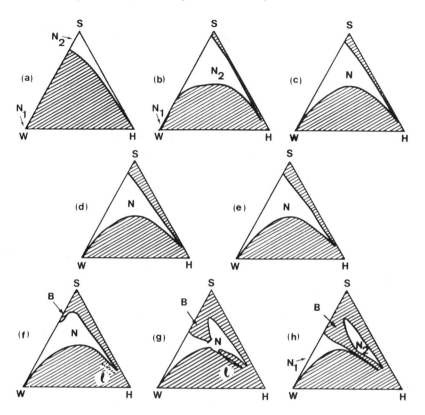

FIG. 8-2 Water/sodium dodecylsulfate/1-butanol/n-dodecane system (T = 25°C). Influence of the surfactant-to-cosurfactant ratio on the configuration of the microemulsion pseudo-ternary domain (blank areas). (a) $k_m = 0$, $k_x = 0$; (b) $k_m = 1/9$, $k_x = 1/35.02$; (c) $k_m = 1/4$, $k_x = 1/15.56$; (d) $k_m = 1/2$, $k_x = 1/7.78$; (e) $k_m = 1/1.68$, $k_x = 1/6.54$; (f) $k_m = 1/1$, $k_x = 1/3.89$; (f) $k_m = 1/0.85$, $k_x = 1/3.31$; (h) $k_m = 1/0.82$, $k_x = 3.19$.

the surfactant-to-cosurfactant mass ratio, corresponding to the different sections of the phase tetrahedron (see Fig. 1) are listed in Figs. 7-1 (1-hexanol) and 8-1 (1-butanol); the corresponding values of k_x, the surfactant-to-cosurfactant molar ratio, are reported in the captions of Figs. 7-2 (1-hexanol) and 8-2 (1-butanol). From the Fig. 7-2, it appears clear that the three-dimensional microemulsion domain of the water/sodium dodecylsulfate/1-hexanol/n-dodecane system consists of two distinct volumes, V_1 and V_2. V_2 leans against and develops from L_2, the water-in-1-hexanol solubilization region

(see Fig. 7-1) and consequently defines in the phase tetrahedron
the range of compositions corresponding to the existence of inverted
microemulsions.* From Fig. 7-2, diagrams d, e, f, g, and h, it
appears clear that the a priori strange lenticular areas l in fact
represent simply sections of the direct microemulsion three-dimensional
region, that is, the volume V_1 whose basis is L_1, the 1-hexanol-in-
water solubilization region (see Fig. 7-1). But for some details,
quite similar results were obtained with the water/sodium dodecyl-
sulfate/1-heptanol/n-dodecane system [33,35]. It is evident from
Fig. 8-2 that the three-dimensional microemulsion domain of the
water/sodium dodecylsulfate/1-butanol/n-dodecane system is an all-
in-one-block volume V whose basis is the vast area of mutual solubilization existing in the ternary diagram of the water/sodium dode-
cylsulfate/1-butanol system (see Fig. 8-1). From a detailed analysis of Fig. 8-2, it turns out that V is a fairly irregular body presenting two dissimilar protusions that both aim at the H apex of the
phase tetrahedron (see Fig. 8-2f, g, and h). A particularly important issue of this analysis is that the small lenticular areas l delimited for $k_m = 1$ and $1/0.85$ (diagrams f and g), appear to belong
to a region of direct microemulsions although they correspond to
n-dodecane-rich compositions; the same remark ought to be made
regarding the tip of the region N_1 delimited for $k_m = 1/0.82$ (diagram h). It may be conjectured that the microemulsions, belonging
by their composition to the l areas and to the tip of N_1, most probably possess an exotic structure. The validity of this statement is
supported by results obtained by Zradba [35,42], on the electroconductive and viscous behavior of microemulsions belonging to l
areas.

It follows from the preceding developments that isolated studies
of the pseudo-ternary microemulsion domain may be insufficient to
ascertain whether a water/ionic surfactant/cosurfactant/hydrocarbon
system belongs to the type S category or the type U one; the only
unbiased criterion is the configuration of the water/surfactant/co-
surfactant mutual solubilization area. It will be shown later that the
type (U or S) of a water/ionic surfactant/cosurfactant/hydrocarbon
system may be determined quite simply, by using the correlations
existing between system type and microemulsion electroconductive
behavior.

---

*As suggested by results obtained by Heil et al. [40,41], the part
of V_2 corresponding to compositions whose water-to-surfactant molar
ratio is less than 12 or so could be in fact a region of existence of
organic solutions of hydrated surfactant aggregates. This aspect
of microemulsion structure will be fully treated later [33] for the
case of the water/sodium dodecylsulfate/1-heptanol/n-dodecane
system.

B. Influence of Hydrocarbon Molecular Structure

Figures 9 and 10 show, for the value $k_X = 1/6.54$ of the sodium dodecylsulfate/n-alkanol molar ratio, how the nature of the hydrocarbon influences the shape and extent of the microemulsion pseudo-ternary domain.

In the case of 1-hexanol (Fig. 9, diagrams a_1 through a_3), the substitution of an aromatic hydrocarbon (benzene or toluene) for n-dodecane does not greatly affect the configuration of the microemulsion domain, which remains type S. Similarly, in the case of 1-butanol (Fig. 9, diagrams c_1 through c_3), the substitution of

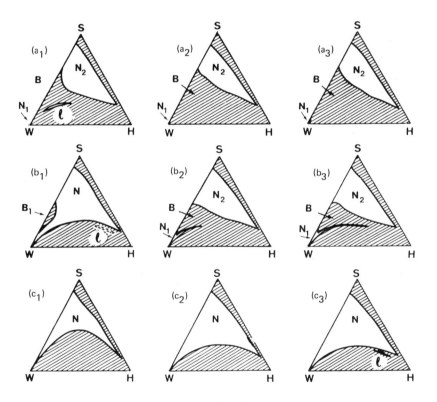

FIG. 9 Water/sodium dodecylsulfate/n-alkanol/hydrocarbon systems ($k_X = 1/6.54$; T = 25°C). Effects of the substitution of benzene or toluene for n-dodecane on the configuration of the microemulsion pseudo-ternary domain (blank areas). (a) 1-Hexanol: (a_1) n-dodecane, (a_2) benzene, (a_3) toluene. (b) 1-Pentanol: (b_1) n-dodecane, (b_2) benzene, (b_3) toluene. (c) 1-Butanol: (c_1) n-dodecane, (c_2) benzene, (c_3) toluene.

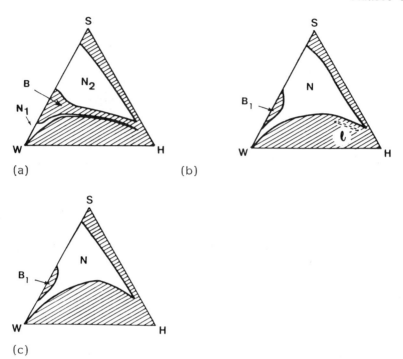

FIG. 10 Water/sodium dodecylsulfate/1-pentanol/n-alkane systems ($k_x = 1/6.54$; T = 25°C). Influence of n_h, the n-alkane number of carbon atoms, on the configuration of the microemulsion pseudo-ternary domain (blank areas). (a) n-octane; (b) n-dodecane; (c) n-hexadecane.

benzene of toluene for n-dodecane has no strong influence on the microemulsion domain, which remains type U. In the case of 1-pentanol, on the other hand (Fig. 9, diagrams b_1 through b_3), the microemulsion domain configuration undergoes drastic changes when n-dodecane is replaced by either benzene or toluene. Be-cause of the greater extension of the region B (i.e., the range of compositions corresponding to the existence of mesomorphous phases), the microemulsion domains of the benzene and toluene systems con-sist of two apparently disjoined areas. The same situation is ob-served in the case of the 1-pentanol/n-octane system (Fig. 10a). In contrast, the microemulsion domain of the 1-pentanol-n-hexade-cane system (Fig. 10c) does not differ fundamentally from that of the 1-pentanol/n-dodecane system (Fig. 10b).

It comes out from the above observations that the nature of the hydrocarbon has little influence on the configuration of the micro-emulsion pseudo-ternary domain of a system whose cosurfactant is either a "long" n-alkanol (1-hexanol in this case) or a typically "short" n-alkanol (1-butanol in this case). This is fully consistent with the main issue of the analysis carried out in the preceding section, that, the surfactant being given, the nature of the n-alkanol is the constitution parameter that entirely determines whether a water/ionic surfactant/cosurfactant/hydrocarbon system belongs to the type S or type U category. Based on the same analysis, it may be taken for granted that, as the 1-pentanol/n-dodecane or n-hexa-decane system does, the 1-pentanol/benzene or toluene or n-octane system belongs to the type U category, although, for $k_x = 1/6.54$, the configuration of the microemulsion pseudo-ternary domain would tend to indicate that it is type S. The validity of this statement has been confirmed by results concerning microemulsion electrocon-ductive behavior [36]. However, in the case of 1-pentanol, which according to the diagrams of Fig. 6, may be considered the n-alkanol corresponding to the transition between "long" n-alkanols and typi-cally "short" ones, the configuration of the microemulsion pseudo-ternary domain is highly sensitive to the hydrocarbon molecular structure. From a comparison of Figs. 9 and 10 with Figs. 5, 7, and 8, it appears at first that the decrease of the n-alkane number of carbon atoms or the substitution of an aromatic hydrocarbon for a long n-alkane has an effect on the microemulsion domain configura-tion somewhat equivalent to that of an increase of the n-alkanol num-ber of carbon atoms. Nicolas-Morgantini [36] has shown that some corrections must be brought to the preceding conclusion to take into account the influence of the surfactant/cosurfactant ratio.

C. Influence of Alkanol Molecular Structure

1. n-Alkanol Number of Carbon Atoms

Comparison of Fig. 11 with Fig. 5 ($k_x = 1/6.54$ in both cases) shows that the hydrocarbon nature does not affect the general pattern of the modifications undergone by the microemulsion pseudo-ternary do-main configuration when the n-alkanol chain length varies. As al-ready explained, the discrepancy, in the case of 1-pentanol, be-tween the diagrams of the benzene and n-dodecane systems is an apparent one only. That the type S → type U transition indeed oc-curs in benzene systems, as in n-dodecane systems, when 1-pentanol is substituted for 1-hexanol, is convincingly illustrated by Fig. 12 ($k_x = 1/9.60$). Although the diagrams correspond to limited ranges of variation of the surfactant/cosurfactant ratio, Figs. 13, 14, and 15 clearly show, on the one hand, that the three-dimensional

Clausse et al.

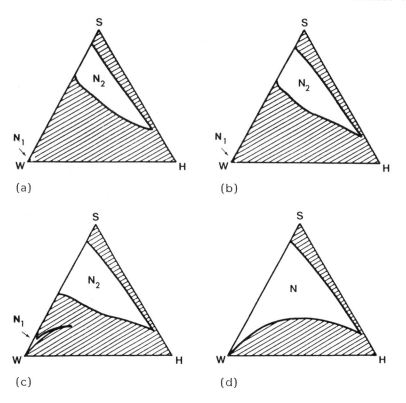

FIG. 11 Water/sodium dodecylsulfate/n-alkanol/benzene systems
($k_X = 1/6.54$; T = 25°C). Influence of n_a, the n-alkanol number of
carbon atoms, on the configuration of the microemulsion pseudo-
ternary domain (blank areas). (a) 1-heptanol; (b) 1-hexanol;
(c) 1-pentanol; (d) 1-butanol.

microemulsion domains of the water/sodium dodecylsulfate/1-hexanol
and 1-heptanol/benzene system consist of two disjoined volumes V_1
and V_2 (type S system). On the other hand, they show that the
three-dimensional microemulsion domain of the water/sodium dodecyl-
sulfate/1-butanol/benzene system forms a vast, all-in-one block
volume V (type U system). This is consistent with the results previously reported concerning the water/sodium dodecylsulfate/n-alka-
nol/n-dodecane systems and confirms that, the ionic surfactant being
given, n_a (i.e., the n-alkanol number of carbon atoms) is the constitution parameter that entirely determines whether a system belongs
to the type S or to the type U category. In the case of sodium

dodecylsulfate, the type S → type U transition occurs when n_a changes from 6 to 5.

2. Alkanol Isomery

Figures 16 and 17 show, for two values of the surfactant/alkanol molar ratio, $k_x = 1/7.78$ and $1/6.54$, the microemulsion pseudo-ternary domains of analogous systems, which differ from one another only in the butylic alcohol used as cosurfactant. It immediately appears from these figures that all of the four water/sodium dodecylsulfate/butylic alcohol/benzene systems belong to the type U category and that the alkanol isomery has no significant influence on

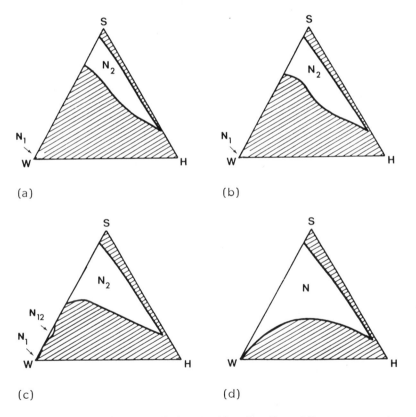

(a)   (b)

(c)   (d)

FIG. 12 Water/sodium dodecylsulfate/n-alkanol/benzene systems ($k_x = 1/9.60$; T = 25°C). Influence of n_a, the n-alkanol number of carbon atoms, on the configuration of the microemulsion pseudo-ternary domain (blank areas). (a) 1-heptanol; (b) 1-hexanol; (c) 1-pentanol; (d) 1-butanol.

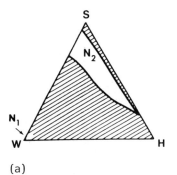

(a)

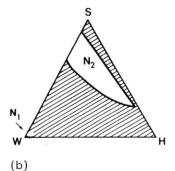

(b)

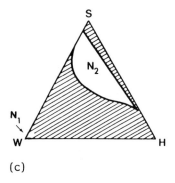

(c)

FIG. 13 Water/sodium dodecylsulfate/1-heptanol/benzene systems
(T = 25°C). Influence of the surfactant-to-cosurfactant ratio on
the configuration of the microemulsion pseudo-ternary domain (blank
areas).   (a) k_m = 1/3.87, k_x = 1/9.60; (b) k_m = 1/2.64, k_x =
1/6.54; (c) k_m = 1/2, k_x = 1.4.97.

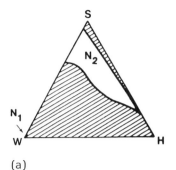

(a)

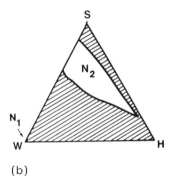

(b)

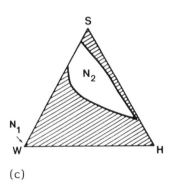

(c)

FIG. 14 Water/sodium dodecylsulfate/1-hexanol/benzene systems ($T = 25°C$). Influence of the surfactant-to-cosurfactant ratio on the configuration of the microemulsion pseudo-ternary domain (blank areas). (a) $k_m = 1/3.40$, $k_x = 1/9.60$; (b) $k_n = 1/2.32$, $k_x = 1/6.54$; (c) $k_m = 1/2$, $k_x = 1/5.64$.

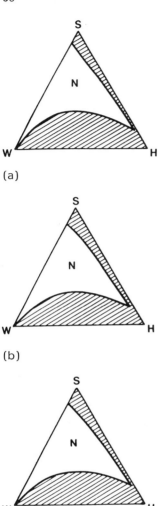

(a)

(b)

(c)

FIG. 15 Water/sodium dodecylsulfate/1-butanol/benzene systems
(T = 25°C). Influence of the surfactant-to-cosurfactant ratio on
the configuration of the microemulsion pseudo-ternary domain (blank
areas). (a) k_m = 1/2.47, k_x = 1/9.60; (b) k_m = 1/2, k_x = 1/7.78;
(c) k_m = 1/1.68, k_x = 1/6.54.

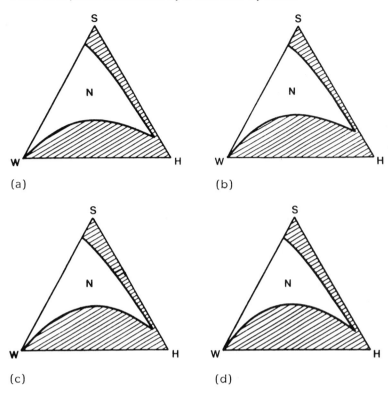

(a) (b)

(c) (d)

FIG. 16 Water/sodium dodecylsulfate/butylic alcohol/benzene systems ($k_X = 1/7.78$; T = 25°C). Influence of alcohol isomery on the configuration of the microemulsion pseudo-ternary domain (blank areas). (a) 1-butanol; (b) 1-propanol-2-methyl; (c) 2-butanol; (d) 2-propanol-2-methyl.

the microemulsion pseudo-ternary domain configuration. It should be noted that the extent of the microemulsion pseudo-ternary domain decreases as the alkanol affinity to water increases.

Figures 18 and 19 show that, in contrast with what is observed in the case of butylic alcohols, the alkanol isomery has a great influence on the configuration of the microemulsion pseudo-ternary domains of the water/sodium dodecylsulfate/amylic alcohol/benzene systems, even though all belong to the type U category. The general trend of the changes induced by the substitution of one amylic alcohol for another one is that the microemulsion pseudo-ternary domain becomes larger as the alkanol-water mutual solubility increases. In this respect, it is to be noted that, for $k_X = 1/6.54$ and $k_X = 1/9.60$,

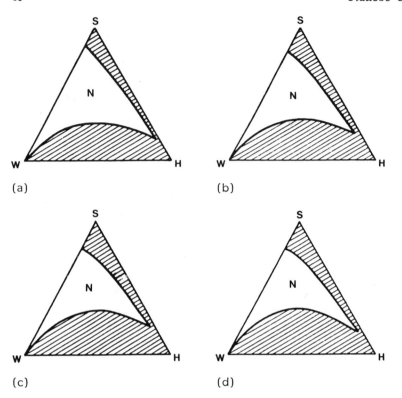

FIG. 17   Water/sodium dodecylsulfate/butylic alcohol/benzene systems (k_x = 1/6.54; T = 25°C). Influence of alcohol isomery on the configuration of the microemulsion pseudo-ternary domain (blank areas). (a) 1-butanol; (b) 1-propanol-2-methyl; (c) 2-butanol; (d) 2-propanol-2-methyl.

the microemulsion domains of the systems 1-butanol/benzene (Fig. 11d and 12d) and 2-butanol-2-methyl/benzene (Figs. 18h and 19c) are fairly similar, which is consistent with the fact that 1-butanol and 2-butanol-2-methyl have almost the same affinity toward water.

It appears from Fig. 20 that, in the case of systems whose cosurfactant is a hexylic alcohol, the replacement of 1-hexanol by one of its isomers, such as 2-hexanol or 3-pentanol-3-methyl, induces a type S → type U transition. The comparison of Figs. 20b and 20c with Figs. 18b and 18c confirms the tight correlation between the configuration of the microemulsion pseudo-ternary domain and the alkanol affinity toward water.

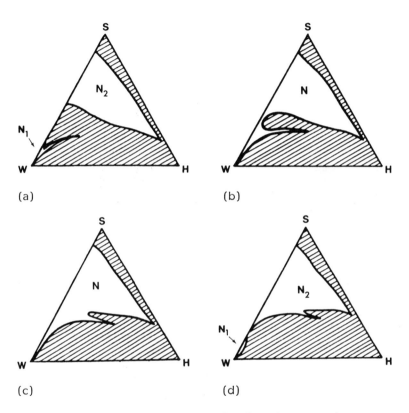

FIG. 18 Water/sodium dodecylsulfate/amylic alcohol/benzene systems (k_X = 1/6.54; T = 25°C). Influence of alcohol isomery on the configuration of the microemulsion pseudo-ternary domain (blank areas). (a) 1-pentanol; (b) 1-butanol-3-methyl; (c) 1-butanol-2-methyl; (d) 1-propanol-2,2-dimethyl; (e) 2-pentanol; (f) 3-pentanol; (g) 2-butanol-3-methyl; (h) 2-butanol-2-methyl.

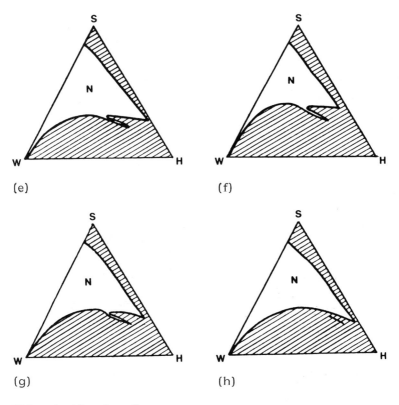

FIG. 18   (Continued)

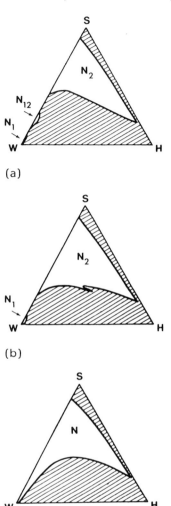

(a)

(b)

(c)

FIG. 19 Water/sodium dodecylsulfate/amylic alcohol/benzene systems ($k_X = 1/9.60$; T = 25°C). Influence of alcohol isomery on the configuration of the microemulsion pseudo-ternary domain (blank areas). (a) 1-pentanol; (b) 1-butanol-3-methyl; (c) 2-butanol-2-methyl.

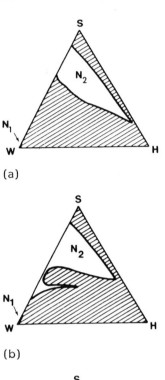

(a)

(b)

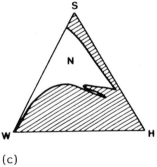

(c)

FIG. 20 Water/sodium dodecylsulfate/hexylic alcohol/benzene systems ($k_X = 1/6.54$; T = 25°C). Influence of alcohol isomery on the configuration of the microemulsion pseudo-ternary domain (blank areas). (a) 1-hexanol; (b) 2-hexanol; (c) 3-pentanol-3-methyl.

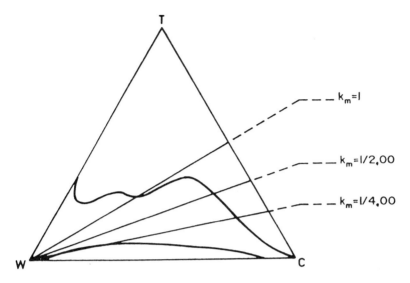

FIG. 21-1 Water/sodium dodecylsulfate/2-pentanol system (T = 25°C).
Sections of the phase diagram corresponding to different values of
the surfactant-to-cosurfactant mass ratio k_m.

The panel of microemulsion pseudo-ternary domain configurations
corresponding to Fig. 18 and Figs. 20b and 20c is particularly informative. It shows that, for all of the eight amylic alcohols and
for the two 1-hexanol isomers considered, the three-dimensional microemulsion domain forms an all-in-one-block volume V that exhibits
two dissimilar protusions, as in the case of the 1-butanol/n-dodecane
system that has been commented upon previously (see Fig. 8-2).
By varying the alkanol ramification, it is possible to obtain detailed
information about the shape and extent of V. This information
matches up with what may be obtained, for some type U system, by
varying the surfactant/cosurfactant ratio. This has been fully
demonstrated elsewhere [35,36], and is just exemplified here by
Fig. 21-2, which shows, in the case of the 2-pentanol/benzene system, how the microemulsion pseudo-ternary domain changes when the
surfactant/cosurfactant ratio varies (see Fig. 21-1). It should be
noted that the patterns of the modifications undergone by the microemulsion pseudo-ternary domains of the 2-pentanol/benzene system
(Fig. 21-2) and of the 1-butanol/n-dodecane system (Fig. 8-2) are
quite similar, which is a consequence of the similarity between the
realms-of-existence of the ternary monophasic solutions of the systems
water/sodium dodecylsulfate/2-pentanol (see Fig. 21-1) and water/
sodium dodecylsulfate/1-butanol (see Fig. 6d).

Clausse et al.

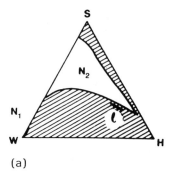

(a)

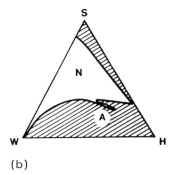

(b)

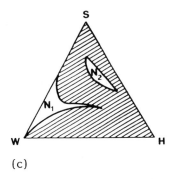

(c)

Figure 21-2 Water/sodium dodecylsulfate/2-pentanol/benzene system
(T = 25°C). Influence of the surfactant-to-cosurfactant ratio on
the configuration of the microemulsion pseudo-ternary domain (blank
areas). (a) k_m = 1/4, k_x = 1/13.09; (b) k_m = 1/2, k_x = 1/6.54;
(c) k_m = 1, k_x = 1/3.27.

IV. CORRELATIONS BETWEEN SYSTEM TYPE AND MICROEMULSION VISCOSITY AND ELECTRICAL CONDUCTIVITY

Ever since the early studies devoted to microemulsions, it was observed that slight variations of system constitution could induce drastic changes in the electroconductive behavior of microemulsion-type media. For instance, Schulman and McRoberts [43] pointed out that "the number and arrangement of the carbon atoms, both in the alcohol and in the oil, decided whether the system was conducting or non-conducting." In a subsequent paper by the same group [44], it was reported that "certain oil-continuous systems containing much soap possessed an intermediate conductivity and are classed as anomalous." They stated that "conditions could be so chosen that the spheres [inverse microdroplets] were squashed, this giving rise to broad X-ray bands and to anomalous electrical conductivity of the oil-continuous micro-emulsion." Winsor [45] also observed that the nature of the alcohol could have a drastic influence on the magnitude of the electrical conductivity of microemulsion-type media and on the trend of its variations with composition. More recently, Shah et al. [8] showed that the mere substitution of 1-pentanol for 1-hexanol induces drastic changes in the electrical conductivity and, correlatively, in other physicochemical properties of water /potassium oleate /n-alkanol /n-hexadecane microemulsions. The same microemulsions also were studied by Clausse et al. [31], who proved that the electroconductive behavior of 1-pentanol microemulsions can be depicted by means of equations derived from the percolation and effective medium theories [46–51], whereas that of 1-hexanol microemulsions cannot.

More systematic investigations into this problem of the influence of constitution parameters on microemulsion physicochemical properties were carried out first by Heil and co-workers [40,41,52], on systems incorporating potassium oleate as surfactant, and later by Zradba, Nicolas-Morgantini, and co-workers [35–37,39,53], on systems incorporating sodium dodecylsulfate as surfactant. From the results thus obtained, it appears that constant correlations exist between system type (U or S) and microemulsion transport properties such as viscosity and electrical conductivity.

That correlations between system type and microemulsion transport properties indeed exist is convincingly exemplified by Figs. 22 and 23. These plots are directly comparable because the variations of the electrical conductivity σ and viscosity η were recorded, for all of the four analogous systems considered ($k_x = 1/6.54$), along the same experimental path ($[R_{h/s}]_x$, the molar ratio of n-dodecane to surfactant-cosurfactant combination, being equal to $1/2.23$ in all cases). It is readily seen from Fig. 22 that the electroconductive

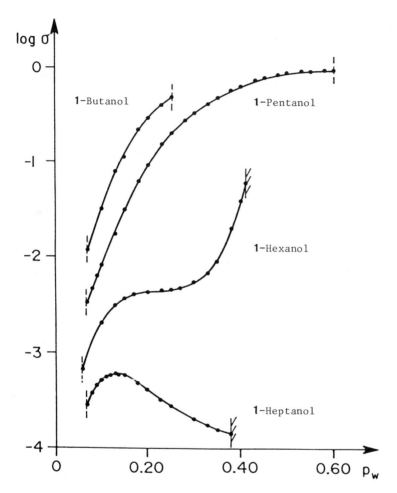

FIG. 22 Water/sodium dodecylsulfate/n-alkanol/n-dodecane micro-emulsions (T = 25°C). Variations of microemulsion electrical con-ductivity σ with p_w, the water mass fraction, along the composition path defined by the value 1/2.23 of $[R_{h/s}]_x$, the molar ratio of hydrocarbon to surfactant/cosurfactant combination. Influence of n_a, the n-alkanol number of carbon atoms (σ is expressed in Sm^{-1}).

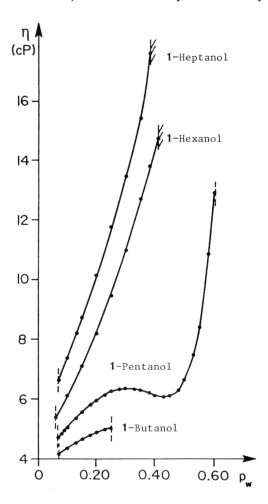

FIG. 23 Water/sodium dodecylsulfate/n-alkanol/n-dodecane micro-emulsions (k_x = 1/6.54; T = 25°C). Variations of microemulsion apparent viscosity η with p_w, the water mass fraction, along the composition path defined by the value 1/2.23 of $[R_{h/s}]_x$, the molar ratio of hydrocarbon to surfactant/cosurfactant combination. Influence of n_a, the n-alkanol number of carbon atoms.

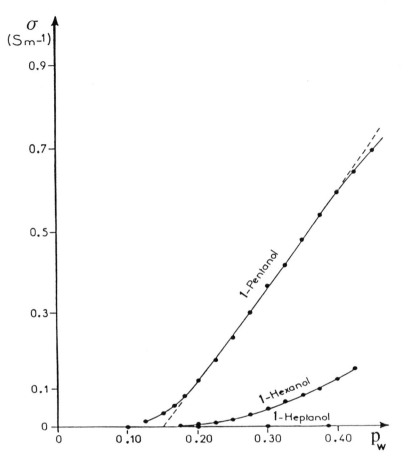

FIG. 24 Water/sodium dodecylsulfate/n-alkanol/n-dodecane microemulsions (k_m = 1/2; T = 25°C). Variations of microemulsion electrical
conductivity σ with p_w, the water mass fraction, along analogous
p_s-type composition paths, (p_s = 0.4). Influence of n_a, the n-alkanol
number of carbon atoms.

behavior of the 1-butanol and 1-pentanol microemulsions (type U,
see Fig. 5) is strikingly different from that of the 1-heptanol microemulsions (type S, see Fig. 5). In the case of the shorter n-alkanols,
the electrical conductivity increases continuously as the water content increases, and its variations may be depicted through the percolation and effective medium theories [51,54,55]. In contrast, no
percolation phenomenon is observed in the case of the 1-heptanol

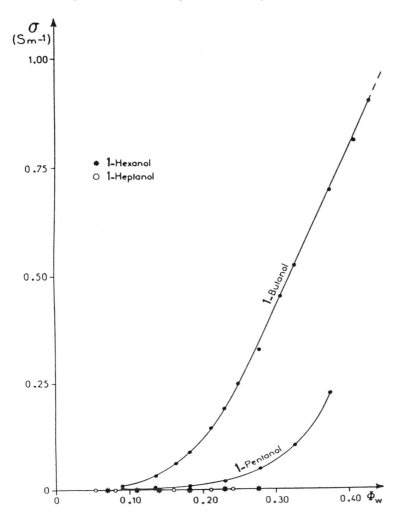

FIG. 25 Water/sodium dodecylsulfate/n-alkanol/benzene microemulsions (k_m = 1/2; T = 25°C). Variations of microemulsion electrical conductivity σ with Φ_w, the water volume fraction, along analogous p_s-type composition paths, (p_s = 0.45). Influence of n_a, the n-alkanol number of carbon atoms.

microemulsions; their electrical conductivity remains quite low (below 10^{-3} Sm^{-1}) and is a nonmonotonic function of the water content. The case of the 1-hexanol microemulsions appears somewhat special, because of the "hydrid" trend of the electrical conductivity variations; nevertheless, the electroconductive behavior of the 1-hexanol microemulsions and that of the 1-heptanol microemulsions have the same general character.

Nearly similar comments may be made with regard to microemulsion viscous behavior (Fig. 23). The variation trend of the apparent viscosity η of the 1-heptanol and 1-hexanol microemulsions differs considerably from that of the 1-butanol microemulsions. As long as the mass fraction of water is less than 0.3 or so, the viscous behavior of the 1-pentanol microemulsions greatly resembles that of the 1-butanol microemulsions; at higher water contents, it has features in common with the viscous behavior of the 1-hexanol and 1-heptanol microemulsions. The additional plots of Figs. 24 and 25 confirm that, along p_s-type experimental paths (see Fig. 2), the microemulsion electrical conductivity variation trend is strikingly different, depending on whether the cosurfactant is a "long" n-alkanol (1-hexanol or 1-heptanol) or a "short" one.

V. CONCLUSIONS

It appears from the phase diagrams data reported and commented upon above that the microemulsion three-dimensional domain of a water/sodium dodecylsulfate/alkanol/hydrocarbon system is the volumic extension of the realm-of-existence of the corresponding water/sodium dodecylsulfate/alkanol monophasic solution. The microemulsion three-dimensional domain configuration is influenced primarily by the alkanol molecular structure and to a much lesser degree by that of the hydrocarbon. Irrespective of the nature of the hydrocarbon, two types of microemulsion domain configuration can be defined. When the cosurfactant is a "long" alkanol, the microemulsion three-dimensional domain consists of two disjoined volumes, V_1 and V_2; V_1 is the volumic extension of the region L_1 corresponding, in the water/sodium dodecylsulfate/alkanol phase diagram, to "direct" solubilization (alkanol in water), and V_2 is the volumic extension of the region L_2 corresponding to "inverse" solubilization (water in alkanol). Systems of this kind are labeled type S. Per contra, the microemulsion domain of systems incorporating "short" alkanols forms in the phase tetrahedron an all-in-one-block volume built on the vast region L that exists in the water/sodium dodecylsulfate/alkanol phase diagram, as the result of the merging of the L_1 and L_2 areas. Systems of this kind are labeled type U. For the systems considered here, all of which incorporate sodium dodecylsulfate as surfactant, the value of n_a, the alkanol number of carbon atoms, corresponding to the type U-to-type S transition is $\bar{n}_a = 6$. Replacing a given alkanol by another alkanol belonging to the same category ("short" or "long") has no effect on system type but may notably affect certain features of the microemulsion three-dimensional domain. Since it is the fact that the alkanol is "long" or "short" that entirely determines whether a water/sodium dodecylsulfate/alkanol/hydrocarbon system belongs to the type S category or the type U one, the nature of the hydrophobic hydrocarbon has no influence on system type.* The hydrocarbon molecular structure, however, may have a strong influence on certain details of the microemulsion three-dimensional domain configuration, especially in the present case, when the cosurfactant is 1-pentanol. Equivalences may be defined regarding the relative influences of system constitution and composition parameters. For instance, in the case of type U systems, decreasing the sodium dodecylsulfate/alkanol molar ratio k_x (alkanol and hydrocarbon given), replacing a "short" n-alkanol by some of its isomers or by a shorter n-alkanol (hydrocarbon and k_x given), increasing the n-alkane number of carbon atoms n_h, or substituting a long n-alkane ($n_h \geqslant 10$) for an aromatic hydrocarbon (alkanol and k_x given) have somewhat similar effects on the microemulsion pseudo-ternary domain configuration.

The existence of two main kinds of water/sodium dodecylsulfate/alkanol/hydrocarbon systems, depending on whether their microemulsion three-dimensional domain consists of two disjoined volumes V_1 and V_2 or is a vast, all-in-one-block body V, can be well correlated with the existence of dissimilarities in certain transport properties of microemulsions, such as viscosity and electrical conductivity. Numerous studies have clearly proved that the electroconductive behavior of microemulsions of type S systems is always strikingly different from that of microemulsions of type U systems. In the first case (type S systems), throughout the pseudo-ternary phase diagram region N_2 that corresponds to inverse solubilization (mass fraction of water $p_w < 0.4$ or so), the microemulsion electrical conductivity remains very low, in the range 10^{-5} to 10^{-2} Sm^{-1}), and varies in a nonmonotonous way as p_w increases. In the second case (type U systems), when the water mass fraction p_w increases up to 0.4 or so, the microemulsion electrical conductivity first rises very drastically according to a power law ("percolation" phenomenon [46–47]) and then increases linearly until it reaches values around 1 Sm^{-1}. At comparable compositions, the electrical conductivity of type U microemulsions is currently 10^2 to 10^3 times higher than that of

---

*At least as far as hydrocarbons in which sodium dodecylsulfate is insoluble (or only very slightly soluble) are considered.

type S microemulsions. The microemulsion viscous behavior also is quite different depending on whether the system belongs to the type U or the type S category. It has been experimentally demonstrated by Zradba et al. [35,39,53] that the dissimilarity existing between the electroconductive behavior of type S microemulsions and that of type U microemulsions is observable whatever the value of the sodium dodecylsulfate/alkanol ratio. This implies that microemulsion electroconductimetry can be used to identify quickly the category to which a water/sodium dodecylsulfate/alkanol/hydrocarbon system belongs, without carrying out multiple phase diagram determinations. Good illustrations of this instrumental consequence of the existence of correlations between system type and microemulsion electroconductive behavior are given in Figs. 26 through 28. It is readily seen from the conductivity plots of Figs. 26 and 27 that, the cosurfactant being given (1-pentanol), the general trend of microemulsion electrical conductivity variations remains the same either when n-dodecane is replaced by n-octane or n-hexadecane, (Fig. 26), or when n-dodecane is replaced by benzene (Fig. 27). This proves that, as the n-dodecane and n-hexadecane systems do, the n-octane and benzene systems indeed belong to the type U category, although for $k_X = 1/6.54$, their microemulsion pseudo-ternary domains consist of two apparently disjoined areas N_1 and N_2 (see Fig. $9b_2$ and Fig. 10a). Similarly, it appears from the conductivity plots of Fig. 28 that the substitution of 2-pentanol for 1-pentanol has no effect on microemulsion electrical conductivity variation trend and, consequently, that the water/sodium dodecylsulfate/1-pentanol or 2-pentanol/benzene systems both belong to the type U category.

All the results reported above concerning microemulsion domain configuration, electrical conductivity, and viscosity are mutually consistent and are in agreement with data reported in the literature by other groups [8,56–63]. They may be interpreted using data obtained on the influence of the chemical structure of the system constituents on the internal organization of and interactions in microemulsions saturated with both water and hydrocarbon [17,18,63–73]. For instance, in the case of water/sodium dodecylsulfate/n-alkanol/n-dodecane saturated microemulsions, it has been shown by Brunetti et al. [64] that the n-alkanol number of carbon atoms has a strong influence on the interactions between microemulsion inverse microdroplets. With 1-heptanol, the value of the second coefficient of the osmotic compressibility virial is 5 or so, which indicates that the interactions are weakly attractive and that the inverse microdroplets behave like hard spheres. In contrast, with 1-pentanol, the value of the second coefficient of the osmotic compressibility virial is -25 or so, which reveals that the interactions are strongly attractive and, consequently, that the inverse microdroplets tend to form extended clusters. These structural data and the concept of "rigidity of

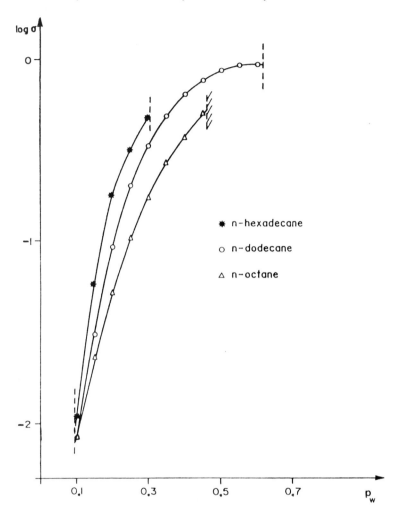

FIG. 26 Water/sodium dodecylsulfate/1-pentanol/n-alkane microemulsions ($k_X = 1/6.54$; $t = 25°C$). Variations of microemulsion electrical conductivity σ with p_w, the water mass fraction, along the composition path defined by the value $1/2.23$ of $[R_{h/s}]_X$, the molar
ratio of hydrocarbon to surfactant/cosurfactant combination. Influence of n_h, the n-alkane number of carbon atoms (σ is expressed
in Sm^{-1}).

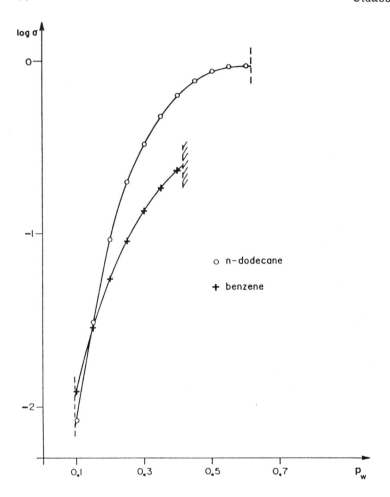

FIG. 27 Water/sodium dodecylsulfate/1-pentanol/n-dodecane or benzene microemulsions ($k_x = 1/6.54$; T = 25°C). Variations of microemulsion electrical conductivity σ with p_W, the water mass fraction, along the composition path defined by the value $1/2.23$ of $[R_{h/s}]_x$, the molar ratio of hydrocarbon to surfactant/cosurfactant combination. Effect of the substitution of benzene for n-dodecane (σ is expressed in Sm^{-1}).

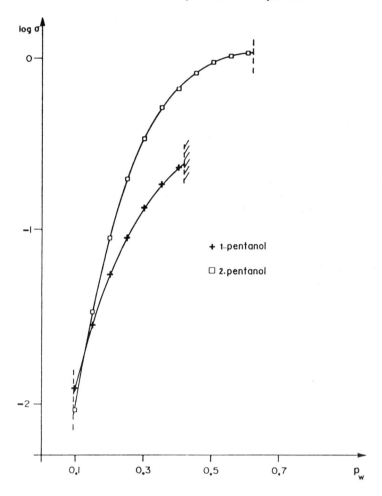

FIG. 28 Water/sodium dodecylsuflate/1-pentanol or 2-pentanol/ben-
zene microemulsions (k_X = 1/6.54; T = 25°C). Variations of microemulsion electrical conductivity σ with p_W, the water mass fraction,
along the composition path defined by the value 1/2.23 of $[R_{h/s}]_X$,
the molar ratio of hydrocarbon to surfactant/cosurfactant combination.
Effect of the substitution of 2-pentanol for 1-pentanol (σ is expressed in Sm^{-1}).

oil-water interfaces" put forward by De Gennes and Taupin [74],
provide a sound basis for interpreting the existence of two types of
systems and two kinds of microemulsion electroconductive behavior.
That microemulsions incorporating a "short" alkanol as cosurfactant
(type U systems) exhibit an electroconductive behavior that is depictable by means of the percolation and effective medium theories is
consistent with the existence of inverse microdroplet clustering and
merging processes that lead to the formation of random structures
("bicontinuous" microemulsions) over a range of compositions located
in the central part of the phase diagram. The clustering and merging processes can develop in type U microemulsions because of the
existence of strongly attractive interactions (formation of clusters of
inverse microdroplets) and of "fluid" interfacial films (merging of
clustered inverse microdroplets). In contrast, when the cosurfactant
is a "long" alkanol (type S system), that inverse microemulsion electrical conductivity is very low and does not vary drastically with water
content indicates that the inverse microdroplets retain their individual-
ity because of poorly attractive interactions (very limited or no clustering process) and because of highly rigid interfacial films (no merging
process). This is consistent with the fact that, in this case, the transition from inverse to direct microemulsions occurs through the formation
of mesomorphous media whose realm-of-existence in the phase diagram
separates the inverse microemulsion area from the direct microemulsion area.

Finally, it must be pointed out that the validity of all the preceding
conclusions is not restricted to the case of systems whose ionic surfactant is sodium dodecylsulfate [40,41]. Only the molecular characteristics
(number of carbon atoms or/and ramification) of the alkanol corresponding to the type U-to-type S transition may possibly change when sodium
dodecylsulfate is replaced by another ionic surfactant of the same kind.

A detailed study of type S microemulsions is reported in Chapter 3.
The case of type U microemulsions is examined in full detail in Chapter 25.

VI. ACKNOWLEDGMENTS

L. Nicolas-Morgantini is grateful to the Ministère de la Recherche
et de l'Industrie (France) for the research scholarship granted to
him from October 1981 to October 1983. A. Zradba wishes to thank
the Ministère des Relations Extérieures (France) for the financial
help provided to him from October 1980 to July 1983. Professor
M. Clausse expresses his deep appreciation for the support granted
to him by the NATO Scientific Affairs Division (Belgium) (Contract
No. 020.82).

The authors wish to express to Mrs. A. M. Schmitt their sincere appreciation for her meticulous typing of this chapter.

VII. REFERENCES

1. Schulman, J. H., Stoeckenius, W., and Prince, L. M., *J. Phys. Chem. 63*, 1677 (1959).
2. Rosano, H. L., Peiser, R. C., and Eydt, A., *Rev. Franc. Corps Gras 16*, 249 (1969).
3. Prince, L. M., *J. Soc. Cosmet. Chem. 21*, 193 (1970).
4. Prince, L. M., in *Emulsions and Emulsion Technology*, Lissant, K. J. (ed.), Marcel Dekker, New York, 1974, part I, Chap. 3, pp. 125–177.
5. Rosano, H. L., *J. Soc. Cosmet. Chem. 25*, 609 (1974).
6. Prince, L. M., in *Microemulsions: Theory and Practice*, Prince, L. M. (ed.), Academic Press, New York, 1977, pp. 1–20.
7. Friberg, S., in *Microemulsions: Theory and Practice*, Prince, L. M. (ed.), Academic Press, New York, 1977, pp. 133–146.
8. Shah, D. O., Bansal, V. K., Chan, K., and Hsieh, W. C., in *Improved Oil Recovery by Surfactant and Polymer Flooding*, Shah, D. O., and Schechter, R. S. (eds.), Academic Press, New York, 1977, pp. 293–337.
9. Rosoff, M., in *Progress in Surface and Membrane Science*, Danielli, J. F., Rosenberg, M. D., and Cadenhead, D. A. (eds.), Academic Press, New York, 1978, vol. XII, pp. 405–477.
10. Taupin, C., in *Physicochimie des Composés Amphiphiles*, Editions du CNRS, Paris, 1979, pp. 255–259.
11. Friberg, S., and Venable, R. L., in *Encyclopedia of Emulsion Technology*, Becher, P. (ed.), Marcel Dekker, New York, 1983, vol. I, pp. 287–336.
12. Clausse, M., Nicolas-Morgantini, L., and Zradba, A., in *Proceedings of the World Surfactants Congress*, Munich, May 6–10, 1984, Kürle Druck, Gelnhausen, West Germany, 1984, vol. III, pp. 209–219.
13. Tadros, Th. F., in *Surfactants in Solution*, Mittal, K. L., and Lindman, B. (eds.), Plenum Press, New York, 1984, vol. III, pp. 1501–1532.
14. Overbeek, J. T. G., De Bruyn, P. L., and Verhoeckx, F., in *Surfactants*, Tadros, Th. F. (ed.), Academic Press, London, 1984, pp. 111–132.
15. Friberg, S. E., *J. Dispersion Sci. Technol. 6*, 317 (1985).
16. Clausse, M., Heil, J., Zradba, A., and Nicolas-Morgantini, L., *J. Com. Esp. Deterg. 16*, 497 (1985).
17. Dvolaitzky, M., Guyot, M., Lagues, M., Le Pesant, J. P., Ober, R., Sauterey, C., and Taupin, C., *J. Chem. Phys. 69*, 3279 (1978).
18. Cazabat, A. M., and Langevin, D., *J. Chem. Phys. 74* 3148 (1981).

19. Graciaa, A., Lachaise, J., Chabrat, P., Letamendia, L., Rouch, J., and Vaucamps, C., *J. Phys. Lett. (Paris) 39*, L-235 (1978).
20. Scriven, L. E., *Nature 263*, 123 (1976).
21. Scriven, L. E., in *Micellization, Solubilization and Microemulsions*, Mittal, K. L. (ed.), Plenum Press, New York, 1977, vol. II, pp. 877–893.
22. Talmon, Y., and Prager, S., *Nature 267*, 333 (1977).
23. Talmon, Y., and Prager, S., *J. Chem. Phys. 69*, 2984 (1978).
24. Talmon, Y., and Prager, S., *J. Chem. Phys. 76*, 1535 (1982).
25. Kaler, E. W., and Prager, S., *J. Colloid Interface Sci. 86*, 359 (1982).
26. Jouffroy, J., Levinson, P., and De Gennes, P. G., *J. Phys. (Paris) 43*, 1241 (1982).
27. Kaler, E. W., Bennett, K. E., Davis, H. T., and Scriven, L. E., *J. Chem. Phys. 79*, 5673 (1983).
28. Auvray, L., *J. Phys. Lett. (Paris) 46*, L-163 (1985).
29. Winsor, P. A., *Trans. Faraday Soc. 44*, 376 (1948).
30. Sjoblom, E., and Friberg, S., *J. Colloid Interface Sci. 67*, 16 (1978).
31. Clausse, M., Boned, C., Peyrelasse, J., Lagourette, B., McClean, V. E. R., and Sheppard, R. J., in *Surface Phenomena in Enhanced Oil Recovery*, Shah, D. O. (ed.), Plenum Press, New York, 1981, pp. 199–228.
32. Bellocq. A. M., and Fourche, G., *J. Colloid Interface Sci. 78*, 275 (1980).
33. Clausse, M., Zradba, A., Rouviere, J., and Sohounhloue, K., *this volume*.
34. Bellocq, A. M., Biais, J., Clin, B., Lalanne, P., and Lemanceau, B., *J. Colloid Interface Sci. 70*, 524 (1979).
35. Zradba, A., Thèse de Doctorat de 3ème Cycle, Université de Pau, France, 1983.
36. Nicolas-Morgantini, L., Thèse de Doctorat de 3ème Cycle, Université de Pau, France, 1984.
37. Clausse, M., Peyrelasse, J., Boned, C., Heil, J., Nicolas-Morgantini, L., and Zradba, A., in *Surfactants in Solution*, Mittal, K. L., and Lindman, B. (eds.), Plenum Press, New York, 1984, vol. III, pp. 2583–1626.
38. Clausse, M., Zradba, A., and Nicolas-Morgantini, L., *this volume*.
39. Zradba, A., and Clausse, M., *Colloid Polymer Sci. 262*, 754 (1984).
40. Heil, J., Thèse de Doctorat de 3ème Cycle, Université de Pau, France, 1981.
41. Heil, J., Clausse, M., Peyrelasse, J., and Boned, C., *Colloid Polymer Sci. 260*, 93 (1982).

42. Clausse, M., Zradba, A., and Nicolas-Morgantini, L., *C. R. Acad. Sci. Paris 296*, Sér. II, 237 (1983).
43. Schulman, J. H., and McRoberts, T. S., *Trans. Faraday Soc. 42B*, 165 (1946).
44. Schulman, J. H., and Riley, D. P., *J. Colloid Sci. 3*, 383 (1948).
45. Winsor, P. A., *Trans. Faraday Soc. 46*, 762 (1950).
46. Kirkpatrick, S., *Phys. Rev. Lett. 27*, 1722 (1971).
47. Kirkpatrick, S., *Rev. Mod. Phys. 45*, 574 (1973).
48. Bruggeman, D. A. G., *Ann. Phys. Leipzig 24*, 636 (1935).
49. Bottcher, C. J. F., *Rec. Trav. Chim. Pays-Bas 64*, 47 (1945).
50. Landauer, R., *J. Appl. Phys. 23*, 779 (1952).
51. Clausse, M., in *Encyclopedia of Emulsion Technology*, Becher, P., (ed.), Marcel Dekker, New York, 1983, vol. I, pp. 481–715.
52. Peyrelasse, J., Boned, C. Heil, J., and Clausse, M., *J. Phys. C: Solid State Phys. 15*, 7099 (1982).
53. Clausse, M., Zradba, A., and Nicolas-Morgantini, L., *Colloid Polymer Sci. 263*, 767 (1985).
54. Lagourette, B., Peyrelasse, J., Boned, C., and Clausse, M., *Nature 281*, 60 (1979).
55. Clausse, M., Peyrelasse, J., Heil, J., Boned, C., and Lagourette, B., *Nature 293*, 636 (1981).
56. Bansal, V. K., Chimnaswamy, K., Ramachandran, C., and Shah, D. O., *J. Colloid Interface Sci. 72*, 524 (1979).
57. Bansal, V. K., Shah, D. O., and O'Connell, J. P., *J. Colloid Interface Sci. 75*, 462 (1980).
58. Lindman, B., Stilbs, P., and Moseley, M. E., *J. Colloid Interface Sci. 83*, 569 (1981).
59. Stilbs, P., *J. Colloid Interface Sci. 94*, 463 (1983).
60. Lindman, B., and Stilbs, P., in *Proceedings of the World Surfactants Congress*, Münich, May 6–10, 1984, Kürle Druck, Gelnhausen, West Germany, 1984, vol. III, pp. 159–167.
61. Sjöblom, E., and Henriksson, U., in *Surfactants in Solution*, Mittal, K. L., and Lindman, B. (eds.), Plenum Press, New York, 1984, vol. III, pp. 1867–1880.
62. Lang, J., Rueff, R., Dinh-Cao, M., and Zana, R., *J. Colloid Interface Sci. 101*, 184 (1984).
63. Roux, D., and Bellocq, A. M., in *Macro- and Microemulsions*, Shah, D. O. (ed.), ACS Symposium Series 272, American Chemical Society, Washington, D.C., 1985, pp. 105–118.
64. Brunetti, S., Roux, D., Bellocq, A. M., Fourche, G., and Bothorel, P., *J. Phys. Chem. 87*, 1028 (1983).
65. Roux, D., Bellocq, A. M., and Bothorel, P., in *Surfactants in Solution*, Mittal, K. L., and Lindman, B. (eds.), Plenum Press, New York, 1984, vol. III, pp. 1843–1863.

66. Lagues, M., Ober, R., and Taupin, C., *J. Phys. Lett. (Paris)* *39*, L-487 (1978).
67. Dvolaitzky, M., Lagues, M., Le Pesant, J. P., Ober, R., Sauterey, C., and Taupin, C., *J. Phys. Chem.* *84*, 1532 (1980).
68. Ober, R., and Taupin, C., *J. Phys. Chem.* *84*, 2418 (1980).
69. Lagues, M., and Sauterey, C., *J. Phys. Chem.* *84*, 3503 (1980).
70. Dvolaitzky, M., Ober, R., and Taupin C., *C. R. Acad. Sci. Paris 293*, Sér. II, 27 (1981).
71. Cazabat, A. M., Chatenay, D., Langevin, D., and Pouchelon, A., *J. Phys. Lett. (Paris) 41*, L-441 (1980).
72. Zana, R., Lang, J., Sorba, O., Cazabat, A. M., and Langevin, D., *J. Phys. Lett. (Paris) 43*, L-829 (1982).
73. Guering, P., and Cazabat, A. M., *C. R. Acad. Sci. Paris 296*, Sér. II, 1129 (1983).
74. De Gennes, P. G., and Taupin, C., *J. Phys. Chem. 86*, 2294 (1982).

3

Structural Transition in Water/Sodium Dodecylsulfate/ 1-Heptanol/n-Dodecane Inverse Microemulsion-type Media

M. CLAUSSE, A. ZRADBA,* and J. HEIL† *Université de Technologie de Compiègne, France*

J. ROUVIERE and K. SOHOUNHLOUE *Université des Sciences et Techniques du Languedoc, Montpellier, France*

I. INTRODUCTION

It has been suggested by different authors that, below a certain water/ionic surfactant ratio, microemulsions incorporating ionic surfactant/alkanol combinations are organic solutions of hydrated surfactant aggregates and not assemblies of water-in-oil microdroplets possessing a well-defined core of "free" water. In view of generalizing former data concerning this problem, a thorough study of the electrical conductivity and refractive index of water/sodium dodecylsulfate/1-heptanol/n-dodecane inverse microemulsion-type media was carried out. From the results thus obtained, it appears possible to define, inside the system phase tetrahedron, a surface Σ that divides the inverse solubilization domain into two regions corresponding respecitively to solutions of surfactant aggregates and to proper inverse microemulsions.

Numerous results gained from studies of microemulsion domain configuration and microemulsion electrical conductivity and viscosity clearly prove that two categories of water/ionic surfactant/alkanol/ hydrocarbon systems can be defined [1–14]. The microemulsion three-dimensional domain of a system belonging to the so-called

Present affiliations:
*Ecole Normale Supérieure de Casablanca, Casablanca, Morocco.
†PUM S.A., Bonneuil sur Marne, France.

type S category [11,14] consists of two disjoined volumes, V_1 and V_2, which are separated by regions of compositions corresponding to the existence of mesomorphous monophases or of polyphasic media of various kinds. This situation is a consequence of the fact that, in the ternary diagram of the water/ionic surfactant/alkanol system, there is a "solubilization gap," that is, a certain range of compositions that separates L_1, the region of "direct" solubilization (aqueous solutions of alkanol), from L_2, the region of "inverse" solubilization (alkanolic solutions of water). Since V_1 and V_2 are the three-dimensional extensions of L_1 and L_2, respectively, V_1 may be a priori considered as the realm-of-existence of water-continuous microemulsion-type media and V_2 as the realm-of-existence of hydrocarbon-continuous microemulsion-type media [12–14].

Initially it may be thought that the problem of the microstructure of type S microemulsions is rather trivial, contrary to that of the microstructure of type U microemulsions, whose realm-of-existence is vast and extends continuously from the region of water-rich compositions to the region of water-poor compositions [15]. However, earlier studies [1,2,9–11,16,17] have shown that type S microemulsions of the inverse kind probably undergo structural transitions as the water content increases. In an attempt to ascertain and generalize this conclusion, a thorough study of the inverse microemulsions of the type S system water/sodium dodecylsulfate/1-heptanol/n-dodecane [14] was undertaken. The microemulsion structural state was probed by means of electroconductimetry and refractometry.

II. EXPERIMENTAL

A. Compounds

The water used was freshly bi-distilled water, (electrical conductivity σ around 10^{-4} Sm^{-1}).

The ionic surfactant was sodium dodecylsulfate (99% purity grade) purchased from Touzart et Matignon, France.

The cosurfactant was 1-heptanol (99.5% purity grade) and the hydrocarbon n-dodecane (98% purity grade), both purchased from Fluka A.G., Switzerland.

B. Methods

The microemulsion pseudo-ternary domains were delimited inside WHS-type triangular diagrams [13,14,18], according to the general method reported previously [14]. All experiments were performed at T = 25°C ± 1°C.

The microemulsion electrical conductivity σ was measured, at T = 25°C ± 0.5°C, by means of Mullard-type cells (Philips PR 9512/00 and PR 9512/01) connected to a semiautomated conductance bridge working at 1.592 kHz (Wayne-Kerr B 331). The relative uncertainty on the values of σ is, on an average, less than 1%

The microemulsion refractive index n was determined, at T = 25°C ± 0.1°C, by means of an O.P.L. Abbe refractometer. The uncertainty on n is ± 0.0002.

III. PHASE DIAGRAMS

A. Water/Sodium Dodecylsulfate/1-Heptanol System

Figure 1-1 shows the realm-of-existence of the water/sodium dodecylsulfate/1-heptanol monophasic solutions. L_1 is the region of "direct" solubilization (aqueous solutions of 1-heptanol), and L_2 is the region of "inverse" solubilization (heptanolic solutions of water). L_1 and L_2 are separated by ranges of compositions corresponding to the existence either of diphasic systems of the L_2/L_1 kind (vicinity of the CW edge of the ternary diagram) or of lamellar mesomorphous

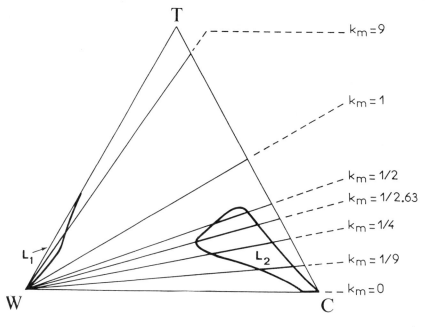

FIG. 1-1 Water/sodium dodecylsulfate/1-heptanol system (T = 25°C). Sections of the phase diagram corresponding to different values of the surfactant-to-cosurfactant mass ratio k_m.

media (either monophasic or belonging to diphasic and triphasic systems). It thus appears that the internal organization of the water/sodium dodecylsulfate/1-heptanol ternary diagram is quite similar to that reported in the literature for other ternary systems incorporating an anionic or cationic surfactant and a reasonably "long" normal alkanol (number of carbon atoms n_a ranging from 5 to 8) [19].

B. Water/Sodium Dodecylsulfate/1-Heptanol/n-Dodecane System

Figure 1-2 shows the evolution of the microemulsion pseudo-ternary domain configuration as a function of the sodium dodecylsulfate/1-heptanol ratio (k_m, mass ratio; k_x, molar ratio). Each value of k_m determines a certain section of the system phase tetrahedron that is represented by one of the straight lines that cut across the ternary diagram of Figure 1-1.

Whatever the value of k_m, the microemulsion pseudo-ternary domain consists of two disjoined regions that are highly asymetrical. The region N_2, which represents a certain section of the phase tetrahedron volume V_2 corresponding to the existence of "inverse" microemulsion-type media, varies in shape and extent with k_m. As k_m increases, N_2 first enlarges regularly, takes its maximum extension at k_m = 1/2.6 or so, then skrinks and eventually vanishes (no N_2 region was found at k_m = 1/1). The region N_1 corresponds to the existence of "direct" microemulsion-type media; it is always restricted to a tiny spot located in the immediate vicinity of the diagram W apex (100% water).*

It is thus confirmed, in the particular case of the water/sodium dodecylsulfate/1-heptanol/n-dodecane system, that the configuration type of the microemulsion domain of a water/ionic surfactant/alkanol/hydrocarbon system is predetermined by the configuration type of the monophosic solution domain of the corresponding water/ionic surfactant/alkanol ternary system [12-14]. If the monophasic solution domain of the ternary system consists of two disjoined areas, L_1 and L_2, the microemulsion domain of the quaternary system (hydrocarbon added) consists of two disjoined volumes, V_1 and V_2 (type S system). V_1 that is the three-dimensional extension of the "direct" solubilization area L_1, is the realm-of-existence of microemulsion-type media of the "direct" kind (hydrocarbon in water). V_2, that is the three-dimensional extension of the "inverse" solubilization area L_2.

---

*Should k_m be fairly high (k_m = 9, for instance, see Fig. 1-1), N_1 probably would be larger, as observed in the case of the analogous system water/sodium dodecylsulfate/1-hexanol/n-dodecane [14].

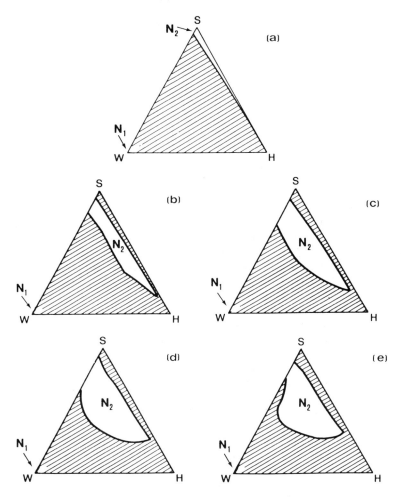

FIG. 1-2 Water/sodium dodecylsulfate/1-heptanol/n-dodecane system
(T = 25°C). Influence of the surfactant-to-cosurfactant ratio on
the configuration of the microemulsion pseudo-ternary domain (blank
areas). (a) $k_m = 0$, $k_x = 0$; (b) $k_m = 1/9$, $k_x = 1/22.34$; (c) $k_m =$
1/4, $k_x = 1/9.93$; (d) $k_m = 1/2.63$, $k_x = 1/6.54$; (e) $k_m = 1/2$,
$k_x = 1/4.96$.

is the realm-of-existence of microemulsion-type media of the "inverse" kind (water in hydrocarbon).

IV. DETECTION OF A STRUCTURAL TRANSITION IN INVERSE MICROEMULSION-TYPE MEDIA

Figure 2 shows the variations of the inverse microemulsion electrical conductivity σ, as recorded along $R_{h/s}$-type experimental paths* cutting across the N_2 region of the $k_m = 1/2.63$ pseudo-ternary phase diagram. At any value of $R_{h/s}$, when the water mass fraction p_w increases, σ first increases, then reaches a local maximum, and eventually decreases as the N_2-region lower boundary is approached. The same phenomenon was observed at the other values of k_m considered ($k_m = 1/9$, $1/4$, and $1/2$). Figure 3 shows that similar conductivity variations may be recorded when the microemulsion composition varies along p_h-type experimental paths. The conductivity maxima composition points located on p_h-type experimental paths are consistent with those determined on $R_{h/s}$-type experimental paths.

The dielectric and electroconductive behavior of type S inverse microemulsions has been analyzed in detail in previous articles [8-10, 16,17]. From this analysis, the electrical conductivity local maxima result from and consequently evidence a structural change that occurs in inverse microemulsion-type media as they are progressively enriched in water. The initial conductivity increase reflects the progressive ionization of the ionic surfactant molecules; at this stage, the "inverse" microemulsion-type media is in fact an organic suspension of aggregates of surfactant molecules to which the water molecules are bonded. Past a certain water content, proper "inverse" microdroplets may form that possess an actual inner pool of "free" water. At this stage, the electrical conductivity of the medium decreases as the water increases, because all the additional water molecules go into the inner parts of the almost electroneutral inverse microdroplets, which grow in both size and number. This interpretation is fully consistent with conclusions arrived at by other groups of authors who used other investigation methods [20-26]. The results reported by Baker et al. [26] are particularly gratifying because they show that the onset of the light-scattering increase, which reveals the formation of supramolecular objects, is well correlated with the existence of a local maximum of the electrical conductivity.

---

*$R_{h/s}$ is the mass ratio of hydrocarbon to the surfactant/cosurfactant combination, and p_h is the mass fraction of hydrocarbon [14].

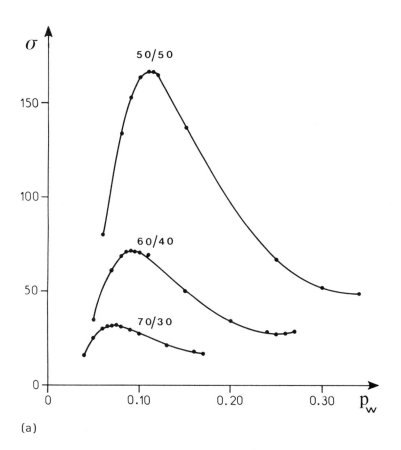

(a)

FIG. 2 Water/sodium dodecylsulfate/1-heptanol/n-dodecane inverse microemulsions (T = 25°C, k_m = 1/2.63). Variations of microemulsion electrical conductivity σ versus p_w, the water mass fraction, along $R_{h/s}$-type experimental paths. (a) σ is expressed in μSm^{-1}; (b) σ is expressed in mSm^{-1}.

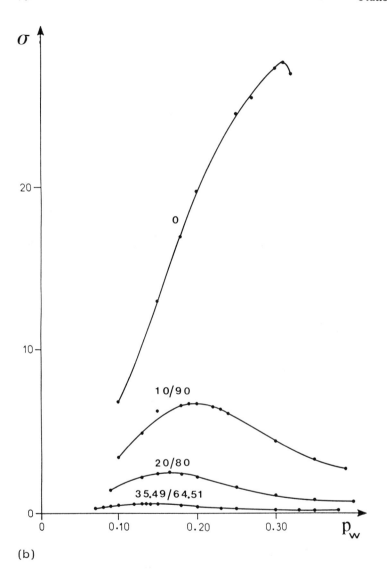

(b)

FIG. 2 (Continued)

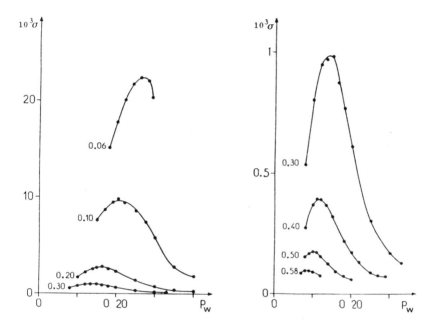

FIG. 3 Water/sodium dodecylsulfate/1-heptanol/n-dodecane inverse microemulsions (T = 25°C; k_m = 1/2). Variations of microemulsion electrical conductivity σ versus p_w, the water mass fraction, along p_h-type experimental paths. σ is expressed in mSm^{-1}.

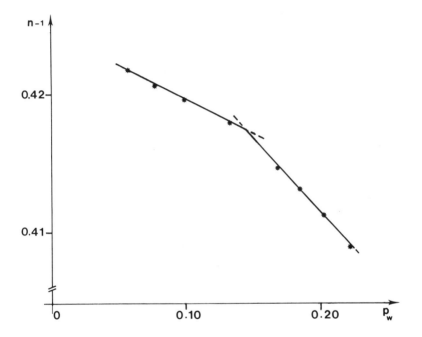

FIG. 4 Water/sodium dodecylsulfate/1-heptanol/n-dodecane inverse microemulsions, (T = 25°C; k_m = 1/2). Variations of microemulsion refractive index n versus p_w, the water mass fraction, along a $R_{h/s}$-type experimental path. $R_{h/s}$ = 50/50.

In an attempt to confirm the existence of the structural transition, as revealed by microemulsion electroconductive behavior, an investigation into microemulsion optical properties was undertaken. The variations of n, the microemulsion refractive index, were recorded versus water content along $R_{h/s}$-type experimental paths, at different values of k_m. Figure 4 shows a typical plot of n versus p_w. It is readily seen from this plot that the n versus p_w variation mode changes abruptly at a certain value of p_w. The compositions determined by the break points in the refractive index plots fit those corresponding to local maxima of microemulsion electrical conductivity, which strongly supports the alleged existence of a structural transition.

V. PARTITION OF THE REALM-OF-EXISTENCE OF INVERSE MICROEMULSION-TYPE MEDIA

On the four diagrams of Fig. 5, the circles represent the composition points corresponding to local maxima of microemulsion electrical conductivity and the stars represent the composition points corresponding to sudden changes in the mode of microemulsion refractive index variations. In any of the four cases considered, these particular composition points unambiguously define a certain line Γ_1 that divides the area N_2, that is, the pseudo-ternary realm-of-existence of inverse microemulsion-type media. On the grounds of the considerations developed in the preceding section, it may be considered that the lines Γ_1 separate regions of proper inverse microemulsions (labeled IM in the diagrams) from regions of organic solutions of hydrated surfactant aggregates (labeled SA in the diagrams). The existence of the lines Γ_1 implies the existence in the phase tetrahedron of a certain surface Σ, which divides V_2, the inverse microemulsion volume, into two subvolumes corresponding to the existence of organic solutions of hydrated surfactant aggregates (in the lower water concentration range) or of proper inverse microemulsions, that is, assemblies of microdroplets possessing an inner core of "free" water (in the upper water concentration range). This surface Σ determines a certain line Γ_0 on the "inverse" solubilization area L_2 of the ternary diagram of the water/sodium dodecylsulfate/1-heptanol system. Figure 6 shows the location of the line Γ_0 resulting from the study of the electrical conductivity (circles) and refractive index (stars) of hydrocarbonless microemulsions (see the σ versus p_w plot labeled $R_{h/s} = 0$ in Fig. 3 and the n versus p_w plot of Fig. 7). Γ_0 divides the inverse solubilization area into two subareas that appear as the realms-of-existence of heptanolic solutions of hydrated surfactant aggregates (subarea labeled SA in Fig. 6) and of solutions of proper inverse micelles in 1-heptanol (subarea labeled IM). This implies that, as already suggested by

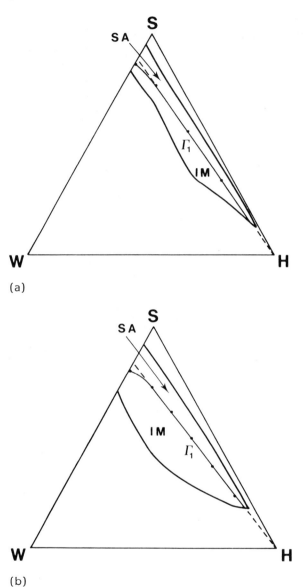

(a)

(b)

FIG. 5 Partition of the inverse microemulsion region N_2. The dividing line Γ_1 is defined by composition points at which the microemulsion electrical conductivity σ reaches a local maximum. (a) $k_m = 1/9$; (b) $k_m = 1/4$; (c) $k_m = 1/2.63$; (d) $k_m = 1/2$.

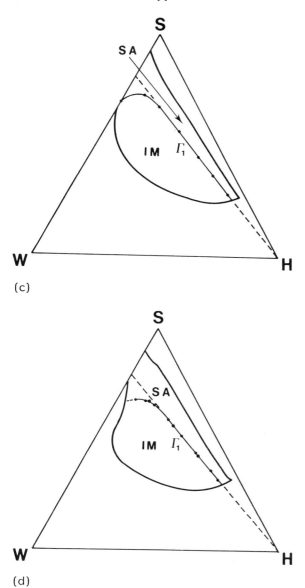

(c)

(d)

FIG. 5 (Continued)

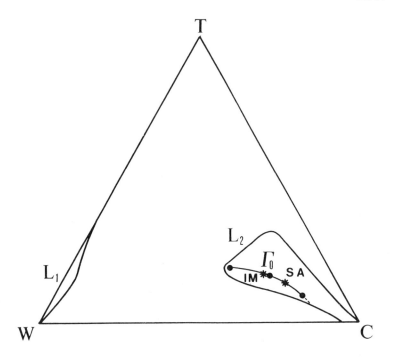

FIG. 6 Partition of the region of water/sodium dodecylsulfate/
1-heptanol monophasic inverse solutions L_2. The dividing line Γ_0
is defined by the composition points at which the electrical conductivity σ reaches a local maximum.

Friberg and co-workers [20,21], the structural state of inverse
microemulsion-type media is significantly predetermined by the structural state of the corresponding hydrocarbon-less inverse monophasic
solutions.

All of the Γ_1 lines determined show the same remarkable feature.
Their major portions are straight segments directed toward the H
apex of the pseudo-ternary phase diagram. This demonstrates that,
at a given value of the sodium dodecylsulfate/1-heptanol mass ratio
k_m, the transition from surfactant aggregates to proper inverse
microdroplets occurs at a certain constant value of water/sodium
dodecylsulfate ratio. It is only in the upper parts of the N_2 areas
(regions of composition simultaneously poor in water and hydrocarbon) that the lines Γ_1 loose their rectilinear character and exhibit
a curvature that becomes more pronounced as k_m increases. By
computing the experimental data [4], it was found that the average
values of $(x_W/x_t)_0$, the water/sodium dodecylsulfate molar ratio corresponding to the straight sections of the Γ_1 lines, are as follows:

k_m	1/9	1/4	1/2.63	1/2
$(x_w/x_t)_0$	15	12.5	12	11.5

As already mentioned, other scientists have used the light-scattering technique to detect the transition between surfactant aggregates and inverse microdroplets [21–23,25,26]. Sjöblom and and Friberg [21] and Fourche and Bellocq [22,23] observed that the scattered light intensity begins to increase rapidly when the inverse microemulsion water content exceeds a certain value corresponding to a water/ionic surfactant molar ratio or 10 or so. This is fully consistent with the data reported above, especially when k_m, the sodium/dodecylsulfate/1-heptanol mass ratio, is equal to or higher than 1/4; the higher value of $(x_w/x_t)_0$ found at $k_m = 1/9$ indicates that the formation of proper inverse microdroplets requires more water when the surface-active mixture is richer in 1-heptanol.

Since the lines Γ_1 are considered as marking the onset of the formation of proper inverse microdroplets, it may be assumed that

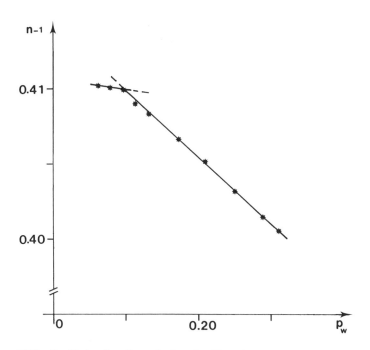

FIG. 7 Water/sodium dodecylsulfate/1-heptanol monophasic inverse solutions (T = 25°C). Variations of solution refractive index n versus p_w, the water mass fraction, along the experimental path defined by $k_m = 1/5.66$.

all of the sodium dodecylsulfate molecules are inserted in the interfacial films that separate the inner water core from the continuous
organic medium. On this ground, it is possible to evaluate $(r_w)_0$,
that is, the average radius of the aquous core of the inverse microdroplets that just form when the water/sodium dodecylsulfate ratio
becomes equal to $(x_w/x_t)_0$. The values of $(r_w)_0$ can be computed
by using the equation

$$(r_w)_0 = \frac{3M_w(x_w/x_t)_0}{N\rho_w\sigma_0} \tag{1}$$

where M_w is the molar mass of water, ρ_w is its specific mass, N is
Avogadro's constant, and σ_0 is the part of the interfacial film area
that corresponds to one molecule of ionic surfactant. σ_0, which is
the only a priori unknown parameter in the above equation, may be
reasonably evaluated to 60 A^2 by taking into account literature data
on the structure of interfacial films incorporating sodium dodecylsulfate and 1-heptanol [27–31]. The $(r_w)_0$ values thus obtained are
as follows:

k_m	1/9	1/4	1/2.63	1/2
$(x_w/x_t)_0$	15	12.5	12	11.5
$(r_w)_0$ A	33	21	22	20

These values are reasonable ones and compare very satisfactorily
with data reported in the literature by other authors [32,33]. The
$(r_w)_0$ values corresponding to the bent sections of the Γ_1 lines are
higher and may rise to 40 A or so when the hydrocarbon content
vanishes [4].

VI. CONCLUSION

It appears from the preceding that the inverse microemulsion-type
media of the so-called type S system [14] exist either as organic
solutions of hydrated surfactant aggregates or as proper microemulsions, that is, assemblies of inverse microdroplets suspended in a
continuous organic phase. The transition between these two main
structural states occurs at certain threshold values of the water/
ionic surfactant molar ratio that define, within the realm-of-existence
V_2 of inverse microemulsion-type media, an internal boundary Σ.
This surface Σ may be determined experimentally by the composition
points at which are observed local maxima of microemulsion electrical
conductivity and sudden changes in the variation mode of microemulsion refractive index. The main feature of Σ is its partially conical

shape, as revealed by the fact that, whatever the value of the ionic surfactant/cosurfactant ratio, the line Γ_1 that, in a WHS pseudoternary diagram, marks the structural transition presents a straight section that is always directed toward the H apex of the diagram. This means that, when the hydrocarbon content is sufficiently high (p_h greater than 0.1 or so), the water/ionic surfactant molar ratio $(x_w/x_t)_0$ at which the structural transition occurs is almost composition independent (the ionic surfactant/cosurfactant ratio being given). $(x_w/x_t)_0$, which is of the order of 12, increases slightly as the ionic surfactant/cosurfactant ratio decreases.

These results confirm and generalize previous data obtained by Heil and al. [1-3, 8-11] in some limited cases (the influence of the ionic surfactant/cosurfactant ratio was not investigated). They are consistent with theoretical predictions derived from the microemulsion geometric model of Biais et al. [29,30], and with experimental results reported by other authors as to the structural state of inverse microemulsion-type media of similar constitution [20-26]. It should also be mentioned that they fit in with findings concerning three-component inverse microemulsion-type media incorporating Aerosol OT as the only surface active compound [1,2,34-38]. The values of the water/AOT molar ratio at which the transition from hydrated surfactant aggregates to proper inverse microdroplets was found to occur, (as detected by different methods) lie in the range 8 to 12 and consequently are consistent with the values of $(x_w/x_t)_0$ reported above.

The existence of two structural states of the inverse microemulsion-type media of so-called type S systems [14] is a fact of more than academic interest. It has practical implications too. This can be well illustrated by considering the case of an inverse microemulsion-type fluid intended to serve as vehicle of some chemically or biologically active substance (a dye, a drug, an enzyme, etc.). The inverse microemulsion capacity of solubilizing the said substance (and, very often, the actual activity potential of the substance) may vary enormously according as the water molecules are bonded to ionic surfactant aggregates or are gathered to form local pools of "free" water. Thus, at least a qualitative knowledge of the physical state of the water incorporated in the inverse microemulsion is required to formulate the contemplated specialty correctly. As demonstrated in this chapter, the qualitative determination of the inverse microemulsion structural state can be achieved by using simple methods such as electrical conductimetry and refractometry.

VII. ACKNOWLEDGMENTS

J. Heil is grateful to the DGRST (France) for the research scholarship granted to thim from October 1979 to October 1981. A. Zradba wishes to thank the Ministère des Relations Extérieures (France) for the financial help provided to him from October 1980 to July 1983. Professor M. Clausse expresses to NATO Scientific Affairs Division (Belgium) his deep appreciation for the support granted by this institution (contract no. 020.82). The authors wish to express to Mrs. A. M. Schmitt their sincere appreciation for her meticulous typing of this chapter.

VIII. REFERENCES

1. Heil, J., Thèse de Doctorat de 3ème Cycle, Université de Pau, France, 1981.
2. Heil, J., Clausse, M., Peyrelasse, J., and Boned, C., *Colloid Polymer Sci.* 260, 93 (1982).
3. Clausse, M., Heil, J., Peyrelasse, J., and Boned, C., *J. Colloid Interface Sci.* 87, 584 (1982).
4. Zradba, A., Thèse de Doctorat de 3ème Cycle, Université de Pau, France, 1983.
5. Zradba, A., and Clausse, M., *Colloid Polymer Sci.* 262, 754 (1984).
6. Nicolas-Morgantini, L., Thèse de Doctorat de 3ème Cycle, Université de Pau, France, 1984.
7. Nicolas-Morgantini, L., Clausse, M., and Poquet, E., *Colloid Polymer Sci.*, 264 (1986).
8. Clausse, M., Report to DGRST (Action concertée "Récupération Assistée du Pétrole"), No. 80-D0850, September 1982.
9. Clausse, M., in *Encylopedia of Emulsion Technology*, Becher, P. (ed.), Marcel Dekker, New York, 1983, vol. I, pp. 481–715.
10. Clausse, M., Boned, C., Peyrelasse, J., Lagourette, B., McClean, V. E. R., and Sheppard, R. J., in *Surface Phenomena in Enhanced Oil Recovery*, Shah, D. O. (ed.), Plenum Press, New York, 1981, pp. 199–228.
11. Clausse, M., Peyrelasse, J., Boned, C., Heil, J., Nicolas-Morgantini, L., and Zradba, A., in *Surfactants in Solution*, Mittal, K. L., and Lindman, B. (eds.), Plenum Press, New York, 1984, vol. III, pp. 1583–1626.
12. Clausse, M., Zradba, A., and Nicolas-Morgantini, L., *Colloid Polymer Sci.* 263, 767 (1985).
13. Clausse, M., Heil, J., Zradba, A., and Nicolas-Morgantini, L., *J. Com. Esp. Deterg.* 16, 497 (1985).
14. Clausse, M., Nicolas-Morgantini, L., Zradba, A., and Touraud, D., *this volume.*
15. Clausse, M., Zradba, A., and Nicolas-Morgantini, L., *this volume.*
16. Boned, C., Clausse, M., Lagourette, B., Peyrelasse, J., McClean, V. E. R., and Sheppard, R. J., *J. Phys. Chem.* 84, 1520 (1980).

17. Boned, C., Peyrelasse, J., Heil, J., Zradba, A., and Clausse, M., *J. Colloid Interface Sci. 88*, 602 (1982).
18. Clausse, M., Nicolas-Morgantini, L., and Zradba, A., in *Proceedings of the World Surfactants Congress*, Munich, May 6–10, 1984, Kürle Druck, Gelnhausen, West Germany, 1984, vol. III, pp. 209–219.
19. Ekwall, P., Mandell, L., and Fontell, K., *Molecular Crystals and Liquid Crystals 8*, 157 (1969).
20. Friberg, S., and Buraczewska, I., *Prog. Colloid Polymer Sci. 63*, 1 (1978).
21. Sjöblom, E., and Friberg, S., *J. Colloid Interface Sci. 67*, 16 (1978).
22. Fourche, G., and Bellocq, A. M., *C. R. Acad. Sci. Paris 289*, Sér. B, 261 (1979).
23. Bellocq, A. M., and Fourche, G., *J. Colloid Interface Sci. 78*, 275 (1980).
24. Danielsson, I., Hakala, M. R., and Jorpes-Friman, M., in *Solution Chemistry of Surfactants*, Mittal, K. L. (ed.), Plenum Press, New York, 1979, vol. II, pp. 659–671.
25. Baker, R. C., Florence, A. T., Tadros, Th. F., and Wood, R. M., *J. Colloid Interface Sci. 100*, 31 (1984).
26. Baker, R. C., Florence, A. T., Ottewill, R. H., and Tadros, Th. F., *J. Colloid Interface Sci. 100*, 332 (1984).
27. Pethica, B. A., and Few, A., *Disc. Faraday Soc. 18*, 258 (1954).
28. Bowcott, J. E., and Schulman, J. H., *Z. Elektrochem. 59*, 283 (1955).
29. Biais, J., Bothorel, P., Clin, B. and Lalanne, P., *J. Colloid Interface Sci. 80*, 136 (1981).
30. Biais, J., Bothorel, P., Clin, B., and Lalanne, P., *J. Dispersion Sci. Technol. 2*, 67 (1981).
31. Brunetti, S., Roux, D., Bellocq, A. M., Fourche, G., and Bothorel, P., *J. Phys. Chem. 87*, 1028 (1983).
32. Dvolaitzky, M., Guyot, M., Lagues, M., Le Pesant, J. P., Ober, R., Sauterey, C., and Taupin, C., *J. Chem. Phys. 69*, 3279 (1978).
33. Cazabat, A. M., Chatenay, D., Langevin, D., and Pouchelon, A., *J. Phys. Lett. (Paris) 41*, L-441 (1980).
34. Eicke, H. F., Kubik, R., and Hammerich, H., *J. Colloid Interface Sci. 90*, 27 (1982).
35. Eicke, H. F., in *Microemulsions*, Robb, I. D. (ed.), Plenum Press, New York, 1982, pp. 17–32.
36. Rouvière, J., Couret, J. M., Lindheimer, M., Dejardin, J. L., and Marrony, R., *J. Chim. Phys.-Phys. Chim. Biol. 76*, 289 (1979).
37. Wong, M., Grätzel, M., and Thomas, J. K., *J. Am. Chem. Soc. 98*, 239 (1976).
38. Bakale, G., Beck, G., and Thomas, J. K., *J. Phys. Chem. 85*, 1062 (1981).

4

Comparison of Emulsions with the Phase Diagrams of the Systems in Which They Form

DUANE H. SMITH* *Phillips Petroleum Company, Bartlesville, Oklahoma*

I. INTRODUCTION

Examination of the phase diagrams of common surfactant/oil/water systems that form three coexisting liquids shows that, theoretically, there are six kinds of emulsions possible, not two as the common "oil-in-water" and "water-in-oil" terminology implies. Two of the six types of emulsions, aqueous phase-in-microemulsion and microemulsion-in-aqueous phase, are theoretically possible in the LCT binodal, which occurs on one of the high-surfactant concentration sides of the three-phase tie-triangle. Two other types, oleic phase-in-microemulsion and microemulsion-in-oleic phase, are theoretically possible for compositions within the UCT binodal, which adjoins the other high-surfactant concentration side of the tie-triangle. (The LCT and UCT binodals are those that terminate at the lower critical end-point temperature and upper critical end-point temperature, respectively.) Oleic phase-in-aqueous phase and aqueous phase-in-oleic phase emulsions form at low surfactant concentrations, on the third side of the tie-triangle at any temperature between the two critical end-point temperatures.

Oleic phase-in-aqueous phase and aqueous phase-in-oleic phase dispersions are often encountered in the literature as "oil-in-water" and "water-in-oil" emulsions. Examination of the literature reveals that the aqueous phase-in-microemulsion and oleic phase-in-microemulsion dispersion types have also been very commonly reported as "water-in-oil" and "oil-in-water" emulsions, respectively. However, only a single study has been found for which the phase diagrams can be constructed and in which it is possible that the other two of the six emulsion types were observed. Thus, further experiments are

Present affiliation: Morgantown Energy Technology Center, Morgantown, West Virginia.

needed to determine whether microemulsion-in-aqueous phase and microemulsion-in-oleic phase emulsions commonly form in the water and oil corners of the LCT and UCT binodals, respectively, or if they are only hypothetical emulsion types.

In 1949 Griffin introduced the hydrophilic-lipophilic balance (HLB) method for choosing a surfactant or mixture of surfactants to prepare an emulsion of the desired type ("oil-in-water," O/W; or "water-in-oil," W/O) [1a]. In this method an HLB number, which depends only on the structure of the surfactant molecule, is assigned to each surfactant. For example, for reaction products of ethylene oxide with linear alcohols, the HLB is often calculated from the equation

$$HLB = \frac{E}{5} \tag{1}$$

where E is the weight percent of oxyethylene in the surfactant molecule [1b]. A scale relates HLB numbers to the applications for which they are suitable [1].

One great advantage of HLB numbers was thought to be their additivity. Thus, it was thought, the HLB of a mixture would be given by

$$HLB = \sum_i X_i (HLB)_i \tag{2}$$

where X_i is the weight fraction of compound or mixture i of hydrophilic-lipophilic balance $(HLB)_i$ [1a].

Equation (1) implicitly suggests the idea of group contributions to HLB. In 1957, Davies introduced a definition of HLB that combined the additivity of Eq. (2) with the group contribution concept of Eq. (1) [2]. His definition can be written as

$$HLB = 7 + \sum_i H_i \tag{3}$$

where H_i is the HLB group number of group i. Davies was apparently the first to give an equation for calculating the HLB of an anionic surfactant.

However, despite the widespread use to this day of the Griffin method, it was apparent almost from the beginning that the method could not account for many effects and thus often made incorrect predictions [2-6]. Griffin realized that the HLB required to form an O/W or W/O emulsion depended on the oil, as well as on the surfactant structure [1]. Moreover, various workers showed that an emulsion could often be made to invert by changes of oil/water ratio, surfactant concentration, or temperature [2-5]. Attempts to rationalize these effects have not been fully satisfactory [7]. Thus, it was apparent when the present work was begun that achievement of the goals of the HLB method would require development of a different method based on a new theoretical understanding of the principles involved.

Griffin's original HLB numbers were based on a rather complicated experimental protocol, and it was these numbers that his equations attempted to codify [1]. This protocol was developed before any complete oil/water/surfactant-temperature phase diagram was reported and thus without benefit of many of the phase principles that are involved [8].

The subsequent work of Shinoda and co-workers on the phase inversion temperature (PIT) concept addressed the role of temperature in "phase"—that is, emulsion—inversion [5,9–23]. These measurements revealed relationships between emulsion types (O/W or W/O) and arbitrary cross sections of the full temperature-composition phase boundary surfaces. However, despite the considerable value of this work, only a fraction of its full potential has been realized, because the data were not plotted as true phase diagrams.*

Recent reviews and books on emulsions do not discuss any comparisons of emulsion behavior to temperature-composition trigonal prismatic diagrams [7,24–28]. Only recently have such comparisons been made in the literature [29].

In this chapter we describe the trigonal prismatic diagram of a typical oil/water/surfactant system that forms three coexisting liquid phases and compare it with the observed emulsion behavior of the system. The comparison reveals several simple relationships between the emulsions and the miscibility gaps of the phase diagram. Examination of the literature on emulsions reveals that these relationships are apparently very general and valid for all liquid-liquid emulsions in surfactant/oil/water systems.

Briefly, the prism contains two binodal surfaces and thus two kinds of tie lines and conjugate phase pairs. These surfaces are called the lower critical end-point temperature (LCT) binodal and upper critical end-point temperature (UCT) binodal, respectively [29]. The various pairs of phases whose compositions fall on the LCT and UCT binodals are called the LCT pair of phases and the UCT pair of phases, respectively. These two two-phase regions border the tie-triangle on two of its sides, but also extend to lower and higher temperatures at which the tie-triangle does not exist.

─────────────

*A PIT plot of temperature versus composition is a true phase diagram only for a two-component system, not for a three-component surfactant/oil/water system.

In addition, there is a third two-phase region that borders the tie-triangle on its low-surfactant concentration side.

Every system composition and temperature that forms two phases can also form an emulsion. Two kinds of emulsions (W/O and O/W) and three kinds of pairs of phases are known. For every temperature and composition that forms a pair of phases and thus an emulsion, we compare the kind of emulsion that forms, as reported by Shinoda and other authors, with the kind of phase pair that forms the emulsion, as shown by the phase diagram of Sec. III.

This comparison (Sec. IV) yields a simple empirical correlation: LCT phases always form W/O emulsions and never the O/W type; UCT phases always form O/W emulsions and never the W/O type. Only a single possible exception to this rule has been found in the literature [16].

Furthermore, Shinoda and co-workers studied a great variety of other oils and surfactants and prepared their emulsions in different ways, always reporting emulsion types that are consistent with this correlation [9-23]. This variety suggests that the correlation is a fundamental and far-reaching property of systems that contain surfactant, oil, and water, and may even obtain for systems that also contain salts and/or cosurfactants.

This general correlation, combined with the known behavior of binodal surfaces and critical lines, allows a simple understanding of why the emulsion type depends on oil and surfactant structure, oil/water ratio, surfactant concentration, and temperature. This understanding in turn allows us to correct interpretational errors in the literature and to establish the foundation for an HLB system based on principles that are correct.

The correlations, simplifications, corrections, and deductions that are made possible by the comparison of phase behavior and emulsion types demonstrate the premise that comparison of emulsion behavior to correctly plotted phase diagrams is a prerequisite for converting the use of emulsions and the solution of emulsion problems from an art to sound engineering practice based on scientific principles.

II. EXPERIMENTAL AND PHASE DIAGRAM CONSTRUCTION PROCEDURES

Although the empirical correlations described in the next sections appear to be very general, they were first discovered for the surfactant NPE-9.2 with water and oil [29]. Details of the experiments and diagram construction for these systems have been reported previously [29].

The surfactant $i\text{-}C_9H_{19}C_6H_4(OCH_2CH_2)_{9.2}OH$ (NPE-9.2) is a common commercial-type polyoligomeric, ethoxylated nonylphenol with an average of 9.2 OCH_2CH_2 groups. Details about the structure and composition of this type of surfactant have been given elsewhere [30,31].

III. PHASE BEHAVIOR AND CRITICAL END-POINT TEMPERATURES

The conventional nomenclature of emulsions and the HLB system of Griffin were developed before the phase behavior of oil/water/surfactant systems was known in more than rudimentary form [1,8]. Thus, the conventional nomenclature lacks terms needed to describe emulsion behavior accurately. For this same reason, the HLB method of Griffin cannot account for important effects; and it often makes wrong predictions because these effects are ignored. These inadequacies can be corrected by consideration of the fact that many amphiphile/oil/water systems can form three coexisting liquid phases.

Thus, the oil/water/surfactant-temperature phase diagram provides the first step for constructing an HLB method based on the true phase and physical behavior of these systems.

In this section, examination of the phase diagram establishes the basis for showing (in Sec. IV) that, theoretically, there are not two kinds of emulsions (W/O and O/W), but six. Comparisons of emulsion types with phase behavior in Sec. IV allows us to infer which of these hypothetical types actually form, and in which two-phase regions of the diagram they are observed. In particular, the temperature-composition regions over which "oil-in-water" and "water-in-oil" emulsions have been reported to form are compared with the phase diagram below, between, and above the two critical end-point temperatures.

Figure 1 shows an isothermal cross section of a simplified (i.e., without liquid crystal) model phase diagram at a temperature between the critical end-point temperatures. This model diagram is characteristic of oil/water/surfactant systems at temperatures between their lower (T_{lc}) and upper T_{uc}) critical end-point temperatures [29, 32-35].

The phase and emulsion behavior below and above the critical end-point temperatures are continuous with the behavior between T_{lc} and T_{uc}, respectively, except for simplifications introduced by the disappearance of one or the other of the binodal curves and emulsion types at a critical end point (cf. Fig. 2).

At the temperature of Fig. 1, the system has one single-phase region, three two-phase regions, and one three-phase region. Compositions within the single-phase region (or most of it) are commonly called microemulsions, and microemulsion compositions along the binodals are sometimes termed limiting microemulsions

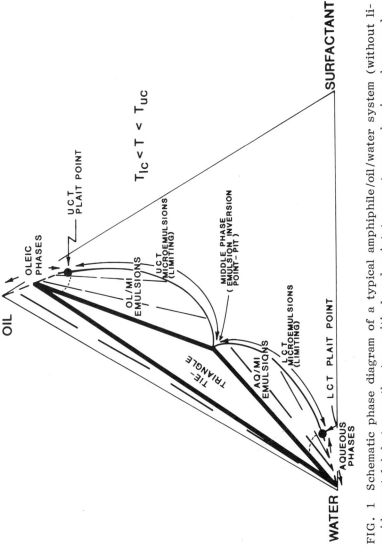

FIG. 1 Schematic phase diagram of a typical amphiphile/oil/water system (without liquid crystals) between the two critical end point temperatures, showing phase and
critical-point nomenclature, and showing regions in which AQ/OL, OL/AQ, AQ/MI,
and OL/MI emulsions form; dotted lines are approximate boundaries at which AQ/MI
↔ MI/AQ and OL/MI ↔ MI/OL inversions may occur.

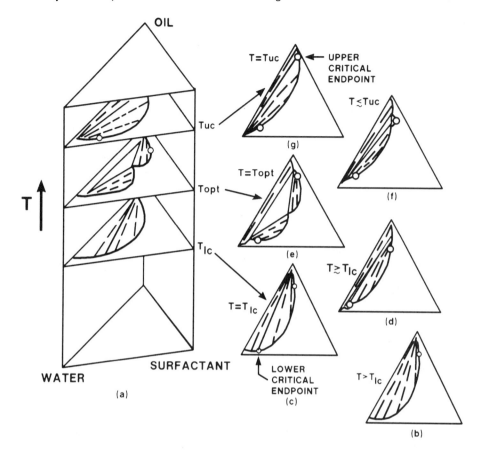

FIG. 2 Temperature-composition prismatic diagram and illustrative isothermal cross sections for the system of Fig. 1.

[36,37].* In this terminology there are two kinds of limiting micro-emulsions, one for each binodal.

For convenience we call the binodal surface with the lower critical end-point temperature the LCT binodal and the binodal surface with the upper critical end-point temperature the UCT

---

*The aqueous and oleic phases are obvious exceptions.

binodal. Conjugate (i.e., coexisting) pairs of phases on the two
binodals are called LCT and UCT phase pairs, respectively.*
   The corner of the tie-triangle where the two binodals meet is
called the middle phase (because it usually has the intermediate density of the three coexisting phases). The other two phases of the
tie-triangle are called the aqueous and oleic phases, because they
are composed largely of water and oil, respectively. Compositions
on the UCT binodal between its line of plait points and the termina-
tion of the binodal on the oil/water side of the diagram may be
called oleic phases, whereas compositions on the LCT binodal between its line of plait points and the oil/water side of the diagram
are termed aqueous phases. This latter terminology for aqueous
and oleic phases is consistent with the "aqueous" and "oleic" terminology for phases of the tie-triangle, with the fact that the corners of the tie-triangle fall on the binodals at surfactant concentrations less than the surfactant concentrations at the plait points,
and with the fact that different names are needed for the two coexisting phases whose compositions fall on opposite sides of a binodal plait point.
   The conjugate phases in the third two-phase region are also
called aqueous and oleic phases, respectively. Although these
phases are often called "water" and "oil" (as in O/W emulsion), the
aqueous and oleic terminology is preferred because it avoids confu-
sion between components (e.g., oil) and phases (e.g., oleic).
   When the lower critical end-point temperature is reached from
higher temperatures, the tie-triangle collapses to a tie-line, and the
aqueous, or LCT, plait line and its binodal disappear (Fig. 2c).
Thus, the division of the compositions on this binodal surface into
aqueous phases and microemulsions also disappears. The limiting
tie-line at which this happens is the lower critical tie-line. Furthermore, the division of limiting microemulsions into LCT and UCT
types by the middle-phase corner of the tie-triangle likewise disappears. Only UCT limiting microemulsions and oleic phases are
left. Figure 2d shows the phase diagram at a temperature slightly

---

*It is possible that each binodal surface has both a lower (T_{lc}) and
an upper (T_{uc}) critical end-point temperature. If so, the T_{lc} of
the UCT binodal surface is less than 0°C and is unobservable, and
the T_{uc} of the LCT binodal is well above 100°C. It seems more likely that the middle phase is produced by the intersection of just two
closed, three-dimensional miscibility-gap surfaces with a single T_{lc}
and a single T_{uc} [cf. Fig. 10 of Ref. 38]. In the literature, UCT
microemulsions are often called "oil-in-water" microemulsions and LCT
microemulsions are often called "water-in-oil" microemulsions. This
terminology is to be avoided because it is incomplete, confusing and
often wrong.

above the lower critical end-point temperature, where the composition of the middle phase is close to the composition of the aqueous phase. In Fig. 2b, the two phases have merged to one ($T < T_{lc}$).

Similarly, when the upper critical end-point temperature is reached from lower temperatures (Fig. 2g), the tie-triangle collapses to a tie-line; the oleic, or UCT, plait line terminates; and its binodal disappears. The division of the compositions on this binodal surface into oleic phases and microemulsions and the division of limiting microemulsions into two types by the middle-phase corner of the tie-triangle likewise disappear. Only LCT conjugate phase pairs remain above the upper critical end-point temperature.

IV. COMPARISON OF EMULSION TYPES WITH PHASE BEHAVIOR

It is generally accepted that emulsions are not really dispersions of components (as implied by the oil/water and water/oil terminology) but, to a much closer approximation, dispersions of the two phases into which the emulsions separate at equilibrium [24–27].

If emulsions are dispersions of phases, then, theoretically, there are four "new" types of liquid-liquid emulsions. This follows from the fact that there are two binodals and two choices of dispersed phase for each binodal (Fig. 1). For the LCT binodal the two emulsion types are aqueous phase-in-LCT microemulsion and LCT microemulsion-in-aqueous phase. For the UCT binodal the two emulsion types are oleic phase-in-UCT microemulsion and UCT microemulsion-in-oleic phase.

There is no provision for these "new" types of emulsions in the Griffin-Davies HLB system, which implicitly assumes only a single two-phase region in the phase diagram.

In this context it is clear that microemulsions should *not* be described as "thermodynamically stable emulsions." A microemulsion is not a type of emulsion. Instead, a limiting microemulsion is one of the phases of a conjugate pair of phases that together form emulsions.

Phrases such as "AQueous phase-in-lower critical end-point temperature limiting MIcroemulsion" emulsion are accurate, but clumsy and confusing. It is convenient and accurate to call them "AQ/MI" emulsions. Likewise, we may call the other emulsion type of the LCT binodal (in which the aqueous phase is the continuous phase) "MI/AQ" emulsions. Similarly, UCT-binodal emulsions in which the oleic phase is the dispersed phase are called "OL/MI" emulsions, and emulsions in which the oleic phase is the continuous phase are "MI/OL" emulsions (cf. Fig. 1). In this terminology, two-phase systems on the low-surfactant concentration side of the tie-triangle form OL/AQ or AQ/OL emulsions.

At any temperature between the two critical end-point temperatures, both binodals exist in the phase diagram. A tie-triangle splits the single two-phase region that exists at temperatures outside the range into three two-phase regions. Since both binodals occur at temperatures between T_{lc} and T_{uc}, in this temperature range all four of the emulsion types AQ/MI, MI/AQ, OL/MI, and MI/OL are theoretically possible within the diagram. On the third side of the tie-triangle, OL/AQ and AQ/OL emulsions form. Therefore, between the critical end-point temperatures there are three two-phase regions, two types of emulsion possible for each region, and six emulsion types theoretically possible in all.

As described in the previous section, there is a lower critical tie-line (Fig. 2c), which is the lower (in temperature) limit of the three-phase region. At and below the temperature of this tie-line, the system has only a single two-phase region and thus only two types of emulsion are possible. These two types are OL/MI and MI/OL emulsions, since it is the UCT binodal that persists below the lower critical end-point temperature. Similarly, there is an upper critical tie-line, which is the upper (in temperature) limit of the three-phase volume in the prismatic diagram (Fig. 2g). Above the upper critical end-point temperature, only the LCT binodal and its theoretical emulsion types AQ/MI and MI/AQ remain.

In summary, below T_{lc} only the UCT binodal surface with its oleic and microemulsion phases exist; therefore, only OL/MI and MI/OL emulsions are possible. Above T_{uc} only the LCT binodal surface with its aqueous and microemulsion phases exist; therefore, only AQ/MI and MI/AQ emulsions are possible. At temperatures between T_{lc} and T_{uc} three two-phase regions occur, and six different liquid-liquid dispersion types are theoretically possible. These facts are listed in Table 1.

The emulsion types AQ/OL and OL/AQ are frequently encountered. The interfacial tension between aqueous and oleic phases passes through a minimum at the temperature T_{opt} (also called the HLB temperature), which is approximately midway between the two critical end-point temperatures [29,39]:

$$T_{opt} \simeq \frac{T_{lc} + T_{uc}}{2} \tag{4}$$

Below T_{opt}, OL/AQ emulsions are more stable than the AQ/OL type; and above T_{opt}, AQ/OL dispersions are the more stable type. However, the free energy differences are very small, and it is often possible to prepare either type at a given temperature and concentration.

The interfacial tension decreases rapidly as tricriticality is approached (i.e., as $T_{uc} - T_{lc}$ approaches zero). Hence, it is likely

TABLE 1 Relationship of Emulsion Types to Critical End-Point Temperatures and Middle-Phase Composition

Temperature	Emulsion type(s)
Surfactant side of tie-triangle	
$T < T_{lc}$	OL/MI, only
$T_{lc} < T < T_{uc}$	OL/MI, oil/water ratio of system > oil/water ratio of middle phase[a]
	AQ/MI, oil/water ratio of system < oil/water ratio of middle phase[a]
$T > T_{uc}$	AQ/MI, only
Oil/Water side of tie-triangle [39]	
$T < \sim(T_{lc} + T_{uc})/2$	OL/MI
$T > \sim(T_{lc} + T_{uc})/2$	AQ/MI

[a]Oil/water ratio of middle phase is a function of T.

that the ease of inversion between the two types increases as the difference between T_{uc} and T_{lc} decreases, although no systematic study of this effect has been found in the literature.

Shinoda and others have published many diagrams of the type shown in Fig. 3 [5,9–15,17–23,39]. These are vertical temperature-composition cross sections of the prismatic diagram shown in Fig. 2. The surfactant concentrations in these vertical cross sections are constant, or nearly constant, and greater than the surfactant concentrations of the lower and upper critical tie-lines. Hence, the vertical cross section cuts through the volume of three phases in the prismatic diagram. Furthermore, a constant-temperature line in Fig. 3 cuts across different tie-lines in the isothermal diagram of that temperature [16]. At temperatures below the three-phase region, the vertical cross section interacts the UCT two-phase region; at temperatures above the three-phase "lens," Fig. 3 interacts the LCT binodal.

In the terminology of this chapter, the emulsions in the upper portion of Fig. 3 must be the AQ/MI or MI/AQ type, since the plane of Fig. 3 intersects the LCT binodal at these temperatures. Emulsions in the lower portion of Fig. 3 must be either the OL/MI or MI/OL type, since Fig. 3 intersects the UCT binodal at these temperatures.

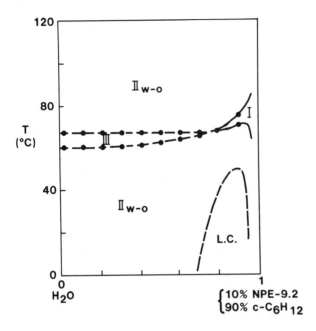

FIG. 3 Typical vertical temperature-composition cross section of the prismatic diagram of an surfactant/oil/water system, showing only "W/O" (AQ/MI) and "O/W" (OL/MI) emulsions (system NPE-9.2/cyclohexane/water). (Redrawn after Ref. 5.)

The designation "O/W" of Fig. 3 is equivalent to OL/MI, and the designation "W/O" is equivalent to AQ/MI. There is no indication in Fig. 3 that MI/OL or MI/AQ emulsions ever formed. Hence, according to Fig. 3, (1) LCT conjugate phases always formed OL/MI emulsions; (2) UCT phases always formed AQ/MI dispersions; and (3) MI/AQ and MI/OL emulsions never formed. Moreover, diagrams similar to Fig. 3 have been reported for many different systems [5, 9-15,17-23,29]. The ubiquity of these reports implies that these empirical correlations are very general, although it was pointed out in a previous paper that MI/AQ and MI/OL emulsions might not have been recognized in these many papers simply because the theoretical possibility of their existence was unknown [29].

However, in an unusual paper, six types of two-phase emulsions were reported for the NPE-9.7/cyclohexane/water system: "W/O," "W/D(O)," "D/W"; and "O/W, "O/D(W)," and "D/O" [16]. (Here D stands for "surfactant phase" in the original paper, i.e., limiting microemulsion in this chapter.) One of the diagrams from this paper is shown in Fig. 4. Partial phase diagrams can be constructed from these temperature-composition cross sections and compared to the reported dispersion types, in an attempt to determine if all six of the theoretical dispersion types were in fact detected in the study of the NPE-9.7/cyclohexane/water system. The comparison yields the following conclusions.

Figure 4 shows a boundary (dotted line) between W/O and W/D(O) emulsions for compositions in the two-phase region at temperatures above the upper critical endpoint temperature (cf. Fig. 2g). This dotted line in Fig. 4 is in a plane of constant surfactant concentration, which cuts across different tie-lines in Fig. 2g. It can be seen from Fig. 2g that the W/O and W/D(O) emulsion types must really be the same. The change of notation between "O" and "D(O)" merely indicates that for the conjugate phase of the greater surfactant concentration (i.e., LCT microemulsion), the concentration of surfactant increases considerably along paths that intersect tie-lines that are increasingly further from the oil/water side of the diagram. As shown by Fig. 2g, this compositional change is continuous, and there is no tie-line at which the surfactant concentration

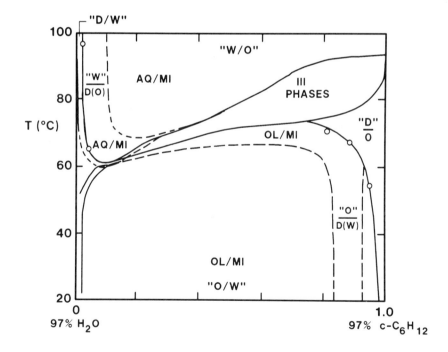

FIG. 4 Temperature-composition cross section of the NPE-9.7/cyclohexane/water prismatic diagram, showing various reported emulsion types. (Redrawn after Ref. 16).

in the microemulsion suddenly increases. In fact, the LCT plait
point is so close to the water corner of the diagram that near the
plait point both of the phases of an emulsion are largely water;
hence it is misleading to refer to this microemulsion phase as either
"O" or "D". Both "W/O" and "W/D(O)" dispersions are AQ/MI
emulsions.

Similarly, in Fig. 4 the designated change from "O/W" to
"O/D(W)" emulsion (lower dotted line) occurred at temperatures
below the lower critical end-point temperature. At these temperatures only the UCT binodal exists. As one goes from the water
side to the oil side of Fig. 4 (a path that cuts across tielines),
the surfactant concentration in the microemulsion phase increases
(Fig. 2c). However, there is no change from "W" to "D(W)" phase
along the UCT microemulsion portion of the binodal line (cf. Figs.
1 and 2c). Hence no change "O/W" to "O/D(W)" of emulsion type
is possible. Both designations are OL/MI emulsions.

In Fig. 4, D/W dispersions correspond to MI/AQ emulsions, and
W/D dispersions correspond to MI/OL emulsions.

Thus, it is seen that the six apparent emulsion types were at
most four (AQ/MI, MI/AQ, OL/MI, and MI/OL), in agreement with
the maximum number theoretically possible on the two higher-
surfactant concentration sides of the three-phase region.

Emulsion types in the NPE-9.7/cyclohexane/water system were
assigned on the basis of volumes and visual observations [16]. And
we now understand that two of the apparent emulsion types were
not really different. Thus, there is some uncertainty about whether AQ/MI ↔ MI/AQ and OL/MI ↔ MI/OL inversions occurred in
the NPE-9.7 system. However, the intriguing possibility remains
that all four of the AQ/MI, MI/AQ, OL/MI, and MI/OL emulsion
types were observed in the study on the NPE-9.7/cyclohexane/water
system [16]. If all four types were observed, emulsion inversion
within the temperature-composition volume of a single binodal can
occur and is almost certainly a general property of surfactant/oil/
water systems. In that case the many diagrams that report only O/W
and W/O emulsions (e.g., Fig. 3) are incomplete [29]. However, as
shown by Fig. 4, the possible occurrence of MI/AQ and MI/OL emulsions is limited to small ranges of composition close to the LCT and
UCT plait lines, respectively.

Regardless of whether MI/AQ and MI/OL emulsions form, direct inversion between OL/MI and AQ/MI emulsions without passage through
a finite intermediate single-phase region or three-phase region is possible only at the middle phase. The middle phase is the cusp where
the two binodals meet. At the cusp, two tie-lines (the aqueous-middle
phase and oleic-middle phase sides of the tie-triangle) have a common
end, the middle phase. The middle-phase line is the only place where
the LCT and UCT microemulsions become identical (Fig. 1).

The two binodals meet along a curved line that begins at the lower critical end-point temperature and composition and ends at the upper critical end point. "Phase" (i.e., emulsion) inversion occurs on any system path that crosses this path traced by the middle phase; the inversion occurs at the point where these two paths intersect. The phrase "phase inversion tempeature" (PIT) has meaning for the AQ/MI $\leftrightarrow$ OL/MI inversion if, and only if, a system composition is specified that is the composition of the middle phase of a three-phase system at some temperature ($T_{lc} < T < T_{uc}$); the PIT is then that temperature.

Since "phase" inversion can be made to occur at any desired temperature between T_{lc} and T_{uc}, there is no unique phase inversion temperature unless $T_{lc} = T_{uc}$. If $T_{lc} = T_{uc}$, this is a tricritical point. Sequences of systems (formed from different members of a homologous series) are known that approach this limit [39].

On the oil/water side of the stack of tie-triangles, the range of inversion temperatures is much narrower [39]. The PIT evidently varies slightly with surfactant concentration between the stack of tie-lines in the face of the prism and the aqueous-oleic tie-lines of the stack of tie-triangles, but by much less than $T_{uc} - T_{lc}$.

As described by critical scaling theory and shown by experiment, the (infinite radius of curvature) interfacial tension between conjugate phases goes to zero as the critical end point is approached [40]. In particular, the aqueous phase-middle phase tension goes to zero from a large value as T_{lc} is approached from higher temperatures, and the oleic phase-middle phase tension goes to zero from much larger values as T_{uc} is approached from below [39]. Thus, these two tensions become equal at T_{opt}.

For $T < T_{opt}$ the aqueous-middle phase tension is lower than the oleic-middle pmase tension, and for $T > T_{opt}$ the reverse inequality holds. The point here is that the aqueous and middle phases always from an AQ/MI (or MI/AQ) emulsion, while the oleic and middle phases always form an OL/MI (or MI/OL) emulsion, regardless of which phase pair has the lower interfacial tension.

Thus, the temperature at which the aqueous phase-oleic phase tensions for a chosen composition reach their minimum is a measure of the PIT only for the stack of tie-lines that contains this composition [39]. The temperature at which the oleic-middle phase and aqueous middle phase tensions become equal (T_{opt}) is *not* "the" phase inversion temperature. It is simply one of an infinite number of composition-dependent emulsion inversion temperatures such that $T_{lc} < T < T_{uc}$.

In summary, one can meaningfully discuss "the" PIT of a system only if a particular composition is specified, or if the lower and upper critical end-point temperatures are identical. A specified composition that is on the surfactant side of the stack of tie-triangles

must be a middle-phase composition. Otherwise, there is a range
of inversion temperatures from T_{lc} to T_{uc}. Most of the phase-
inversion temperatures reported in the literature do not meet either
criterion and are therefore ambiguous [39,41,42].

As more components and/or intensive (i.e., field) degrees of
freedom are added to a surfactant/oil/water-temperature diagram,
the emulsion-inversion line traced by the middle phase successively
becomes a surface, volume, and so forth. In this common case,
ambiguities associated with reported phase-inversion temperatures
become even greater [41,42].

It is now possible to summarize the dependence of emulsion types
on temperature and composition, and to understand the fundamental
limitations of conventional HLB methods. Six—not two—types of
liquid-liquid emulsions are theoretically possible; the conventional
O/W and W/O designations are not rigorous, but are reasonably
descriptive in the OL/AQ and AQ/OL regions, respectively. OL/MI
emulsions can occur only when $T < T_{uc}$, and AQ/MI emulsions form
only when $T > T_{lc}$. For $T_{lc} < T < T_{uc}$, either type can form,
depending on composition, as shown in Fig. 4 and Table 1. In the
temperature range where either type is possible, OL/MI emulsions
can form only when the oil/water ratio of the sample is greater than
the oil/water ratio of the middle phase, and AQ/MI emulsions are
possisle only when the oil/water ratio of the system is less than the
oil/water ratio of the middle phase (compositions on the surfactant
side of the stack of tie-triangles).

Hence, there is always a temperature (T_{uc}) above which a conventional HLB prediction of O/W emulsions will be wrong for all compositions: The system can form only AQ/MI dispersions. Similarly,
if the Griffin-Davies method predicts water-in-oil emulsions, there is
a temperature (T_{lc}) below which this prediction cannot be true:
Any liquid-liquid emulsion that forms must be the OL/MI type. At
temperatures between T_{lc} and T_{uc}, either HLB prediction will be
correct—if the experimenter is fortunate to choose a composition and
temperature that place the system within the two-phase region required by the Griffin-Davies prediction. This HLB method now appears to be a remarkably successful method, for its time, of pre-
dicting which surfactant/oil combinations have approximately the
same critical end-point temperatures. Recent work has shown that
the Griffin-Davies HLB numbers are linear functions of T_{opt} for
common nonionic surfactants [43].

To a first approximation, the oil/water ratio of the middle phase
($T_{lc} < T < T_{uc}$) is simply

$$\frac{[oil]}{[water]} \cong \frac{T - T_{lc}}{T_{uc} - T} \tag{5}$$

This equation is based on the approximations that the fractional middle-phase oil and water concentrations, [oil] and [water], change linearly with temperature and range from zero to one between the two critical end-point temperatures. The first of these approximations is reasonably accurate except near T_{lc} and T_{uc}. However, these exceptions are not a severe limitation. Near a critical end-point temperature the compositional area bounded by the smaller of the two binodals is so limited that emulsions of this type are unlikely to be encountered in most practical cases. The assumption that the oil and water concentrations of the middle phase vary from zero to one is accurate for all but the lowest-molecular-weight proto-surfactants [33,34]. But these exceptions are unimportant, because proto-surfactants seldom form stable emulsions. Of course, for an emulsion to form, the surfactant concentration must be chosen such that the composition falls within a two-liquid region.

Thus, as suggested by Eq. (5), the first requirement for a satisfactory HLB method is the accurate prediction of the critical end-point temperatures. And to be fully satisfactory, an HLB method must predict with reasonable accuracy the complete binodal surfaces and their dependence on oil and surfactant structure. Definitive new experiments, based on knowledge of the phase diagrams, are needed to determine whether and where AQ/MI → MI/AQ and OL/MI → MI/OL inversions occur.

V. REFERENCES

1. Griffin, W. C., (a) *J. Soc. Cosmetic Chem.* 1, 311 (1949); (b) *J. Soc. Cosmetic Chem.* 5, 249 (1954).
2. Davies, J. T., in *Proceedings of the Second International Congress on Surface Activity*, Butterworths, Longon, 1959, vol. 1, p. 426.
3. Davies, J. T., in *Proceedings of the Third International Congress on Surface Activity*, Universitaetsdruckerei, Mainz, 1960, vol. 2, p. 585.
4. Sherman, P., in *Rheology of Emulsions*, Sherman, P. (ed.), Pergamon Press, London, 1963, p. 93.
5. Shinoda, K., in *Solvent Properties of Surfactant Solutions*, Shinoda, K., (ed.), Marcel Dekker, New York, 1967, p. 27.
6. Blakey, B. C., and Lawrence, A. S. C., *Disc. Faraday Soc.* 18, 268 (1954).
7. Becher, P., in *Surfactants in Solution*, Mittal, K. L., and Lindman, B. (eds.), Plenum Press, New York, 1984, vol. 3, pp. 1925–1946.
8. See Marsden, S. S., and McBain, J. W., *J. Phys. Colloid Chem.* 52, 110 (1948) for the phase diagram available to Griffin.

9. Shinoda, K., Hanrin, M., Kunieda, H., and Saito, H., *Colloids Surfaces 2*, 301 (1981).
10. Shinoda, K., and Kunieda, H., *J. Colloid Interface Sci. 42*, 381 (1973).
11. Shinoda, K., *J. Colloid Interface Sci. 24*, 4 (1967).
12. Shinoda, K., and Sagitani, H., *J. Colloid Interface Sci. 64*, 68 (1978).
13. Shinoda, K., and Ogawa, T., *J. Colloid Interface Sci. 24*, 56 (1967).
14. Arai, H., and Shinoda, K., *J. Colloid Interface Sci. 25*, 396 (1967).
15. Shinoda, K., and Arai, H., *J. Colloid Interface Sci. 25*, 429 (1967).
16. Shinoda, K., and Saito, H., *J. Colloid Interface Sci. 26*, 70 (1968).
17. Shinoda, K., and Saito, H., *J. Colloid Interface Sci. 30*, 258 (1969).
18. Shinoda, K., Moriyama, N., and Hattori, K., *J. Colloid Interface Sci. 30*, 270 (1969).
19. Shinoda, K., and Takeda, H., *J. Colloid Interface Sci. 32*, 642 (1970).
20. Tsutsumi, H., Nakayama, H., and Shinoda, K., *J. Am. Oil Chem. Soc. 55*, 363 (1978).
21. Saito, H., and Shinoda, K., *J. Colloid Interface Sci. 35*, 359 (1971).
22. Shinoda, K., Saito, H., and Arai, H., *J. Colloid Interface Sci. 35*, 624 (1971).
23. Shinoda, K., and Yoneyama, T., *J. Disp. Sci. Technol. 1*, 1 (1980).
24. Becher, P., *Emulsions: Theory and Practice*, Reinhold, New York, 1957.
25. Sherman, P. (ed.), *Emulsion Science*, Academic Press, New York, 1968.
26. Lissant, K. J. (ed.), *Emulsions and Emulsion Technology*, parts I and II, Marcel Dekker, New York, 1974.
27. Lissant, K. J., *Demulsification*, Marcel Dekker, New York, 1983.
28. Becher, P. (ed.), *Encyclopedia of Emulsion Technology*, vol. 1, Marcel Dekker, New York, 1983.
29. Smith, Duane H., (a) *J. Colloid Interface Sci., 102*, 435 (1984); (b) *108*, 471 (1985).
30. Enyeart, C. R., in *Nonionic Surfactants*, Shick, M. J. (ed.), Marcel Dekker, New York, 1967, pp. 44–85.
31. Smith, Duane H., and Fleming P. D., *J. Colloid Interface Sci. 105*, 80 (1985).
32. Kunieda, H., and Friberg, S. E., *Bull. Chem. Soc. Jpn. 54*, 1010 (1981).

33. Kunieda, H., *Bull. Chem. Soc. Jpn. 56*, 625 (1983).
34. Kahlweit, M., Lessner, E., and Strey, R., *J. Phys. Chem. 87*, 5032 (1983).
35. Kunieda, H., and Shinoda, K., *J. Dispersion Sci. Technol. 3*, 233 (1982).
36. Lindman, B., and Stilbs, P., in *Surfactants in Solution*, Mittal, K. L., and Lindman, B. (eds.), Plenum Press, New York, 1984, vol. 3, p. 1651.
37. Biais, J., Bothorel, P., Clin, B., and Lalanne, P., *J. Dispersion Sci. Technol. 2*, 67 (1981).
38. Knickerbocker, B. M., et al., *J. Phys. Chem. 86*, 393 (1982).
39. Kunieda, H., and Shinoda, K., *Bull. Chem. Soc. Jpn. 55*, 1777 (1982).
40. Fleming, P. D., and Vinatieri, J. E., *J. Colloid Interface Sci. 81*, 319 (1981).
41. Belzer, D., and Kosswig, K., (a) *Tenside Detergents 16*, 256 (1979); (b) West German patents EP88206 (DE3208206), EP73894, EP58371 (DE3105913), and EP47370 (DE3134530).
42. Kraft, H. R., and Pusch, G., in *Proceedings of the Third Joint SPE/DOE Symposium on Enhanced Oil Recovery*. Tulsa Oklahoma, April 4–7, 1982 (SPE/DOE 10714), and pertinent references therein.
43. Kunieda, H., and Shinoda, K., *J. Colloid Interface Sci. 107*, 107 (1985).

5

Phase Equilibria in the Glycerol-Aerosol OT System with Decanol or Hydrocarbon

STIG E. FRIBERG and YUH-CHIRN LIANG *University of Missouri at Rolla, Rolla, Missouri*

I. INTRODUCTION

Phase diagrams were determined for the glycerol-Aerosol OT system combination with decanol as well as with the hydrocarbons decane and p-xylene.

A comparison with the corresponding aqueous systems showed the glycerol systems to be less ordered. The regions for the liquid crystals with lamellar and inverse hexagonal structure were strongly reduced. The isotropic solution region was extended toward lower Aerosol OT content, showing the importance of the enhanced interactions between the hydrophobic component and glycerol in comparison with that in the aqueous system.

Microemulsion systems are usually stabilized by a combination of a water-soluble ionic surfactant combined with a medium-chain-length alcohol as a cosurfactant [1-8]. Other possibilities are the use of a single surfactant of more hydrophobic character such as a nonionic one [9-13], or a hydrophobic ionic surfactant such as Aerosol OT [14-18].

The microemulsion region has recently been extended to include nonaqueous systems [19-21], following the studies on micellization in nonaqueous systems by Evans and collaborators [22-24] and on liquid crystals by Friberg and collaborators [25-28]. Our introductory studies on glycerol-hydrocarbon systems [29] led to the conclusion that although the phase regions in the glycerol/surfactant/cosurfactant/hydrocarbon systems appeared similar to those in aqueous systems, the presence of the plaitpoint, also at high hydrocarbon content, indicated the microemulsions to be solutions with critical behavior rather than well-defined aqueous microdroplets as found in corresponding water in oil (W/O) microemulsion systems [3,4]. This conclusion has

been confirmed by NMR determinations of the diffusion coefficients
of individual components by Lindman et al. [30].

With this difference in mind and the fact that the aqueous microemulsion stabilized by Aerosol OT has been extensively investigated
[14-18], we found phase diagrams with glycerol/Aerosol OT systems
to be a necessary foundation for continued research into nonaqueous
microemulsions. In this chapter we have combined the glycerol-
Aerosol OT pair with decanol and with hydrocarbons.

II. EXPERIMENTAL

A. Materials

Decyl alcohol (Aldrich, 99%), Decane (Aldrich, 99+%, gold label),
and p-xylene (Fisher, 99.9 mol%) were used as received. Glycerol
(Aldrich 99.5%, gold label) was dried by successive treatment of
molecular sieves (Davison Chemical, 8-12 mesh beads, 3 Å effective
pore size) to less than 0.05% to water content as determined by Karl
Fischer titration. Aerosol OT (Aldrich, 96%) was purified according
to Kunieda and Shinoda [31]. The water-soluble impurities (i.e.,
Na_2SO_4) were extracted from the Aerosol OT-benzene solutions by
water, and the organic impurities were extracted from Aerosol OT-
aqueous methanol solution by petroleum ether. The purified Aerosol
OT was dried in a high-vacuum oven at 40°C for a week and kept in
P_2O_5 desiccator.

B. Phase Diagrams

The ternary phase diagrams of glycerol, Aerosol OT, and hydrocarbon systems were determined at 25°C. Solutions of Aerosol OT in
hydrocarbons with various weight ratio were prepared in glass vials
and then glycerol was added. The phase boundary was noted by
observing the transition from clear to turbid or from isotropic to
anisotropic. For the determination of lamellar liquid crystal region,
hydrocarbons were added to the mixtures of Aerosol OT and glycerol
and microscopic observation in polarized light was used to establish
the borders of the liquid crystals.

C. Small-Angle X-Ray Diffraction

The interlayer distances of lamellar liquid crystals were determined
by a Siemens (Crystalloflex 4) X-ray diffraction equipment at 25°C.
The radiation (λ = 1.54 Å) was the Ni-filtered K_α line of copper.

III. RESULTS

The results describe the phase diagrams in the three systems, followed by a comparison between the interlayer spacings in the part

of the lamellar liquid crystal that was common to the aqueous and the glycerol systems.

A. Glycerol-Aerosol OT Systems

The system contains a glycerol solution of Aerosol OT with a maximum solubility of 35% of the surfactant. The solubilization of decanol in this solution was limited; a maximum of approximately 4% was found. The Aerosol OT was soluble in the decanol to a high degree; the equilibrium conditions at high surfactant concentrations were difficult to determinep and the maximum value of 78% Aerosol OT should be considered approximate. The decanol/Aerosol OT solution solubilized huge amounts of glycerol; Fig. 1 shows a maximum solubilization of 74%. The decanol solution and the glycerol solution were separated by a lamellar liquid crystal reaching 87% glycarol.

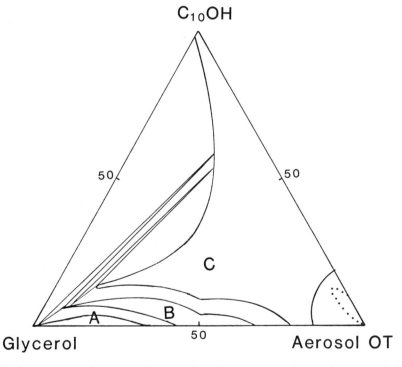

FIG. 1 Partial phase diagram for the system glycerol, Aerosol OT (AOT), and decanol ($C_{10}OH$). A = isotropic glycerol solution; B = lamellar liquid crystal, C = decanol/Aerosol OT solution with solubilized glycerol.

The system with decane, Fig. 2, is characterized by the fact
that the region of the lamellar liquid crystal was reduced. The
narrow, crescent-shaped channel region found for the decanol system, Fig. 1, was now reduced to include an insignificant solubilization of hydrocarbon, reaching only 2% from the glycerol-Aerosol OT
axis. The isotropic liquid solution regions were extended to unite
the glycerol and decanol solutions in a broad channel. The equilibrium conditions at high-Aerosol OT content were also difficult to
ascertain, and the dashed lines in the diagram represent approximate values.

The system with p-xylene, Fig. 3, on the other hand, displayed a strongly enhanced isotropic solubility region toward the
glycerol-p-xylene axis. The maximum surfactant concentration for
the left-hand borders of the solubility region was now 9%, compared

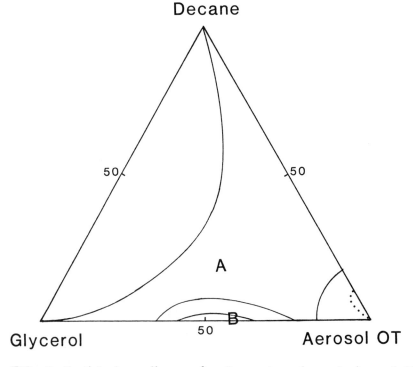

FIG. 2 Partial phase diagram for the system glycerol, Aerosol OT
(AOT), and decane. A = isotropic liquid solution, B = lamellar
liquid crystal.

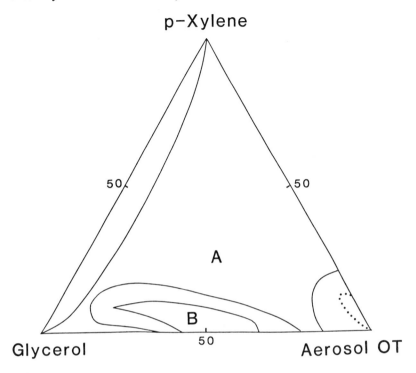

FIG. 3 Partial phase diagram for the system glycerol, Aerosol OT (AOT), and p-xylene. A = isotropic liquid solution, B = lamellar liquid crystal.

to 33% for the decanol system and 34% for the decane combination. The lamellar liquid crystal displayed a region similar to the one found in the decanol system.

B. Small-Angle X-Ray Diffraction

The interlayer spacing in the lamellar liquid crystal along the Aerosol OT-polar solvent axis is given in Fig. 4. The interlayer spacing in the glycerol system had lower values than the ones in the aqueous system; the extrapolated values for zero concentration of solvent were identical.

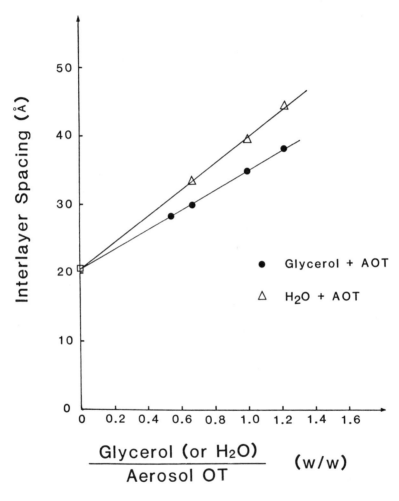

FIG. 4 Interlayer spacing in the lamellar liquid crystal from Aerosol OT (AOT) plus water (△) or glycerol (●).

IV. DISCUSSION

The value of this study is in the comparison between the aqueous systems [32] and the present glycerol systems. The differences in phase regions are illustrated in Figs. 5 and 6. In the decanol system we found an extended solubilization of glycerol in the Aerosol OT-decanol solution of 74% glycerol, compared to 28% in the aqueous system. In addition, the large aqueous regions of the inverse hexagonal phase and of the isotropic liquid crystal in the glycerol system reached only insignificant size and were replaced with an isotropic solution. The extension of solubility regions toward lower Aerosol OT content for low decanol content was another conspicuous feature.

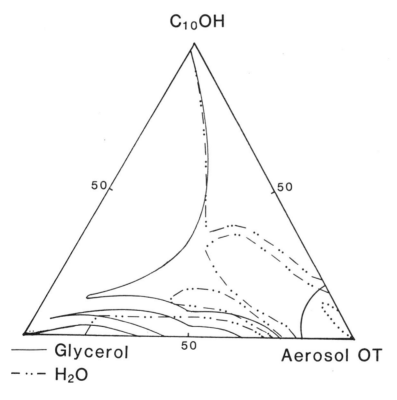

FIG. 5 A comparison of the phase regions in systems of Aerosol OT and decanol combined with glycerol (———) or water (−·· −) [32].

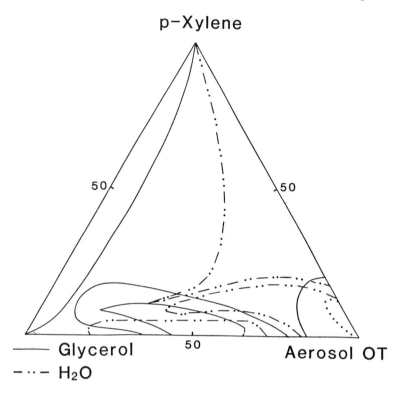

FIG. 6 A comparison of the phase regions in systems of Aerosol OT and p-xylene combined with glycerol (———) or water (–···–) [32].

The tendency of these changes was strongly exaggerated for the p-xylene system, Fig. 6. The isotropic liquid solution of the glycerol system not only covered the regions for the inverse hexagonal and the isotropic liquid crystal, but extended far toward the glycerol-p-xylene axis.

It appears that all these changes are a consequence of one factor. The glycerol has a more enhanced molecular interaction with the hydrophobic compounds than water. This difference leads to a higher penetration of glycerol molecules in the hydrocarbon part of the surfactant. The higher penetration of glycerol means an increase of the cross-sectional area of the surfactant, which in turn lowers the amount of polar solvent to change inverse structures over lamellar liquid crystals to normal micelles. Such a change should lead to destabilization of all liquid crystals at a lower percentage of glycerol than of water. This was found in our results.

On the other hand, a low glycerol content in the lamellar liquid crystal is higher than the corresponding value in the aqueous system. The result of the glycerol content for the start of the lamellar liquid crystalline region along the Aerosol OT-glycerol axis is not real. A diagram drawn on volume fraction instead of weight fraction would show the upper limit of the lamellar phase as being identical for the two systems.

The second feature is the exterior to higher glycerol content of the lamellar liquid crystal when decanol is added. This phenomenon may be understood by realizing the competition between glycerol and decanol for sites in the lamellar layer. Decanol is known to remove water from sites between the surfactant polar groups in a lamellar liquid crystal with water [32]. A similar interaction in the glycerol system appears reasonable.

The reduced penetration of glycerol means an alleviation of the disorder caused by the penetrating glycerol, and an enhanced capacity of the lamellar liquid crystal to accept a greater amount of glycerol without destabilization.

Another factor to be considered is the influence on the exterior of the electric double layer by the changed solvent. Unfortunately, information about the electric double layer properties in different solvents is not yet complete. Lyklema and collaborators have reviewed the factors involved [33-35], but no final conclusion has been reached concerning the influence on colloidal stability. Hence, the reason why the higher stability toward polar solvent addition in the glycerol system occurs must await further development in the basic theory for colloidal stability in different media.

V. ACKNOWLEDGMENT

This research was supported by U.S. Department of Energy, Office of Basic Sciences Grant #DE-AC02-83ER13083.

VI. REFERENCES

1. Hoar, T. P., and Schulman, J. H., *Nature 152*, 102 (1943).
2. Shah, D. O., and Hamlin, R. M., Jr., *Science 171*, 483 (1971).
3. Gillberg, G., Lehtinen, H., and Friberg, S., *J. Colloid Interface Sci. 33*, 40 (1970).
4. Sjöblom, E., and Friberg, S., *J. Colloid Interface Sci. 67*, 16 (1978).
5. Rosano, H. L., Schiff, H., and Schulman, J. H., *J. Phys. Chem. 66*, 1928 (1962).
6. Adamson, A. W., *J. Colloid Interfate Sci. 29*, 261 (1969).

7. Lindman, B., Stilbs, P., and Moseley, E., *J. Colloid Interface Sci. 83*, 569 (1981).

8. Ruckenstein, E., in *Micellization, Solubilization and Microemulsions*, Mittal, K. L. (ed.), Plenum Press, New York, 1977, p. 755.

9. Shinoda, K., *J. Colloid Interface Sci. 24*, 4 (1967).

10. Nilsson, P.-G., Wennerström, H., and Lindman, B., *J. Phys. Chem. 87*, 1377 (1983).

11. Friberg, S., Lapczynska, I., and Gillberg, G., *J. Colloid Interface Sci. 56*, 19 (1976).

12. Shinoda, K., and Friberg, S., *Adv. Colloid Interface Sci. 4*, 281 (1975).

13. Shinoda, K., and Ogawa, T., *J. Colloid Interface Sci. 24*, 56 (1967).

14. Eicke, H.-F., *J. Colloid Interface Sci. 52*, 65 (1975).

15. Eicke, H.-F., Shepherd, J. C. W., and Steinmann, A., *J. Colloid Interface Sci. 56*, 168 (1976).

16. Eicke, H.-F., and Shepherd, J. C. W., *Helv. Chim. Acta 57*, 1951 (1974).

17. Magid, L. J., in *Solution Chemistry of Surfactants*, Mittal, K. L. (ed.), Plenum Press, New York, 1979, vol. 1, p. 427.

18. Martin, C. A., and Magid, L. J., *J. Phys. Chem. 85*, 3938 (1981).

19. Friberg, S. E., and Podzimek, M., *Colloid Polymer Sci. 262*, 252 (1984).

20. Rico, I., and Lattes, A., *N. J. Chim. 8*, 424 (1984).

21. Fletcher, P. D. I., Galal, M. F., and Robinson, B. H., *J. Chem. Soc. Faraday Trans. I 80*, 3307 (1984).

22. Evans, D. F., Chen, S. H., Schriver, G. W., and Arnett, E. M., *J. Am. Chem. Soc. 103*, 481 (1981).

23. Evans, D. F., Yamauchi, A., Roman, R., and Casassa, E. Z., *J. Colloid Interface Sci. 88*, 89 (1982).

24. Ramadan, M., Lumry, R., Evans, D. F., and Philson, S., *J. Phys. Chem. 87*, 4538 (1983).

25. Moucharafieh, N., and Friberg, S. E., *Mol. Cryst. Liq. Cryst. 49*, 231 (1979).

26. El Nokaly, M. A., Ford, L. D., Friberg, S. E., and Larsen, D. W., *J. Colloid Interface Sci. 84*, 228 (1981).

27. Ganzuo, L., El Nokaly, M., and Friberg, S. E., *Mol. Cryst. Liq. Cryst. 72*, 183 (1982).

28. Frieberg, S. E., Solans, C., and Ganzuo, L., *Mol. Cryst. Liq. Cryst. 109*, 159 (1984).

29. Friberg, S. E., and Liang, Y.-C., *in press*.

30. Lindman, B., et al. *in preparation*.

31. Kunieda, H., and Shinoda, K., *J. Colloid Interface Sci.* 70, 577 (1979).
32. Ekwall, P., in *Advanced Liquid Crystals*, Brown, G. H., (ed.), Academic Press, New York, 1975, p. 1.
33. De Wit, J. N., and Lyklema, J., *Electroanal. Chem Interfacial Electrochem.* 47, 259 (1973).
34. Lyklema, J., and De Wit, J. N., *J. Electroanal. Chem.* 65, 443 (1975).
35. Lyklema, J., and De Wit, J. N., *Colloid Polymer Sci.* 256, 1110 (1978).

6

The Nature of the Middle Phase in Brine-Carbon Tetrachloride – Microemulsion Three-phase System at Optimal Salinity

LI ZHI-PING and HU MEI-LONG   *Xinjiang Institute of Chemistry, Academy of Science of China, Urumuqui, Xinjiang, People's Republic of China*

I. INTRODUCTION

Microphotographs of hydrophilic and hydrophobic micelles in various locations of the middle phase have been obtained with Sudan III as a dye.

We have observed the nonhomogeneity of the middle phase of two- and three-phase systems, including heptane-sodium petroleum sulfonate, 2-phenoxyethanol-brine, and heptane-linear dodecylbenzene-sulfonate, 2-phenoxyethanol-brine, at optimal salinity and provide that the nonhomogeneity's related to the property of interface [1]. Obviously, further investigation of this phenomenon is needed.

K. Shinoda et al. have made some inferences about the middle phase. They think that the two ends of the middle phase are the micelle with different structures. The end near the hydrophilic phase is a hydrophilic micelle, and the end near the hydrophobic phase is a hydrophobic micelle [2]. This inference is made to explain the fact that interfacial tension is lower as the nature of the solution on both sides of the interface is more similar. Most researchers can accept this opinion at present, but up to now experimental evidence has not been obtained. This chapter addresses this problem, and attempts to find the distribution of different micelles and to provide further experimental evidence of nonhomogeneity in the middle phase.

II. REAGENTS AND APPARATUS

The surfactant used was sodium dodecylbenzenesulfonate (DDBS) (BDH Chemicals, Ltd.), which is suitable for use as the stationary

phase in gas-liquid chromatography. The DDBS was further purified before use by the following method: 20 g of DDBS were dissolved in 600 ml of a 1:1 (v/v) mixture of distilled water and 95%
ethanol. The solution containing DDBS was transferred to a sep-
aratory funnel and extracted five times with a 200-ml portions of
petroleum ether (b.p. 60–90°C). The aqueous phase was separated
and evaporated to dryness in a steam bath. After drying in a
vacuum dessicator, 200 ml of 95% ethanol were added to the residue.
After the DDBS was dissolved by heating, the alcoholic solution was
immediately filtered and then washed three times with 95% ethanol.
The filtrate was evaporated to dryness in a steam bath. Finally,
the residue was dried to constant weight in a vacuum dessicator
over phosphorous pentoxide. The purity of DDBS was 98.5% after
purifying [3].

The cosurfactant for this study was isopropanol (A. R. E.
Merck). Also used were sodium chloride (A. R. Chemical Plant of
Beijing) and carbon tetrachloride (A. R. Changjiang Chemical Plant
of Shanghai).

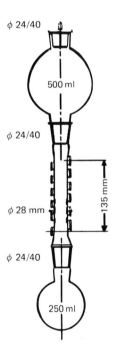

FIG. 1 Apparatus for determining the composition at various locations in the middle phase.

Other chemicals used in this experiment are all analytical reagents, and the aqueous phase was prepared with redistilled water.

The glass apparatus used in this study is shown in Fig. 1.
Its middle portion is a cylindrical tube with 12 oil-proof, rubber-
capped ports through which the samples of the middle phase can
be withdrawn. Before being used, the oil-proof rubber caps were
immersed in CCl_4 for more than 30 days to remove any CCl_4-
soluble impurities. Other equipment used included a gas chroma-
tograph (Shimadzu Seisakusho, Ltd., Japan, model GC-5A), an ultra-
centrifuge Hitachi, Japan, model 80P-7, and an Olympus (Japan)
universal research microscope.

III. PROCEDURES

A. Measurement of the Optimal Salinity of the System

The same volume of CCl_4 and an aqueous solution containing 1.7%
(w/v) DDBS, 9% (w/v) isopropanol, and 2.4% (w/v) sodium chloride were mixed together. The result showed that the WOR of the
middle phase was 1:1 in this case. Therefore, we infer that the
optimal salinity of the system is 2.4% at the above-mentioned conditions.

B. Preparing the Three-phase System

According to the above-determined condition, 225 ml of CCl_4 and
225 ml of aqueous solution containing 2.4% (w/v) NaCl, 9% (w/v)
isopropanol, and 1.7% (w/v) DDBS were mixed in a 1000-ml volumetric flask and shaken vigorously for 30 min. The solution was
then transfered into the glass apparatus shown in Fig. 1.

The apparatus was immersed in a water bath maintained at
25°C. The height of the middle phase was 130 mm after equilibrium for 60 days. Then a series of the properties of the middle
phase were measured.

C. Measurement of DDBS Concentration by Two-phase
Titration

In this procedure, 80 μl of the microemulsion in various locations
of the middle phase were withdrawn by a 100-μl microsyringe. We
transferred it into a 25-ml volumetric flask and diluted it to the
mark with distilled water. The content of DDBS was measured by
two-phase titration, using the diluted solution. The results are
shown in Table 1 and Fig. 2.

Figure 2 shows that the concentration gradients of DDBS regularly exist in the middle phase from high to low.

TABLE 1 Distribution of DDBS Concentration in the Middle Phase
After Equilibrating at 25°C for 60 Days

Distance below upper interface (%)	2	19	45	66	94
DDBS concentration $\times 10^{-1}$ M	1.927	1.870	1.846	1.846	1.768

D. Measurement of the Amount of Water, Isopropanol, and CCl₄ in the Middle Phase by GC

In this procedure, 1 µl of the samples were withdrawn by 5-µl
microsyringe at various locations of the middle phase, and the
relative contents of isopropanol, water, and CCl_4 were measured
by gas chromatography, (GC). Experimental GC conditions were
as follows:

Thermal conductivity detector
Stainless column, 3 m × 3 mm, packed with 60-80 mesh GDX-403
Carrier gas: H_2 at 50 ml/min
Column temperature: 170°C

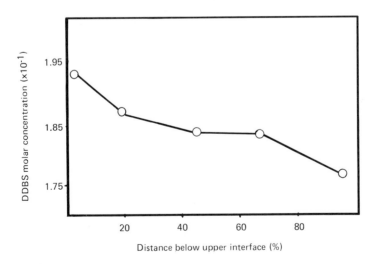

FIG. 2 Molar concentration of DDSS in the middle phase as a function of distance below the upper interface (%).

Injection port temperature: 200°C
Detector current: 100 mA
Chart speed: 2.5 mm/min

Figure 3 shows the gas chromatogram of three components in the middle phase. The results in Table 2 and Fig. 4 show the relative amounts of CCl_4, water, and isopropanol measured by GC. The percentage of the three components is obtained from an auto-numerical integrator.

The results obtained from GC show that there is more water and isopropanol in the locations of the middle phase near the brine phase and more CCl_4 in the location of the middle phase near the

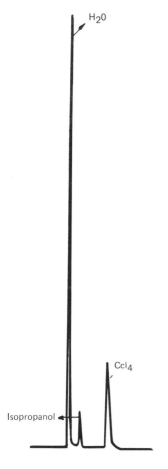

FIG. 3   Gas chromatogram of three components in the middle phase.

TABLE 2   Distribution of Components at Various Locations in the Middle Phase

Distance below upper interface (%)	2	20	44	68	97
H_2O %		59.8	50.2	49.4	46.6	46.6
CCl_4 %		33.6	43.3	44.6	48.2	48.9
Isopropanol %		6.61	6.41	6.04	5.27	4.74

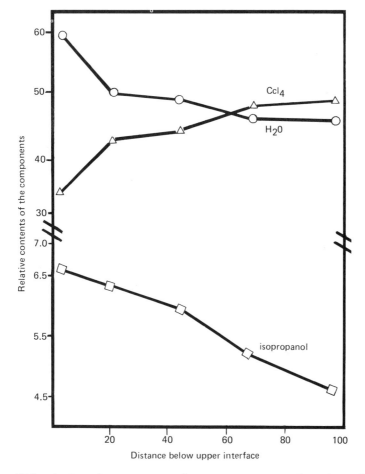

FIG. 4   Relative amounts of components as a function of distance below the upper interface (%).

CCl$_4$ phase. We have also measured the absolute amount of water by the Karl Fisher method [4] to check the results from GC.

E. Measurement of Water Content at Various Locations of the Middle Phase by the Karl Fisher Method

Preparation of Karl Fisher reagent 5: 110 ml of anhydrous pyridine dissolvee in 63 g of pure iodine was cooled outside by an ice-salt bath and the weight of the solution was increased to 205 g by adsorbing SO$_2$. The product was allowed to stand for 30 min and then was diluted to 500 ml with anhydrous methanol. The resulting reagent was standardized by standard water solution. 100 ml of this reagent were equal to 0.2861 g of water.

Measurement of water: 60 µl of microemulsion from various locations of the middle phase were transferred into a 100-ml glass-stopped conical flask 25 ml of anhydrous methanol was added as a solvent, and then the solution was titrated to a bright gold-yellow color with the standardized Karl Fisher reagent. The results are shown in Table 3 and Fig. 5.

F. The Effect of Centrifugation

Following these measurements, the middle phase was separated from the three-phase system. 10 ml of homogeneous microem'lsion was mixed by shaking and placed into a plastic centrifuge tube, where it remained for 30 min at 10,000 rpm and a temperature of 25°C. Samples were taken from different levels of the microemulsion in the tube. The components in the microemulsion were measured by GC within 4 hr after centrifuging. The data from the gas chromatography are shown in Table 4.

From the centrifuge experiment, we found that no concentration gradient can be found if the centrifuge speed is less than 10,000 rpm. The microemulsion was destroyed above 40,000 rpm.

TABLE 3 Distribution of Water in the Middle Phase After Equilibrating for 60 Days

Distance below upper inter-face (%)	2	12	48	81	98
Grams of water per ml of microemulsion	0.4378	0.3910	0.3443	0.3147	0.2851

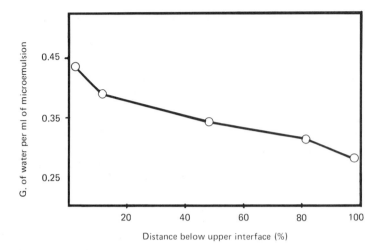

FIG. 5   Water content in middle phase as a function of distance below upper interface (%).

G.   Test of Micelle Type at Various Locations in the Middle Phase

A three-phase system prepared by Procedure B was equilibrated at
25°C for 60 days.   Then 5 µl of the samples was withdrawn with a
microsyringe at various locations of the middle phase.   One or two
drops of the samples were dropped on an object carrier, and one
drop of 0.5% Sudan III in CCl_4 solution was added.   They were
mixed in a glass roll covered with glass, and then surveyed and
photographed with an Olympus universal research microscope.   We
clearly observed that the middle phase near the CCl_4 phase was very

TABLE 4   Concentration Gradient in Microemulsion After
Centrifuging

Distance below upper inter-face (%)	2	31	67	98
Water %	51.5	48.4	48.2	46.1
Isopropanol %	5.4	5.2	5.2	5.0
CCl_4 %	43.5	46.4	46.4	48.9

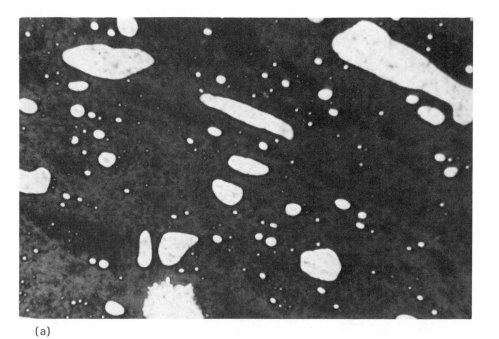

(a)

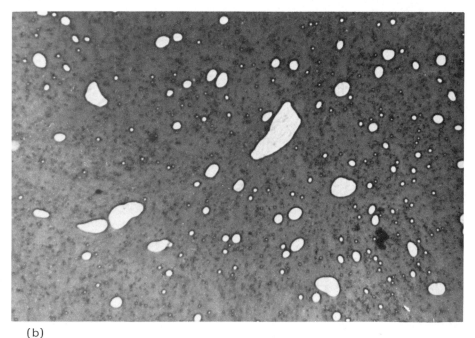

(b)

FIG. 6 (a) Photomicrograph of middle phase of 2% distance to upper interface; (b) Photomicrograph of the middle phase of 20% distance to upper interface. (c) Photomicrograph of the middle phase of 20% distance
to lower interface. (d) Photomicrograph of the middle phase of 20% distance to lower interface.

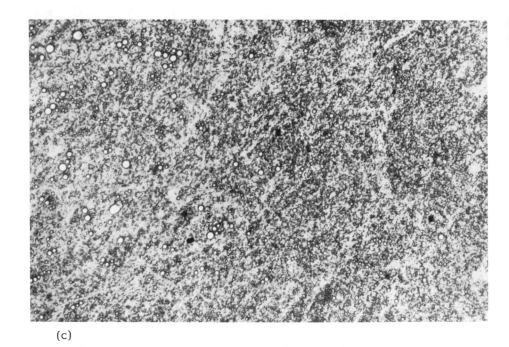

(c)

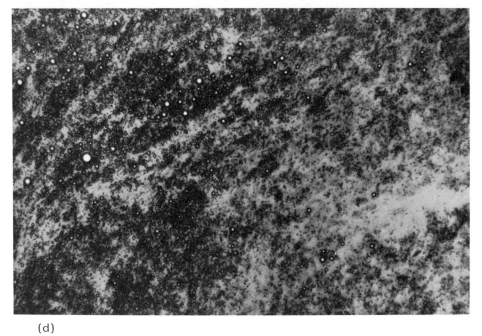

(d)

FIG. 6 (Continued)

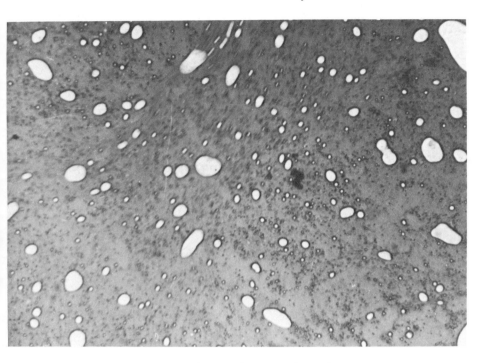

FIG. 7 Photomicrograph of the microemulsion without upper and lower phase.

miscible with dye. In this case, the red-brown picture could be observed from the field of vision in the microscope. It proved that the micelles are hydrophobic. On the other hand, the middle phase near the brine phase was not miscible with dye, at least in the first half-hour. In this case, blue spots could be observed on a red-brown background, indicating hydrophilic micelles.

Figure 6A and Fig. 6B show, respectively, the microphotographs of 2% and 20% distance to the upper interface. Figure 6C and Fig. 6D are microphotographs of 2% and 20% distance to the lower interface. These pictures illustrate that there are less hydrophilic micelles in the location of the middle phase near the lower interface and less hydrophobic micelles in the location of the middle phase near the upper interface. To separate the whole middle phase from the upper and lower phases, we allowed the microemulsion to stand for 24 hours. The picture obtained from this microemulsion and 0.5% Sudan III solution is shown in Fig. 7. Figure 7 is very similar to Fig. 6A and Fig. 6B, showing that the microemulsion in this case is mainly hydrophilic micelle.

II. DISCUSSION

The middle phase of the three-phase system at optimal salinity shows
that the surfactant, cosurfactant, carbon tetrachloride, and water
appear in obvious concentration gradients after equilibration for 60
days at 25°C. The amount of surfactant, cosurfactant, and water
increase from high to low, but the amount of carbon tetrachloride
decreases from high to low. In other words, the characteristics of
the concentration gradients are that the middle phase near the brine
phase contains more surfactant, cosurfactant, and water, and the
middle phase near the organic phase contains more carbon tetrachloride. The results of this study conform to the conclusions of
Rosen and Li [1].

Following the centrifuging of the homogeneous microemulsion,
the concentration gradients are similar to those found in the middle
phase of the three-phase system, they were formed when the centrifuge speed was above 10,000 rpm and lower than 40,000 rpm.
This result shows that droplets or different size and density may
exist in the microemulsion. Under the effect of gravitation, the
microemulsion is homogeneous, due to the diffusion caused by molecular thermal motion. However, the strong centrifuge effect causes
the different sized droplets to produce a regular distribution as a
result of the different density.

We have proved the presence of both hydrophilic and hydrophobic micelles in the middle phase. The hydrophilic micelles exist in
the middle phase near the brine phase, and the hydrophobic micelles
exist in the end of the middle phase near the organic phase. According to classical thermodynamics, the nonhomogeneity exists only
in a region of a few molecular layers near the interface. However,
we discovered that different types of micelles and concentration gradients exist in the middle phase of the three-phase system. The
microemulsion separated from the upper and lower phases, equilibrated for 24 hours at room temperature, shows the main properties
of hydrophilic micelles. This result suggests that the different
types of micelle in the middle phase are related to the properties
of interface. According to the condition of thermodynamics for
equilibrium systems, the chemical potentials of various locations in
the same phase must be equal to each other: that is, the concentration of materials is homogeneous in the same phase. The nonhomogeneity of the middle phase in the three-phase system studied
can increase the free energy of the system, but on the other hand,
the nonhomogeneity in the middle phase can also decrease the interface free energy of w/m and m/o two interfaces. The result of two
opposite effects can finally lead the three-phase system to a minimum energy, that is, the equilibrium state of the system.

Based on our experiment, we suggest that it is necessary to distinguish the middle phase from the microemulsion. The middle phase not only possesses the general properties of a microemulsion but also some nonhomogeneities caused by w/m and m/o, two interfaces that have a large interfacial property difference. The microemulsion can be understood as a system or a phase of some two-phase system.

We suggest a model of micelle distribution in the middle phase. From the pattern of the figure, hydrophilic micelles exist in the upper part of the middle phase and hydrophobic micelles in the lower part of the middle phase. Hydrophilic and hydrophobic micelles occupy at least 20% of the middle phase, and there may also be lamella micelles in the middle part of the middle phase. This inference is in accordance with the concentration gradient. The distribution of micelles in this pattern is evidently beneficial in decreasing the different properties of the solution on the two sides of the interface, that is, in decreasing the interfacial tension.

V. ACKNOWLEDGMENT

Special thanks go to Prof. M. J. Rosen, who provided the authors with valuable advice.

VI. REFERENCES

1. Rosen, M. J., and Li, Z.-P., *J. Colloid Interface Sci. 97*, 456–464 (1984).
2. Shinoda, K., Hanrin, M., Kunieda, H., and Siato, H., *Colloid Surfaces 2*, 301–314 (1981).
3. Li., Z.-P., and Tong, J.-S. (trans.), *Systematic Analysis of Surface-Active Agents*, Xinjiang Peoples Press, 1982, p. 33.
4. Li, Z.-P., and Rosen, M. J., *Anal. Chem. 53*, 1516–1519 (1981).
5. Li, Z.-P., and Tong, J.-S. (trans.), *Systematic Analysis of Surface-Active Agents*, Xinjiang Peoples Press, 1982, p. 229.
6. Liu, Z.-J., Jiang, Z.-J., Shen, Z.-H., and Sheng, X.-M. (trans.), *The Analysis of Detergents and Detergent Products*, Light Industry Press, 1983, pp. 293–296.

7

Microemulsion Structure from Molecular Self-diffusion: A Brief Overview

BJÖRN LINDMAN   *Lund University, Lund, Sweden*

PETER STILBS   *Uppsala University, Uppsala, Sweden*

I. INTRODUCTION

Homogeneous phases of small amounts of surfactant and large amounts of water and oil are examples of organized solutions [1,2] that are of widespread theoretical, biological, and technical interest and are currently being intensely studied using a range of experimental and theoretical approaches. Although phase behavior is necessarily the initial characterization of any system, the second step of investigation is naturally that of microstructure. Some years ago it occurred to us that basic structural information on isotropic surfactant solution systems was contained in the long-range (lateral) mobilities of the individual molecules and that, therefore, molecular self-diffusion studies should offer a convenient route to microemulsion microstructure [3,4]. With the advent of the Fourier transform (FT) version of the pulsed-gradient spin-echo (PGSE) NMR technique [5–7], it became feasible to measure rapidly and simultaneously the self-diffusion coefficients of several constituents in a complex mixture with high precision and over a broad range of molecular mobilities and composition. Therefore, we started [8] a rather broad program for investigating the structure of microemulsion phases appearing in different surfactant systems. Here we will present a brief review of our studies and exemplify that the microstructure of microemulsions can be very different depending on composition, salinity, cosurfactant, temperature, etc. A comprehensive review of principles and applications of the approach is published elsewhere [9].

Self-diffusion work on microemulsion structure has recently been started also by other groups, and at the end of this chapter a bibliography concerning these studies is given.

II. DETERMINATION OF SELF-DIFFUSION COEFFICIENTS

In the determination of self-diffusion coefficients by the FT PGSE
NMR technique, a spin echo is produced by a 90–180° pulse sequence and the second half of the echo is Fourier-transformed to
resolve individual contributions of the echo in the frequency domain.
From the variations of the signal amplitudes with the time during
which diffusion is monitored, the self-diffusion coefficients of a
large number of components can generally be determined simultaneously. A detailed account [5] of the technique has been prepared,
and here we only illustrate the potential of the technique with
studies of different microemulsion systems. Note in Figs. 1–3
how differently the signal amplitudes of the different components
can decay. (Rapid decay corresponds to rapid diffusion.)

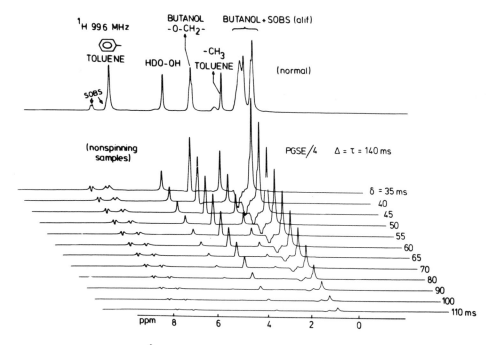

FIG. 1 FT PGSE ^{1}H NMR self-diffusion studies of a microemulsion
composed of sodium octylbenzenesulfonate (SOBS), n-butanol, toluene, and water (D$_2$O) at 100 MHz. Spin echo spectra shown as
a function of duration (δ) of magnetic field gradient pulse. (From
Ref. 5.)

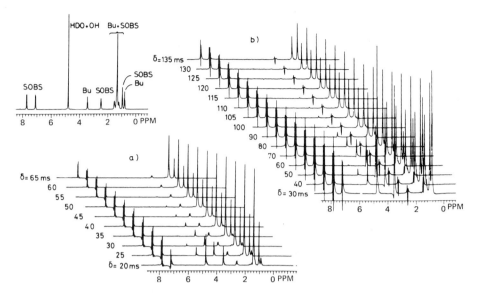

FIG. 2 FT PGSE ^{1}H NMR self-diffusion studies of a solution composed of sodium octylbenzenesulfonate (SOBS), n-butanol, and water at 300 MHz. Spin echo spectra shown as a function of duration (δ) of magnetic field gradient pulse. (From Ref. 5.)

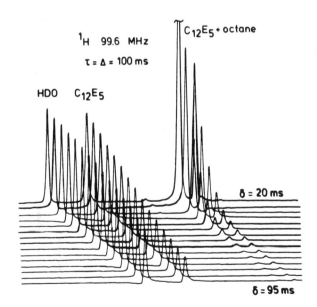

FIG. 3 FT PGSE ^{1}H NMR self-diffusion studies of a microemulsion composed of pentaethyleneglycol dodecylether, n-octane, and water at 100 MHz. Spin echo spectra shown as a function of duration (δ) of magnetic field gradient pulse. (From Ref. 5.)

III. MOLECULAR SELF-DIFFUSION AND SOLUTION STRUCTURE

The use of self-diffusion coefficients for extracting information on solution structure is based on the relation between diffusivity and size of the diffusing objects, as expressed for example by the Stokes-Einstein equation. For a spherical case, there is an inverse proportionality between self-diffusion coefficient and radius. A molecule that is part of a big aggregate and has a long lifetime in the aggregate will diffuse slowly over macroscopic distances, as its translational mobility is governed by that of the entire aggregate. (On the other hand, the molecule may move rapidly *within* the aggregate and have a high, short-time self-diffusion coefficient, but this is not monitored by the present type of experiments.) If there is a certain probability for the studied molecule to exist in a nonaggregated state in the interaggregate solution, the self-diffusion coefficient will become higher than if it is confined entirely to the aggregates. In fact, the effective observable self-diffusion coefficient is the weighted average over the different environments the molecule samples as a function of time.

Based on such simple reasoning, molecular self-diffusion over macroscopic distances is very sensitive to confinement into closed domains. Therefore, multicomponent self-diffusion studies, as obtained most conveniently in the Fourier transform spin-echo NMR work, can easily distinguish between different structural models in many cases. The following limiting cases form a sufficient basis of the present discussion.

Water-in-oil droplet structure: With confinement in closed domains, water diffusion will be slow. If surfactant molecules occur only at the interfaces, their diffusion will also be the same as that of the droplets themselves. Oil diffusion will be high as it forms continuous domains. It will only be retarded with respect to the neat oil by an obstruction effect due to the droplets and by penetration between surfactant alkyl chains. Therefore, we predict that $D_{oil} \gg D_{water} \cong D_{surfactant} \cong D_{droplet}$.

Oil-in-water droplet structure: In this case water diffusion occurs in a continuous medium and is retarded compared to neat water only by surfactant solvation and obstruction. Thus, $D_{water} \gg D_{oil} \cong D_{surfactant} \cong D_{droplet}$.

Bicontinuous structure: A large number of structures are possible in which both water and oil form domains that are continuous over macroscopic distances and where surfactant molecules are located at the interfaces between these domains. In particular, we may think of layered or channel-type structures; however, we will not in the present treatment consider any distinction between different possible bicontinuous structures. D_{water} and D_{oil} will both be high for bicontinuous microemulsions, lowered from the values of the neat liquids only by obstruction and solvation/penetration effects. Surfactant diffusion will be hindered by the location at interfacial films but will be unrestricted. We expect $D_{surfactant}$ to be intermediate between the value of nonassociated surfactant molecules (of the order of 10^{-9} m^{2}s^{-1}) and its value for droplet-type structures (of the order of 10^{-11} m^{2}s^{-1}).

Molecule disperse solutions: When the different components diffuse as single molecules, all D values will be high ($5 \times 10^{-9} - 5 \times 10^{-10}$ m^{2}s^{-1}). However, this case is clearly inapplicable for microemulsions with low surfactant contents, and the presence of surfactant in an aggregated state for other cases is borne out in observations of field-dependent NMR relaxation [10,11].

The concentration dependences can also be informative, as both medium and droplet molecules are expected to diffuse more slowly with increasing volume fraction (ϕ) of the dispersed phase. For example, for spherical droplets, $D_{medium}/D^0_{medium} = 1/(1 + \phi/2)$; $D_{droplet}/D^0_{droplet} = 1 - 2\phi$. For other geometries see, for example, Ref. 12.

There exists a broad range of different isotropic solutions of surfactants with different characteristics, and a number of different structures have been suggested. However, it is often not easy to test different structures experimentally in an unambiguous way. Here we will, by way of examples, demonstrate that the self-diffusion behavior of different isotropic surfactant solutions can be very different and thus that a broad range of microemulsion structures is possible.

IV. SOLUBILIZED IONIC MICELLAR SOLUTIONS

Solubilized ionic micellar systems, which are "precursors" of oil/water (O/W) microemulsions, illustrate well the great effect of confinement, since at higher concentrations $D_{water} \gg D_{surfactant} \cong D_{oil} \cong D_{micelle}$. Depending on the system (micelle size effects), D_{water}/D_{oil} ratios lie in the range 40–400 for high concentrations. An example is shown in Fig. 4. Studies of this type have been frequently used to investigate phenomena such as surfactant self-association, micelle size and intermicellar interactions, counterion binding, and solubilization partitioning, all factors that are essential to understanding microemulsion phenomena.

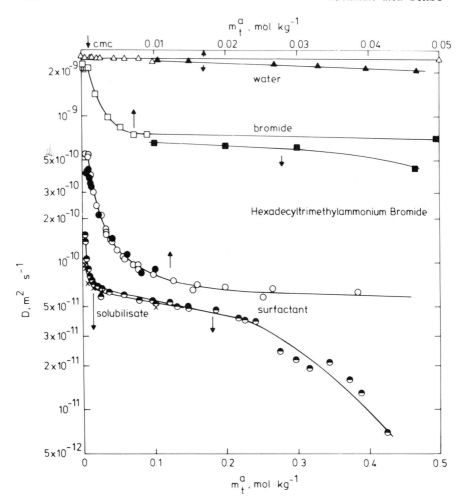

FIG. 4 Self-diffusion behavior of aqueous solutions of hexadecyl-
trimethylammonium bromide at 33°C. Note the difference by more
than two orders of magnitude between water diffusion and surfactant or solubilizate diffusion at high concentrations. (From Ref. 13.)

V. DOUBLE-CHAIN IONIC SURFACTANT-HYDROCARBON-WATER SYSTEMS

The most studied synthetic double-chain surfactant is Aerosol OT (sodium di-2-ethylhexylsulfosuccinate). Aerosol OT is highly soluble in many hydrocarbons, and these solutions can take up large amounts of water. The self-diffusion behavior of these systems is, as illustrated in Fig. 5, effectively opposite to that of solubilized micellar

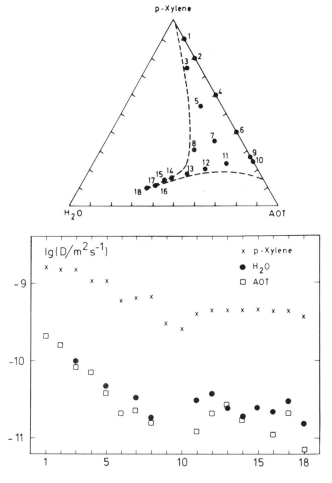

FIG. 5 In Aerosol OT-hydrocarbon-water systems, hydrocarbon self-diffusion may be one to two orders of magnitude more rapid than water and surfactant self-diffusion. (From Ref. 2.)

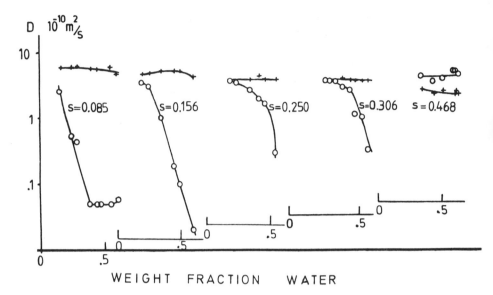

FIG. 6 Self-diffusion of water (o) and oil (+) versus the water content for the system didodecyldimethylammonium bromide-water-decane.
Five series with constant molar ratio (s) between surfactant and
(surfactant + oil) were studied. (From Ref. 23.)

solutions. Thus, $D_{oil} \gg D_{water} \cong D_{surfactant} \cong D_{droplets}$, corresponding to a water-in-oil droplet situation. Water diffusion stays
slow up to the highest water content studied, demonstrating that
rapid interdroplet exchange of water is relatively insignificant. The
self-diffusion behavior of Aerosol OT systems may be very temperature-dependent and, therefore, microemulsion structure may be quite
different at different temperatures [14–17].

A number of cationic double-chain surfactant-hydrocarbon-water
systems have recently been studied in detail by Ninham and Evans
[18–20] with interesting results. In the didodecyldimethylammonium
bromide-hydrocarbon-water system the surfactant is not soluble in
hydrocarbon but there is an extensive isotropic solution region in
the presence of water (ca. 15–60% water for decane). Hydrocarbon
self-diffusion is rapid throughout the microemulsion phase, whereas
surfactant diffusion is 10–100 times slower [21,22]. Water diffusion
is strongly dependent on water content and hydrocarbon. Over parts
of the microemulsion regions, water diffusion is very rapid (often
more rapid than hydrocarbon diffusion); whereas with increasing water
contents it may become very slow [21]. This is also illustrated in a
recent study by Fontell et al. [23] (see Fig. 6). It is evident that

in these systems one has in part the same pattern as for Aerosol OT, with a clear-cut W/O structure, but in part a bicontinuous structure. It appears that surfactant diffusion is more rapid in the bicontinuous concentration regions that in the W/O region (and of the order of 10^{-10} m^2/s, as expected for a surfactant continuous case). The observation that upon increasing the water content one introduces a transition from connected to disconnected water domains is unusual and suggests a different bicontinuous structure than for other systems studied (cf. below).

VI. FOUR-COMPONENT, SURFACTANT-COSURFACTANT-HYDROCARBON-WATER SYSTEMS

With butanol or pentanol as cosurfactant, D_{oil} and D_{water} are both high and of the same order of magnitude and differ only by a factor 2-3 over wide concentration ranges (Fig. 7). $D_{surfactant}$ is of the

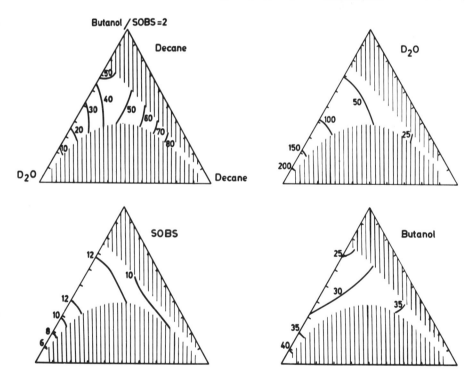

FIG. 7 Isodiffusion lines within the isotropic solution region of the system sodium octylbenzenesulfonate-butanol-decane-water (D_2O) (self-diffusion coefficients displayed in units of 10^{-11} m^{2}s^{-1}). (From Ref. 2.)

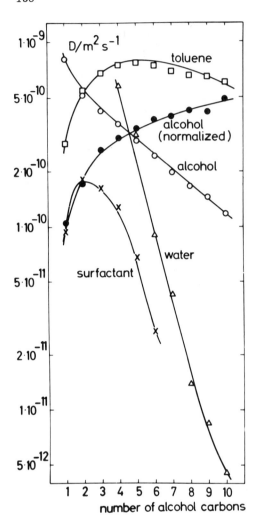

FIG. 8   The water self-diffusion coefficient decreases by more than
two orders of magnitude as the cosurfactant is changed from butanol
to decanol.   (Sodium dodecylsulfate-alcohol-toluene-water system
with weight ratios 17.5:35:12.5:35.)   (From Ref. 11.)

order of 10^{-10} m^2s^{-1}, as expected for a surfactant continuous structure, and is remarkedly constant over wide ranges of water-to-oil ratio. D_{water} decreases at the highest hydrocarbon contents and D_{oil} at the highest water contents, but these changes are not dramatic. Apparently characteristic of these microemulsion systems is that a bicontinuous structure persists over very wide concentration ranges [3,8,11].

The structure of four-component cosurfactant systems is dramatically dependent on the cosurfactant [24]. Changing the cosurfactant from butanol to decanol at constant composition changes D_{water} by two orders of magnitude, while D_{oil} remains quite constant (Fig. 8). With a long-chain alcohol as cosurfactant, the structure is of a distinct W/O droplet type.

Ceglie et al. [25] have further documented the cosurfactant effect and in particular demonstrated that a change of oil has little or no influence on the microemulsion structure.

VII. EFFECT OF SALINITY ON MICROEMULSION STRUCTURE

Several examples have demonstrated the great effect of salinity on phase diagrams and microemulsion structure. Increasing the salinity of a system of water, salt, cosurfactant, surfactant, and oil may change the phase behavior Winsor I → Winsor III → Winsor II, and this is paralleled by dramatic changes in self-diffusion behavior [26] (Fig. 9) from a region where $D_{water} \gg D_{oil}$, via a region where $D_{water} \cong D_{oil}$, both being high, to a region where $D_{oil} \gg D_{water}$. The change in the ratio D_{water}/D_{oil} is by a factor of ca. 10^4. $D_{surfactant}$ is around 10^{-11} m^2s^{-1} or somewhat above in the low- and high-salinity regions, whereas it rises to 10^{-10} m^2s^{-1} in the intermediate region.

Increasing addition of salt thus induces a transition from an oil-in-water structure to a bicontinuous structure (also surfactant continuous) and then to a water-in-oil structure [26].

VIII. NONIONIC SURFACTANT SYSTEMS

Systems studied [27] include tetraethyleneglycol dodecylether ($C_{12}E_4$)-hexadecane-water, $C_{12}E_4$-decane-water, and $C_{10}E_4$-hexadecane-water. The temperature-dependent phase behavior is accompanied by a self-diffusion behavior that is very sensitive to temperature. Since much more work is needed before we will have any reliable picture of these systems, we give here only a few pertinent observations.

In a wide concentration region from the water corner and down to ca. 60% water, D_{water} may be two orders of magnitude or more larger than D_{oil} and $D_{surfactant}$, these two being approximately equal. Then the structure is of the O/W type, with water diffusion decreasing with increasing droplet volume fraction at a rate corresponding to obstruction and hydration of spherical particles.

For the water-rich systems, $D_{surfactant}$ and $D_{hydrocarbon}$ in some cases decrease steadily up to high-volume fractions of surfactant + hydrocarbon, which can be ascribed to repulsive interactions between droplets slowing down droplet motion. At higher concentrations, $D_{hydrocarbon}$ and $D_{surfactant}$ have a minimum and then start to increase, which is referred to an exchange of molecules between droplets. A striking observation is that for some systems this exchange starts to become dominant at very low volume fractions (cf. Fig. 10). For the $C_{10}E_4$-hexadecane-water system, exchange of hydrocarbon molecules between micelles occurs more rapidly than

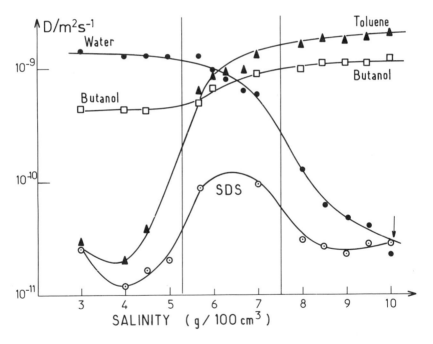

FIG. 9 Salinity may dramatically change the self-diffusion behavior of microemulsions, as shown for the system sodium dodecylsulfate-butanol-toluene-brine. For details (composition, etc.), see Ref. 26, from which the figure is taken. Between vertical lines there is a "middle-phase" microemulsion.

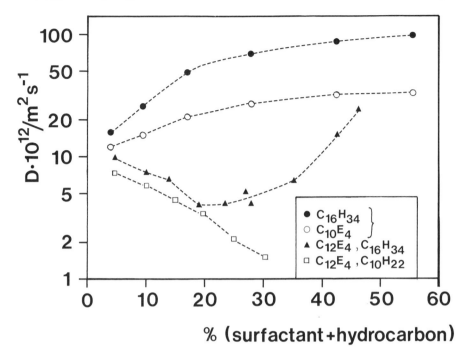

FIG. 10 Surfactant and hydrocarbon self-diffusion coefficients in the water-rich solution phase of various three-component nonionic surfactant systems. The experimental temperatures are 30°C ($C_{10}E_4$-$C_{16}H_{34}$), 19.6°C ($C_{12}E_4$-$C_{16}H_{34}$), and 4.7°C ($C_{12}E_4$-$C_{10}H_{22}$). (From Ref. 28.)

surfactant exchange even at low concentrations. These observations can be correlated with the distance from the so-called phase inversion or HLB temperature. The observations demonstrate attractive droplet-droplet interactions and rapid fusion-fission processes.

At high hydrocarbon contents, $D_{hydrocarbon} \gg D_{water}$, demonstrating W/O structure.

In the surfactant phase, which is found in the center of the ternary phase diagram, the self-diffusion of all three components is rapid, showing that no closed aggregates exist above the nanosecond range. Shinoda [29] has suggested the structure of the surfactant phase to be both oil- and water-continuous and to have a layered structure.

IX. DIFFERENT BICONTINUOUS STRUCTURES

In conclusion, it appears that the self-diffusion behavior, as obtained most conveniently in FT PGSE NMR studies, offers a direct and often detailed insight into microemulsion structure. The technique is very sensitive, as illustrated in our observations of changes in the D_{water}/D_{oil} ratio by a factor of 10^5 between different (macroscopically very similar) isotropic surfactant solutions at the same time as the individual D values can normally be obtained with a precision of 1% (90% confidence interval).

The technique has become a standard approach to distinguish between water-in-oil droplet, oil-in-water droplet, and bicontinuous structures. Bicontinuous structures have previously been somewhat overlooked but have now been demonstrated for a range of systems. It is also clear from these observations that basically different types of bicontinuous structures are possible. It is an important aspect of our current and future work to characterize the different bicontinuous structures.

X. REFERENCES

1. Shinoda, K. J., *Phys. Chem. 89*, 2429 (1985).
2. Stilbs, P., and Lindman, B., *Progr. Colloid Polymer Sci. 69*, 39 (1984).
3. Lindman, B., Kamenka, N., Kathopoulis, T. M., Brun, B., Nilsson, P.-G., *J. Phys. Chem. 84*, 2485 (1980).
4. Stilbs, P., Moseley, M. E., and Lindman, B., *J. Magn. Resonance 40*, 401 (1980).
5. Stilbs, P., *Progr. NMR Spectrosc.*, in press.
6. Stilbs, P., and Moseley, M. E., *Chem. Scr. 15*, 176 (1980).
7. Stilbs, P., and Moseley, M. E., *Chem. Scr. 15*, 215 (1980).
8. Lindman, B., Stilbs, P., and Moseley, M. E., *J. Colloid Interface Sci. 83*, 569 (1981).
9. Lindman, B., Stilbs, P., in *Microemulsions*, Friberg, S. E., and Bothorel, P., (eds.), CRC Press, Cleveland, Ohio, *in press*.
10. See Söderman, O., et al., *this volume*, and several other papers from our group.
11. Lindman, B., Ahlonäs, T., Söderman, O., Walderhaug, H., Rapacki, K., and Stilbs, P., *Faraday Disc. Chem. Soc. 76*, 317 (1983).
12. Jönsson, B., Wennerström, H., Nilsson, P.-G., and Linse, P., *Colloid Polymer Sci., 264*, 77 (1986).
13. Lindman, B., Kamenka, N., Puyal, M. C., Rymdén, R., and Stilbs, P., *J. Phys. Chem. 88*, 5048 (1984).

14. Eicke, H.-F., Hilfiker, R., and Holz, M., *Helv. Chim. Acta* 67, 361 (1984).
15. Eicke, H.-F., and Kubik, R., *Faraday Disc. Chem. Soc. 76*, 305 (1983).
16. Carnali, J. O., Ceglie, A., Lindman, B., and Shinoda, K., *Langmuir 2*, 417 (1986).
17. Stilbs, P., and Lindman, B., *unpublished observations*.
18. Chen. S. J., Evans, D. F., and Ninham, B. W., *J. Phys. Chem. 88*, 1631 (1984).
19. Ninham, B. W., Chen, S. J., and Evans, D. F., *J. Phys. Chem. 88*, 5855 (1984).
20. Evans, D. F., and Ninham, B. W., *J. Phys. Chem. 90*, 226 (1986).
21. Blum, F. D., Pickup, S., Ninham, B. W., Chen, S. J., and Evans, D. F., *J. Phys. Chem. 89*, 711 (1985).
22. Chen, S. J., Evans, D. F., Ninham, B. W., Mitchell, D. J., Blum, F. D., and Pickup, S., *J. Phys. Chem. 90*, 842 (1986).
23. Fontell, K., Ceglie, A., Lindman, B., and Ninham, B. W., *Acta Chem. Scand. A40*, 247 (1986).
24. Stilbs, P., Rapacki, K., and Lindman, B., *J. Colloid Interface Sci. 95*, 583 (1983).
25. Ceglie, A., Das, K. P., and Lindman, B., *J. Colloid Interface Sci., in press*.
26. Guéring, P., and Lindman, B., *Langmuir 1*, 464 (1985).
27. Nilsson, P.-G., and Lindman, B., *J. Phys. Chem. 86*, 271 (1982).
28. Nilsson, P.-G., Wennerström, H., and Lindman, B., *Chem. Scr. 25*, 67 (1985).
29. Shinoda, K., *Progr. Colloid Polymer Sci. 68*, 1 (1983).

XI. BIBLIOGRAPHY

Wärnheim, T., Sjöblom, E., Henriksson, U., and Stilbs, P., *J. Phys. Chem. 88*, 5420 (1984).

Stilbs, P., and Lindman, B., *J. Colloid Interface Sci. 99*, 290 (1984).

Sjöblom, E., Henriksson, U., and Stilbs, P., in *Reverse Micelles*, Luisi, P. L., and Straub, B. E., (eds.), Plenum Press, New York, 1984, p. 131.

Klose, G., Bayerl, P., Stilbs, P., Brückner, S. Zirwer, D., and Gast, K., *Colloid Polymer Sci. 263*, 81 (1985).

Sjöblom, E., Wärnheim, T., Henriksson, U., and Stenius, P., *Tenside Deterg. 21*, 303 (1984).

Cheever, E., Blum, F. D., Foster, K. R., and Mackay, R. A., *J. Colloid Interface Sci. 104*, 121 (1985).

Foster, K. R., Cheever, E., Leonard, J. B., Blum, F. D., and Mackay, R. A., *ACS Symp. Ser. 272*, 275 (1985).

Clarkson, M. T., Beaglehole, D., and Callaghan, P. T., *Phys. Rev. Lett.* *54*, 1722 (1985).

Ahlnäs, T., and Söderman, O., *Colloids Surf.* *12*, 125 (1984).

Lindman, B., and Stilbs, P., in *Surfactants in Solution*, Mittal, K. L., and Lindman, B. (eds.), Plenum Press, New York, 1984, vol. 3, p. 1651.

Lindman, B., and Stilbs, P., in *Surfactants in our World—Today and Tommorrow*, Proc. World Surfactants Congress, Kürle Druck, Gelnhausen, 1984, vol. III, p. 159.

Chatenay, D., Guéring, O., Urbach, W., Cazabat, A. M., Langevin, D., Meunier, J., Léger, L., and Lindman, B., *Proceedings of the 5th International Symposium on Surfactants in Solution*, Bordeaux, 1984, in press.

Sjöblom, E., and Henriksson, U., in *Surfactants in Solution*, Mittal, K. L., and Lindman, B. (eds.), Plenum Press, New York, 1984, vol. 3, p. 1867.

Lindman, B., Kamenka, N., Brun, B., and Nilsson, P.-G., in *Microemulsions*, Robb, I. D. (ed), Plenum Press, New York, 1982, p. 115.

8

The Interpretation of Nuclear Magnetic Resonance Relaxation Data from Micellar and Microemulsion Systems

O. SÖDERMAN   *Lund University, Lund, Sweden*

D. CANET   *Université de Nancy, Nancy, France*

J. CARNALI   *Lund University, Lund, Sweden*

U. HENRIKSSON   *Royal Institute of Technology, Stockholm, Sweden*

H. NERY   *Université de Nancy, Nancy, France*

H. WALDERHAUG   *Lund University, Lund, Sweden*

T. WÄRNHEIM   *Royal Institute of Technology, Stockholm, Sweden*

I. INTRODUCTION

Microemulsions have been a topic of intense research during the past decade. This is not surprising, since it has been realized that there are several potentially very interesting technical applications of these systems, in enhanced oil recovery, among others. Furthermore, some crucial aspects regarding solution structure and dynamics of these systems are still unknown. This is a situation unlike other surfactant systems, such as ordinary micelles, for which we have rather detailed knowledge about aggregate structure and molecular dynamics [1,2].

Most studies of microemulsions have employed various scattering techniques [3-6], which have undoubtedly shed much light on the microstructure in microemulsion systems. However, these studies are limited in one particular way, in that they can provide information only on global properties of the aggregates present in microemulsions. That is, the scattering techniques provide little, if any, information on the state of the different molecular species in the microemulsion systems. Such information can be obtained from spin resonance techniques, either electron (ESR) [7] or nuclear (NMR) [8,9]. Of these techniques, NMR has the advantage over ESR in that the unperturbed systems can be studied [10]. In particular, NMR relaxation studies have proven very valuable in the study of aggregated surfactant systems [11].

One problem with such studies is that the interpretation of the NMR relaxation data is complicated by the mere fact that the surfactants are aggregated, which renders their motion complex. A further complication in microemulsions as compared to, for instance, ordinary micellar systems is that the structure of the aggregates is often uncertain.

The study reported in this chapter shows how NMR relaxation data can be used to give information on both local and global properties of surfactant aggregates. We will exemplify with two systems: one containing spherical and rod-shaped micelles, where detailed knowledge about aggregate structure is available; and another, a microemulsion system with different cosurfactants, where the aggregate structure is uncertain. The local information obtained pertains to the amplitude (order parameters) and rate of the motions (correlation times) that the aggregated surfactants undergo within the aggregates, whereas the global information is in terms of correlation time(s) for the motion of the aggregates or interfaces. More specifically, extensive relaxation data for ^{2}H at several magnetic field strengths are presented for the decylammonium chloride/water and sodium dodecylsulfate/alcohol/toluene/water (where the alcohol is varied from propanol to decanol) systems. The relaxation data are interpreted with a relaxation model developed by Halle and Wennerström [12].

II. MATERIALS AND METHODS

Decylammonium chloride (DAC), selectively deuterated in the α-position, was synthesized according to the method suggested by Amudsen and Nelson [13], whereas α-deuterated sodium dodecylsulfate (SDS) was obtained from Syntestjänst, Lund, Sweden.

Two solutions of DAC, 0.5 M and 2.0 M, were prepared from triply distilled water. These samples were made by weighing directly into NMR tubes, which were then sealed. Six microemulsion samples in the SDS/alcohol/toluene/water system were prepared by weighing the constituents into glass ampules that were then flame-sealed. The weight ratios among the different components were 17.5:35.0:12.5: 35.0, and this composition was the same for all six alcohols used, propan-1-ol, butan-1-ol, pentan-1-ol, hexan-1-ol, octan-1-ol, and decan-1-ol. The water was triply distilled and the other chemicals were of p.a. grade.

The NMR experiments at and below 2.11 T were performed on a Bruker CXP 100 spectrometer (the microemulsion systems), equipped

with a field-variable flux stabilizer HS 90 var., and on a modified
Bruker HX 90 (the DAC system). Experiments at higher field
strengths were performed at 2.31 T with a Varian XL-100, at 4.70 T
with a modified Bruker WP 200, at 6.0 T with a home-built spectrometer using a wide-bore superconducting magnet, and finally, at 8.45 T
with a Nicolet NTC-360. The deuterium T_1's were measured with the
standard inversion recovery method, whereas T_2's were obtained with
the Carr-Purcell-Meiboom-Gill method or deduced from the line widths
after suitable corrections for contributions from inhomogeneities in
the static magnetic field.

The temperature in all experiments was 27°C. We estimate the
difference in temperature to be less than 1°C between measurements
performed at different field strengths. The evaluation of the various
parameters (see below) from the relaxation data was done as described
in Ref. 14.

III. THEORY

For simple liquids or solutions of small, nonaggregated molecules at
not too extreme temperatures or pressures, the NMR spin-lattice
(R_1) and spin-spin (R_2) relaxation rates for deuterium are given by

$$R_1 = R_2 = \frac{3\pi^2}{2} \chi^2 \tau_c \tag{1}$$

In Eq. (1), χ is the quadrupole coupling constant and τ_c is the
(effective) correlation time for the motion responsible for the NMR
relaxation [in Eq. (1), it is assumed that the asymmetry parameter
of the electric field gradient tensor is zero]. As an example, Eq.
(1) would hold in the case of a specifically deuterated amphiphile
dissolved in water at a concentration well below its cmc. However,
for associated surfactant systems, micellar systems among others,
one finds that R_1 is no longer equal to R_2 and, furthermore, that
R_1 depends on the magnetic field strength [12] (see also Figs. 1
and 2). Obviously, Eq. (1) is no longer sufficient to account for
the NMR relaxation data. Instances where this has not been realized
are abundant in the literature. Instead, the general expressions
for R_1 and R_2 have to be used [15]:

$$R_1 = \frac{3\pi^2}{40} \chi^2 [2\tilde{J}(\omega_0) + 8\tilde{J}(2\omega_0)] \tag{2}$$

$$R_2 = \frac{3\pi^2}{40} \chi^2 [3\tilde{J}(0) + 5\tilde{J}(\omega_0) + 2\tilde{J}(2\omega_0)] \tag{3}$$

Here, $\tilde{J}(\omega_0)$ is the reduced spectral density, evaluated at the Larmor frequency. To be more specific, $\tilde{J}(\omega_0)$ is the Fourier transform of the autocorrelation function for the irreducible field gradient tensor χ expressed in the laboratory frame.

The autocorrelation function contains the relevant molecular information, but the experiments give information on the Fourier transform of this quantity at only a few frequencies. To solve this problem, one usually constructs models for the molecular motions and tests these against the obtained relaxation data. Needless to say, such models have to be chosen as realistically as possible. On the other hand, important information can be obtained from the frequency-dependent relaxation rates, and in favorable cases spectral density functions can be obtained without the use of any motional model [11b].

Such a model, termed the "two-step model," has been developed by Wennerström and co-workers [12,16] for isotropic surfactant systems. (In the present context, the term isotropic system implies a system that shows no static effects in the NMR spectrum, except for the chemical shift and J couplings). The basic feature of the model is a division of the molecular motions giving relaxation into fast, local motions (trans-gauche isomerizations, torsions, etc.), which, due to the "anchoring" of the polar end of the amphiphile, are rendered slightly anisotropic; and slower motions, which, when combined, are isotropic. Within the framework of the two-step model, the spectral density is given by

$$\tilde{J}(\omega_0) = (1 - S^2)\tilde{J}^f(\omega_0) + S^2\tilde{J}^s(\omega_0) \tag{4}$$

where $\tilde{J}^f(\omega_0)$ and $\tilde{J}^s(\omega_0)$ are the spectral densities for the fast and show motions, respectively. S is defined as

$$S = \frac{1}{2} <3 \cos^2 \theta - 1>_f \tag{5}$$

where θ is the angle between the C$-$D bond and the normal (director) to the aggregate surface, and the subscript "f" denotes that the average should be taken over a time that is long compared to the time scale of the fast motion but short compared to the time scale of the slow motion(s).

Given functional forms for the different $\tilde{J}$'s, we are now in a position to apply the model to relaxation data. For the fast motions we shall assume that they are so rapid that they fulfill the criteria for extreme narrowing, which means that they produce no field dependence in the relaxation rates. Thus, we have

$$\tilde{J}^f(\omega) = 2\tau_c^f \tag{6}$$

where $\tau_c{}^f$ is an effective correlation time for the rapid motions. For the slow motions, we will treat three cases. First, consider the case of spherical micelles. In this instance, Wennerström et al. [16] suggested that the slow motion is given by the rotational tumbling of the micelle. Since this motion can be described by an exponential correlation function, we have

$$\tilde{J}(\omega_0) = \frac{2\tau_c{}^s}{1 + (\omega_0\tau_c{}^s)^2} \tag{7}$$

where $\tau_c{}^s$ is the correlation time for the rotational tumbling. Second, we have the case of rod-shaped micelles. If we, as in the spherical case, assume that the slow motions are the rotational diffusion of the micelle around the long axis, and end-over-end tumbling of the rod, we have (again with exponential correlation functions and ignoring any end effects)

$$\tilde{J}^s(\omega_0) = \left(\frac{1}{4}\right) \left[\frac{6\tau_c{}^{im}}{1 + (\omega_0\tau_c{}^{im})^2} + \frac{2\tau_c{}^{\perp}}{1 + (\omega_0\tau_c{}^{\perp})^2} \right] \tag{8}$$

where

$$\tau_c{}^{im} = \frac{1}{2\theta_\perp + 4\theta_\parallel} \tag{9}$$

and

$$\tau_c{}^{\perp} = \frac{1}{6\theta_\perp} \tag{10}$$

$\theta_\parallel$ and $\theta_\perp$ are the rotational diffusion coefficients for diffusion around the long axis and end-over-end tumbling, respectively. For the case where $\theta_\perp \ll \theta_\parallel$, which would be the case for rods with large axial ratios, Eq. (9) can be written

$$\tau_c{}^{im} = \frac{1}{4\theta_\parallel} \tag{11}$$

Thus, we have effectively performed yet another time-scale separation of motions, this time between the rotational diffusion around the long axis and the end-over-end micellar tumbling; we have arrived at a "three-step" model.

At this point we should mention one relevant detail, namely, the influence of surfactant diffusion over the curved surface of the spherical micelle or around the long-axis of the rod-shaped micelle. These motions will contribute to the relaxation, but since they average the *same* part of the interaction as rotational tumbling of the spherical micelle and micellar rotational diffusion around the long axis, respectively, we are permitted to use an effective correlation time for the surfactant and aggregate diffusion:

$$(\tau_c)^{-1} = (\tau_c)_{rot}^{-1} + (\tau_c)_{diff}^{-1} \tag{12}$$

Finally, we turn to microemulsions. In this case we do not know the microstructure of the solution a priori, and therefore, we do not know any details about the aggregate motions responsible for the NMR relaxation. As a starting point we will assume that the slow motions can be described with exponential correlation functions. Thus, we are going to attempt to use Eq. (7) when interpreting data from the microemulsions.

IV. RESULTS AND DISCUSSION

The outline of this discussion will be as follows: First we will discuss the DAC/water system; then we will treat the microemulsion systems; and finally, we will draw some general conclusions.

A. The DAC/Water System

In Fig. 1a we represent R_1 and R_2 as functions of the Larmor frequency for a 0.5 M DAC/water system; the R_1 data for the 2 M system are presented in Fig. 1b. In addition, Table 1 shows R_2 data for the latter system.

The following features are immediately noticeable: For both systems, R_1 depends on the measuring frequency and R_1 is different from R_2. For the 0.5 M system, R_1 levels off to a constant value, within the experimental uncertainty, and R_2 at 2.31 T is equal to the R_1 value at the plateau. For the 2 M system, on the other hand, R_1 levels off at intermediate frequencies, but as the frequency is further lowered, R_1 begins to increase again. Furthermore, R_2 at 2.31 T is much larger than R_1 at the lowest frequency used. Taken together, these findings clearly point to a situation where, for the 0.5 M system, there is but one slow motion affecting the NMR relaxation, e.g., a system consisting of spherical (or nearly spherical) micelles; whereas for the 2 M system, there are at least two slow motions contributing to the NMR relaxation, which in line with the discussion above, indicates the presence of rod-shaped micelles. This does not mean that these are the *only* slow motions

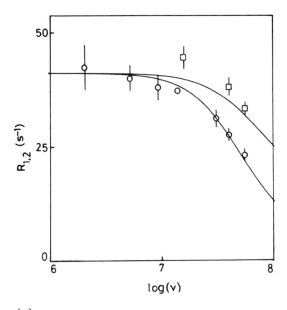

(a)

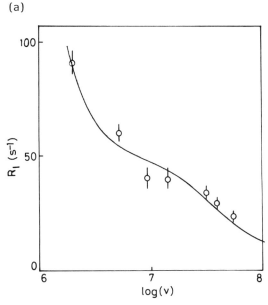

(b)

FIG. 1 ^{2}H spin-lattice (○) and spin-spin (□) relaxation rates for
0.5 M (a) and 2 M (b) samples of DAC/water as a function of the
logarithm of frequency, expressed in hertz. Solid lines are results
of fitting the relaxation expressions to the data (see text for details).

TABLE 1 ^{2}H Spin-Spin Relaxation Rates for a
2 M DAC/Water Sample at Three Field Strengths

B_0 (T)	ω (MHz)	R_2 (s^{-1})
2.31	15.1	550 ± 25
6.00	39.1	490 ± 25
8.45	55.3	510 ± 30

present. It simply means that these are the fastest *isotropic* motions [11,17].

The next step is then to fit Eqs. (6), (7), and (8) together with Eqs. (2) and (3) to the relaxation data in Figs 1a and 1b. These fits produce the solid lines in Figs. 1a and 1b. The agreement is satisfactory. In fact, we have also included extensive multifield ^{13}C relaxation data for all resolved carbons along the chain in the fits (for ^{13}C, completely analogous expressions hold for the relaxation). Thus, we are able to rationalize a large body of experimental data from different nuclei for the case of spherical and rod-shaped micelles. The resulting parameters of the fits are given in Table 2. The values of S and τ_c^f in Table 2 refer to the

TABLE 2 Parameters Obtained from Fits of the Relaxation Expressions to the Data in Figs. 1a and 1b[a]

Parameter	Concentration	
	0.5 M	2.0 M
S	0.224 ± 0.005[b]	0.234 ± 0.005
τ_c^f (ps)	13.0 ± 1.0	14.7 ± 1.0
τ_c^s (ns)	1.7 ± 0.1	–
τ_c^{im} (ns)	–	2.6 ± 0.2
$\tau_c^\perp$ (ns)	–	300 ± 30

[a]Plus extensive ^{13}C data; see text for details.
[b]Uncertainties correspond to an approximately 80% level of confidence, taking only random errors into account.

C-D bond of the α-methylene group. Corresponding values of S
and $\tau_c{}^f$ for the other carbons along the chain are also obtained,
but these will be presented in a separate publication [18]. Turning to the individual values in Table 2, we note that the values for
both S and $\tau_c{}^f$ are the same in spherical and rod-shaped micelles.
The value of S can be compared with that measured directly from
the deuterium quadrupolar splitting in the hexagonal phase, viz.,
S = 0.204. Thus, the value of S depends marginally on the structure of the aggregates present or whether one has an isotropic or
anisotropic phase. For the fast correlation time, a similar situation
holds, i.e., there is no difference between spherical and cylindrical
micelles. To put the values of $\tau_c{}^f$ into perspective, it is interesting to compare it with the correlation time found for DAC in aqueous solution at a concentration below the cmc, which is 11.6 ±
0.1 ps [18]. The correlation time for the monomer is thus slightly
shorter, but if we take into account the fact that motions of the
molecule as a whole contributes to the monomeric correlation time,
it would appear that the influence of micellization is marginal on
the rapid, local motions that the C_α-D bond undergo.

It remains to interpret the values of the correlation times for
the slow motions. Starting with the case of spherical micelles, we
suggested above that we can write an effective correlation time for
the combined effects of micellar tumbling and surfactant diffusion;
viz., Eq. (12) can be used. In Eq. (12), $(\tau_c)_{rot}$ can be calculated
from the Debye-Stokes-Einstein equation,

$$\left(\tau_c\right)^{-1}_{rot} = \frac{3kT}{4\pi r_M{}^3 \eta} \tag{13}$$

and $(\tau_c)_{diff}$ is obtained from the diffusion equation,

$$\left(\tau_c\right)^{-1}_{diff} = \frac{6D}{r_M{}^2} \tag{14}$$

Here, r_M is the radius of the micelle, D is the lateral diffusion coefficient of the surfactant, and the rest of the quantities have their
usual meaning. In previous work we have computed r_M from the
experimental value for $\tau_c{}^s$. This presupposes a knowledge of the
surfactant lateral diffusion coefficient, which was obtained from
measurements in cubic or anisotropic phases. However, since this
quantity has not yet been measured for DAC, we shall instead compute D from the obtained $\tau_c{}^s$. To do this, we need a value for
$(\tau_c)_{rot}$ that can be calculated, given a value for r_M. Now, theoretical considerations imply that the radius of a spherical micelle should
be equal to the length of the all-trans chain [19], which for a DAC

molecule is equal to 16.9 Å [20], giving a value of 4.1 ns for $(\tau_c)_{rot}$. Equations (12) and (14) then give a diffusion coefficient $D = (1.6 \pm 0.2) \times 10^{-10}$ m^2s^{-1} for surfactant lateral diffusion over the curved micellar surface. As a comparison, we note that the corresponding value for the C_8 analog in the lamellar phase is 3.3×10^{-10} m^2s^{-1} [21]. Since surfactant lateral diffusion depends marginally on the chain length for a given head group [21], this indicates that the diffusion rate for a given surfactant is not very dependent on the aggregate geometry. Finally, for the 2 M sample we shall perform an analogous calculation. To do this we first need to know the length of the rod-shaped micelle. Assuming that the micelle can be treated as a stiff ellipsoid with a minor axis equal to 16.9 Å, we obtain, from $\tau_c^\perp$ and Perrin's equations [22], a major axis equal to 150 Å. The next step is to compute, again using the appropriate equation from Perrin's work [22], the correlation time for the micellar rotational diffusion around its long axis, viz., $\tau_c^\parallel = 30.6$ ns. Finally, the value for τ_c^{im} from Table 2, together with Eq. (11), gives $D = (2.5 \pm 0.5) \times 10^{-10}$ m^2s^{-1}. Within the uncertainty, this value is essentially the same as that found for the spherical case, providing further support for the statement given above that the lateral diffusion depends marginally on the aggregate geometry. As one more test of the validity of the three-step model in interpreting data from the 2 M system, we have, using the data in Table 2, predicted the value for R_2 at 2.31 T. The result is $R_2 = 544$ s^{-1}, which is in agreement with the experimental value in Table 1.

B. The Microemulsion System

The reason for choosing the SDS/toluene/water/alcohol microemulsion system to investigate the influence of the cosurfactant is the existence of extensive self-diffusion data due to Lindman et al. [23] for that system. In short, these workers found (for a constant weight composition equal to the one used in the present work) that the hydrocarbon self-diffusion is rapid for all investigated alcohols, but that the water self-diffusion is critically dependent on the alcohol chain length. For C_4-C_5 alcohols the water and hydrocarbon diffusion rates are not very different, indicating that the structure of the microemulsion is bicontinuous on the time scale of the diffusion measurement (100 ms), thus excluding both closed water and oil droplets. For the longer alcohols and in particular for decan-1-ol, the water diffusion is roughly two decades slower than the hydrocarbon diffusion, clearly indicating the existence of closed water droplets in a hydrocarbon medium.

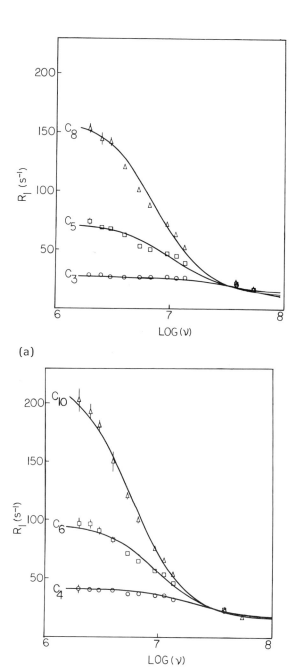

(a)

(b)

FIG. 2 ^{2}H spin-lattice relaxation rates for the surfactant α-CD$_2$ group in the SDS/water/toluene/alcohol (with six different alcohols) system. The alcohols are indicated in the figure. Solid lines are results of fitting the relaxation expressions to the data (see text for details).

TABLE 3 ^{2}H Spin-Spin Relaxation Rates in s^{-1} for the Surfactant α-CD$_2$ Group in Microemulsions Containing Different Alcohols

| | B_o | | | |
| | 2.11 T | | 6.00 T | |
Alcohol	Obs	Predicted[a]	Obs	Predicted[a]
C$_3$OH	33.0 ± 1.7	27.3	28.6 ± 1.0	25.0
C$_4$OH	45.2 ± 2.3	37.2	34.5 ± 1.0	30.0
C$_5$OH	69.0 ± 2.8	54.0	47.8 ± 1.5	38.8
C$_6$OH	95.2 ± 3.0	69.5	74.1 ± 1.5	47.3
C$_8$OH	188 ± 5	97.6	149.3 ± 2.0	65.9
C$_{10}$OH	417 ± 10	119.5	303 ± 3	84.3

[a]R_2's predicted from Eqs. (3), (4), (6), and (7) and data in Table 4.

Presented in Figs. 2a and 2b are the field-dependent deuterium R_1's for the six straight-chain alcohols investigated, whereas Table 3 gives the corresponding R_2's measured at two field strengths. With reference to Figs. 2a and 2b, and Table 3, the following features are immediately noticeable. First, there is a marked field dependence in the R_1's at lower field strengths (B_o 2.1 T). Regardless of the model used to interpret the relaxation data, this indicates the presence of slower motions contributing to the NMR relaxation, or in other words, the presence of surfactants aggregated into some sort of interface.

Second, all the R_1's converge toward a common value at higher field strengths. In the framework of the two-step model, this is a consequence of the fact that at these high field strengths, the R_1's are dominated by the fast, local motions. For the case of $(\omega_0 \tau_c^s) > 1$ in Eq. (7), the second term on the right-hand side in Eq. (4) can be neglected, reducing the expression for R_1 to the following equation:

$$R_1 = \left(\frac{3\pi^2}{2}\right) \chi^2 (1 - S^2) \, \tau_c^f \tag{15}$$

Now, since S typically is much smaller than 1, S^2 can be neglected in Eq. (15). Thus the data in Figs. 2a and 2b indicate that the rate of the fast, local motions for the C_α-D bond does not depend on the type of cosurfactant used. Finally, we note that for propan-1-ol and butan-1-ol the R_1's at the lower fields are almost equal to the R_2's at 2.11 T presented in Table 3; whereas for the longer-chain alcohols, and in particular for decan-1-ol, R_2 is significantly larger at 2.11 T than R_1 at the lowest field.

We now proceed to fit Eqs. (2) combined with Eqs. (6) and (7) to the R_1 data in Figs. 2 a and 2b. These fits produce the solid lines in Figs. 2a and 2b and the parameters presented in Table 4. With the data in Table 4 and Eqs. (3), (6), and (7), it is possible to predict the outcome of measurements of R_2, *provided* that the motion causing the frequency dependence in R_1 is the fastest iso-tropic motion responsible for the NMR relaxation. Presented in Table 3 are such predicted values of R_2 together with experimental R_2's. As noted above, the difference between these quantities is minor for propan-1-ol and butan-1-ol but, starting with pentan-1-ol, the discrepancy becomes significant and increases with increasing chain length of the alcohol. Thus, the assumption made above that the motion causing the frequency dependence in R_1 is the fastest isotropic motion does not hold for alcohols longer than butan-1-ol. Rather, we have a situation analogous to the case of the 2 M DAC/water system presented above, i.e., there is a motion on intermediate time scales that is anisotropic and leaves a residual anisotropy that is averaged to zero by still slower motion(s). In the case of DAC/water we could be rather specific about the nature of this motion, since we knew in advance the structure of the solution. In the microemulsion case, such independent information is not at hand, and we can therefore not be as specific for these systems (vide supra).

TABLE 4 Parameters Obtained from Fits of the Relaxation Expressions to the Data in Figs. 2a and 2b

Alcohol	S	τ_c^f (ps)	τ_c^s (ns)
C_3OH	0.157 ± 0.045[a]	25.0 ± 11.4	1.7 ± 0.8
C_4OH	0.117 ± 0.017	38.8 ± 2.5	4.4 ± 0.6
C_5OH	0.124 ± 0.007	42.2 ± 3.8	8.4 ± 1.3
C_6OH	0.147 ± 0.004	38.4 ± 2.1	9.0 ± 0.6
C_8OH	0.171 ± 0.002	36.8 ± 1.4	11.9 ± 0.3
$C_{10}OH$	0.181 ± 0.001	36.9 ± 1.2	15.0 ± 0.6

[a]Uncertainties as in Table 2.

Turning to the individual parameter values in Table 4, it is
clear that for propan-1-ol and butan-1-ol the correlation time, τ_c^S,
is comparatively small (compare, for instance, with the value of 1.7
as found for the spherical DAC micelles), pointing to a situation
where one has predominantly small aggregates present (whose dimensions do not exceed the length of a surfactant molecule) and/or a
very flexible hydrophobic-hydrophilic interface that changes direction extensively within a few nanoseconds. Since these motions are
almost isotropic, it is necessary that their amplitude is large, so
that the interaction is effectively averaged to zero. For the longer-
chain alcohols, τ_c^S is getting progressively longer and reaches the
value of 15 ns for the decan-1-ol case. As stated in the introduc-
tion of this section, the self-diffusion data for the systems containing longer-chain alcohols imply a situation with closed water droplets.
In such a case, candidates for the intermediate and slow motions are
shape fluctuations of the droplet, which are not isotropic, and tumbling of the whole droplet, respectively. Such shape fluctuations
would in principle give rise to nonexponential correlation functions
[24]. As is evident from Figs. 2a and 2b, the fits to exponential
correlation functions (note also that quoted uncertainties of the
parameters in Table 4) are reasonably good. Therefore, we will
not attempt to fit any other functional form of the spectral density
to the data in the present work, but defer such an analysis to a
more detailed publication, in which we will also include ^{13}C measurements and additional [2]H R_1's at a few more frequencies. To conclude, for the shorter-chain alcohols there are motions that are comparatively rapid (on a time scale of a few nanoseconds) and of large
amplitudes. This picture is gradually changed when the chain length
of the cosurfactant is increased and we have a complex situation with
motions occurring over different time scales that are effective in
producing NMR relaxation.

Concerning the fast correlation times in Table 4, they are, in
complete agreement with the statement made above, independent of cosurfactant chain length. They are roughly a factor of 3 longer than
for the DAC case, a difference that can be ascribed to the difference
in head group between the two systems. The SDS head group is
larger and also heavier, presumably influencing the local dynamics of
the C_α–D bond. This is in agreement with the result found for another
surfactant with a bulky head group, namely, dodecyltrimethylammonium
chloride, where the fast correlation time was 37 ps [11].

Finally, we turn to the order parameters. First, we note that
we have used a two-step model in the fits, producing the solid lines
in Table 4. This is justified for the shorter-chain alcohols, where
the motions causing the dispersion in R_1 are practically isotropic, but
not for the longer-chain alcohols, where this is not the case. For
these latter systems the order parameters, given in Table 4, are

then lower limits (see the discussions in Ref. 11). The magnitude
of S for all alcohols falls between 0.1 and 0.2. Unfortunately, to
our knowledge, at present there are no reported values from lamellar
phases of SDS that can serve as comparisons (we are presently
collecting such data). Usually, one finds in lamellar phases of other
surfactants values of around 0.2 (see, for instance, values for the
DAC system above). Thus, the conformational state in the interfacial film in the microemulsions studied does not seem to be very
different from that in lamellar phases (perhaps with the exception
of butan-1-ol). This is partly in contradiction to the conclusion
reached by De Gennes and Taupin [26], who state that the "fluidity,"
as measured with order parameters, is considerably lower in microemulsions as compared to lamellar phases.

Disregarding propan-1-ol, where the uncertainty is rather large,
it is clear that the value of S increases as the cosurfactant is made
longer. This is an effect that has also been observed in three-
component surfactant/alcohol/water lamellar phases [25]. In this
work the effect was discussed in terms of the difference in chain
lengths between the surfactant and the cosurfactant, which, for the
shorter-chain alcohols, would tend to compress the bilayers, leaving
more conformational freedom for the surfactant chain.

Another factor that may contribute to the effect is suggested by
the fact that butan-1-ol is soluble in water up to some 10%, which
renders the water domains "more hydrophobic," leading to an increased concentration of monomeric surfactant. Since the observed
S is an average over the different sites where SDS is distributed,
and since S for the monomers is zero, this would lead to a lower S
for butan-1-ol in comparison with decan-1-ol, which has a much lower
solubility in water. In this context, it is of interest to mention a
study by Stilbs [27], who concludes from solubilization data of short-
chain alcohols in SDS micelles that the alcohols cause a breakdown
of the micellar aggregates.

In conclusion, there are differences in the surfactant conformational order as the cosurfactant is made longer, but the differences
are slight. Furthermore, the conformational state is not expected to
be very different form that found in corresponding anisotropic phases
of SDS.

V. CONCLUDING REMARKS

The intention of this chapter has been to show how NMR relaxation
data can be used to draw conclusions not only about global properties,
such as aggregate size, but also about local properties, such as the
conformational and motional state of the individual surfactant molecules, in aggregated surfactant systems.

For the micellar case, we were able to rationalize a large body of relaxation data for both spherical and elongated micelles in terms of fast, local motions and slower motions associated with aggregate tumbling and/or surfactant diffusion over the curved aggregate surface. Among other things, we found that the state of the surfactants in the aggregates were marginally dependent on the geometry of the aggregates or the type of phase present.

For the microemulsions, because of lack of independent structural data, we could not be so specific. It was, however, possible to fit the two-step model to the field-dependent R_1's, and the data thus obtained indicated that the rate of the local motions was independent of the chain length of the cosurfactant. The surfactant C_α–D order parameter was dependent on cosurfactant chain length, but the magnitude of S was not expected to differ dramatically from those found in lamellar phases. Finally, the NMR relaxation data indicate that the aggregates, whatever their shape may be, is in a very dynamic state for the shorter-chain alcohols; i.e., the ordering imposed on the surfactants by the interface is very short-lived (a few nanoseconds). For the longer-chain alcohols, there are motions over different time scales that produce NMR relaxation, possibly related to shape fluctuations and aggregate tumbling, respectively.

VI. REFERENCES

1. Lindman, B., and Wennerström, H., *Top. Curr. Chem. 87*, 1 (1980).
2. Wennerström, H., and Lindman, B., *Phys. Rep. 52*, 1 (1979).
3. Dvolaitzky, M., Lagues, M., Le Peasant, J. P., Ober, R., Sauterey, C., and Taupin, C., *J. Phys. Chem. 84*, 1532 (1980).
4. Ober, R., and Taupin, C., *J. Phys. Chem. 84*, 2418 (1980).
5. (a) Auvray, L., Cotton, J.-P., Ober, R., and Taupin, C., *J. Phys. Chem. 88*, 4586 (1984); (b) *J. Physique 45*, 913 (1984).
6. Kaler, E., Bennet, K. E., Davis, H. T., and Scriven, L. E., *J. Chem. Phys. 79*, 5673 (1983).
7. Paz, L., Di Meglio, J. M., Dvolaitzky, M., Ober, R., and Taupin, C., *J. Phys. Chem. 88*, 3415 (1984).
8. (a) Stilbs, P., and Lindman, B., in *Physics of Amphiphiles*, Corti, M., and Degiorgio, V. (eds.), North Holland, Amsterdam, 1984; (b) Söderman, O., and Walderhaug, H., *Langmuir 2*, 57 (1986).
9. Tricot, Y., Kiwi, J., Niederberger, W., and Grätzel, M., *J. Phys. Chem. 85*, 862 (1981).
10. Taylor, M. G., and Smith, I. C. P., *Biochim. Biophys. Acta 85*, 862 (1981).

11. (a) Söderman, O., Walderhaug, H., Henriksson, U., and Stilbs, P., *J. Phys. Chem.* 89, 3693 (1985); (b) Henriksson, U., Stilbs, P., Söderman, O., and Walderhaug, H., in "Magnetic Resonance and Scattering in Surfactant Systems," Magdid, L. (ed.), Plenum, N.Y., *in press.*

12. Halle, B., and Wennerström, H., *J. Chem. Phys.* 75, 1928 (1981).

13. Amudsen, L. H., and Nelson, L. S., *J. Am. Chem. Soc.* 73, 242 (1951).

14. Walderhaug, Söderman, O., and Stilbs, P., *J. Phys. Chem.* 88, 1655 (1984).

15. Abragam, A., *The Principles of Nuclear Magnetism*, Clarendon Press, Oxford, 1961.

16. Wennerström, H., Lindman, B., Söderman, O., Drakenberg, T., and Rosenholm, J. B., *J. Am. Chem. Soc.* 101, 6860 (1979).

17. (a) Lipari, G., and Szabo, A., *J. Am. Chem. Soc.* 104, 4546 (1982); (b) *J. Am. Chem. Soc.* 104, 4559 (1982).

18. Nery, H., Söderman, O., Canet, D., Walderhaug, H., and Lindman, B., *J. Phys. Chem.*, *in press.*

19. Jönsson, B., *personal communication.*

20. Tanford, C. J., *J. Phys. Chem.* 76, 3020 (1972).

21. Lindblom, G., and Wennerström, H., *Biophys. Chem.* 6, 107 (1971).

22. Perrin, F., *J. Phys. Radium 10*, 33 (1934).

23. Lindman, B., Ahlnäs, T., Söderman, O., Walderhaug, H., Rapacki, K., and Stilbs, P., *Faraday Disc. Chem. Soc. 76*, 317 (1983).

24. (a) Brown, M. F., *J. Chem. Phys.* 80, 2808 (1984); (b) Brown, M. F., *J. Chem. Phys.* 80, 2832 (1984); (c) Marqusee, J. A., Warner, M., and Dill, K. A., *J. Chem. Phys.* 81, 6404 (1984).

25. Söderman, O., Walderhaug, H., and Lindman, B., *J. Phys. Chem.* 89, 1795 (1985).

26. De Gennes, P. G., and Taupin, C., *J. Phys. Chem.* 86, 2294 (1982).

27. Stilbs, P., *J. Colloid Interface Sci.* 89, 547 (1982).

9

Phase Behavior and Structures by Neutron Scattering of Aqueous Hydrocarbon/Water/Nonionic Surfactant Systems

JEAN-CLAUDE RAVEY, MARTINE BUZIER, and GASTON DUPONT
Université de Nancy, Nancy, France

I. INTRODUCTION

Oil solubilization in aqueous nonionic surfactant systems has been investigated by a correlative study of phase behavior and the structure of the micellar aggregates (by small-angle neutron scattering). In this chapter the respective influence of the temperature and the oil/surfactant and water/surfactant ratios is described. According to these experimental conditions, oil-swollen micelles and microemulsions can be defined. Their respective structure is specified in terms of area per polar head and oil penetration in the surfactant film. This study has shown the existence of a particular critical end point, and analyzes the formation and influence of liquid crystal phases.

   A nonionic surfactant is hydrophilic below a certain temperature, which is related to the balance between the hydrophobic and hydrophilic parts of the molecule. Accordingly, the ternary oil/water/surfactant phase diagrams exhibit a succession of various outlooks that depend on the chemical nature of both the oil and the emulsifiant [1–12]. In the case of n-alkyl polyglycol ethers, C_nEO_m, systematic work on the isotropic phases of $C_{12}EO_m$-water-oil was initiated by Friberg and the Swedish school [2,3]. They showed that this evolution follows a unique mean scheme. At low temperature, the aqueous one-phase realm is a sector-like domain along the water-surfactant interface. When the temperature increases, there must be a splitting of that region for temperatures in the proximity of the lower critical point of the binary water-surfactant solution. One of these realms becomes an isolated domain, which then moves toward the oil-surfactant interface.

As far as the two- and three-isotropic phase domains are concerned, these real diagrams seem to bear some resemblance to the classical (Winsor) diagrams only for the higher temperatures. Indeed, for these "long" surfactants the phase diagrams appear much more complex than for the shorter compounds [10], which include the alcohols as the first terms of these series; we note the existence of large domains with liquid crystal phases (mainly lamellar), which express the strong tendency to associate into large "palissade" layers. Hence, there must be multiple equilibria among several isotropic surfactant-water-oil phases and the liquid crystals. At the present time, full experimental diagram studies (including plait points and tie-lines) of these complex nonionic systems remain quite scarce: Recent results from Kunieda and Shinoda constitute an impressive example of such phase behaviors [4].

In an attempt to generalize, and to give a "simple" explanation for the formation of three-phase systems, Kahlweit et al. [13] claim that the phase behavior is qualitatively very similar, irrespective of the chemical nature of the surfactant, i.e., whether the nonionic C_nEO_m emulsificant is a short- or long-chain molecule. For that purpose he suggests considering these systems as being "near tricritical." The behavior is quite clear for shorter nonionics [10]: When the critical (end) point of the aqueous (oil + surfactant) solution touches the oil/water solubility gap and when the plait point of this gap still faces the oil-surfactant interface, then the three-phase triangle emerges. Similarly, both the oil-surfactant and water-surfactant phases form critical systems.

For longer surfactants, and perhaps because of the presence of anisotropic phases, it is not yet evident whether or not such "simple" behavior still holds, at least for the lower temperatures. Hence, more detailed experimental investigations are necessary in order to understand the more or less progressive change from the water phase (containing oil-swollen micelles) to the surfactant phase (with "microemulsion" structures); such studies should also try to correlate the phase behavior with structural results.

More specifically, this study is interested in the splitting process of the aqueous isotropic phase region when temperature increases, and in the progressive formation and evolution of the isolated microemulsion (or surfactant-phase) realm. Incidentally, the actual dependence of the hydrocarbon solubilization on the cloud point of the system, i.e., the sudden onset of turbidity of the solution on raising the temperature, can be observed. We have to show that simple statements such as "the hydrocarbon solubilization hinders the cloud point" may be insufficient. And we will see that the onset of turbidity may not always be taken as indicative of the true limit of oil incorporation [14–16].

In order to understand this behavior more thoroughly, we have performed a detailed phase study of the ternary mixture $C_{12}EO_6$ +

decane + water for temperatures in the range 10–60°C, but with special interest in the systems containing more than 70% water. Similarly, we have tentatively determined the structures of the micellar aggregates by using small-angle neutron scattering, taking advantage of the contrast variation that such a technique affords. In this chapter we give only a general survey of our structural results, summarizing the method we used for the interpretation of the scattering data and emphasizing the actual correlation between phase behavior and structure.

II. EXPERIMENTAL

The decane used was a Merck product and applied without further purification. The $C_{12}EO_6$ surfactant was purchased from Nikko Chemicals, Tokyo, and was used as received (purity specified 98%). The critical temperature of its aqueous solution was found to be about 47.5 ± 0.5°C. This value was subjected to aging, mainly as a result of long storage at high temperatures ($t > 40°C$). Since the critical temperature also depends on the manufacturer's batch, the same sample was used throughout our study.

Phase behavior determination was performed in two steps. First a "quick" span of the whole domain of interest was carried out using an automatic apparatus. To each sample of given water/surfactant concentration ratio, small aliquots of decane were successively added, the oil concentration increased by steps of 1% w/w, and with the mixture being continuously stirred. The temperature of each batch of samples was continuously raised and then decreased to the same rate of 0.1°C/min. Using such temperature loops allowed eventual hysteresis effects to be noted, indicating a deviation from the static true equilibrium. The change of the phases was recorded by classical turbidimetry. The eventual anisotropic character of the phases in the sample was periodically checked with the help of an automatic device that inserts polarizers into the laser beam for a short time. The whole experimental setting was monitored by a microordinator whose internal slots are equipped with a programmable timer module and AD/DA boards. After preset values were given to the timer and while the pulses of this clock monitored the thermostatic bath, the temperature and turbidity of the samples were continuously measured and displayed on a scope; data were stored on floppy disks from which a BASIC program determined the points of phase change. Although it was quite inexpensive, this apparatus appeared highly versatile. Automatic, continuous scanning proved useful with the present system; in many cases phases were of very limited stability in temperature, and we could have missed them by a discrete scanning in steps of a few degrees.

Nevertheless, this scanning method should be considered as only a preliminary study that helps in the choice of the samples which must be stored for a long time at fixed and given temperatures. Indeed, many equilibria required several days. This was particularly true for systems at low temperatures which correspond to the oil solubilization limit. In that case, the kinetics of the phase separation may be extremely slow: Upon cooling, initially one isotropic phase system may seem metastable for many hours. It appeared that the molecular aggregates in these samples, after they were formed at higher temperatures, had a structure that was not easily broken by lowering the temperature. As will be seen below, this metastability could be related to the large penetration rate of the decane into the surfactant layer of the oil/water globules.

It should be noted that the apparent temperature of phase change measured upon continuous heating was always much closer to the true equilibrium value than when it was measured upon continuous cooling. At the end of the bath storage, the volume of each phase in the sample was measured. These volumetric data allow us to state more precisely the phase domain boundaries and the tie-lines in the ternary diagram.

III. PHASE BEHAVIOR RESULTS

For temperatures below about 32°C, the one-isotropic phase realm keeps a simple sector-like shape along the surfactant-water interface, connected to the water corner and limited to water concentrations above 60% w/w. Liquid crystals (hexagonal and/or lamellar phases) exist in the more water concentrated systems. Hence, when the temperature is raised to 32°C, there is a gradual and uniform increase of the maximum decane solubilization, which is independent of the water/surfactant ratio (Fig. 1A). Typical maximum uptakes are 0.8, 1.1, and 2 oil molecules per surfactant at 20°C, 25°C, and 32°C, respectively. In raising the temperature, this limit also increases for some time; this ratio is 3.4 at 40°C. But a continuous oil incorporation is no more possible for the higher surfactant concentrations: A fingerlike two-phase region gradually moves toward the water corner for compositions with about one oil molecule per surfactant. There we can also note the progressive formation of a liquid crystal phase (Fig. 1).

Interestingly, at 40°C and for this particular oil/surfactant ratio, the samples containing 4–6% surfactant exhibit a very pronounced opalescence. Therefore, there is a critical system for a temperature noticeably lower than the critical point of the purely aqueous solution. In this respect, the incorporation of decane *depresses* the cloud point of this nonionic, a result that is at variance with previous statements [16]. As a matter of fact, this particular system

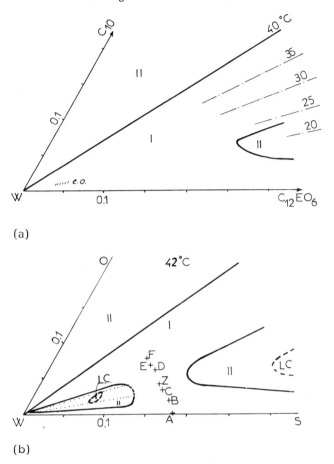

(a)

(b)

FIG. 1 Phase diagram of the aqueous system water/decane/C$_{12}$EO$_6$.
a) Temperatures between 20 and 40°C, c.o. stands for critical opalescence. b) t = 42°C. I, II, III represent the domains with one,
two, and three isotropic phases. LC stands for liquid crystal phase.
Dotted lines suggest the three-phase regions involving the anisotropic phase.

appears to have a *multicritical* end point, which can be seen by
raising the temperature 1-2 degrees (Fig. 1b): Right in the center
of this (quasi) *isolated* multiphase region, we notice the presence
of a small domain that cannot contain anything but than a *one-
anisotropic* phase. Hence, in the median part of the multiphase
region there are binary and ternary equilibria involving that anisotropic phase. The external parts consist of two-isotropic-phase
realms. In Figs. 1-3, the tiny triphasic regions are represented
schematically by dotted lines at their most propable location. The
exact composition of the water phases in these equilibria is not
known, but they certainly consist of almost pure water. When the
temperature increases a few more degrees, there is a considerable
extension of these anisotropic phase domains toward *higher* oil/sur-
factant ratios. (Since we were not able to determine precisely their
respective delineations, they are only suggested by broken lines in
the Figs. 1-3.)

Correlatively, the former (fingerlike) multiphase region also
grows larger until it merges with the other one: The splitting is
completed at about 45.5°C, that is, just before the critical point of
the purely aqueous surfactant solution is attained. As a first conclusion, the oil solubilization is also clearly dependent on the water/
surfactant ratio: The most stable systems are those that contain
about 80% water (Figs. 1b and 2a).

At 47°C, the isotropic phase with a high oil/surfactant ratio is
largely disconnected from the water corner, as the result of the
rapid extension of the two-phase region adjacent to the upper limit
of the crystalline phase sector. Instead, it is connected to a large
three-isotropic-phase triangle (Fig. 2b). One of the other apices
of this triangle corresponds to decane with a small percent of surfactant, and the third corresponds to almost pure water. From a
practical point of view, everything happens as if there were an invariant stage (eutectoid) leading, within an infinitely small temperature range, to the formation of a three-phase triangle of finite extent.
This (apparent) discontinuity between limiting water phase (almost
pure water) and surfactant phase (containing noticeable amounts of
both oil and water) has already been noted in our study of a three-
phase behavior with salt added to the nonionic.

On the other hand, this point marks the maximum extension for
the liquid crystal domain; further increase of the temperature reduces its area in such a way that its maximum oil/surfactant ratio
gradually tends toward zero. Correlatively, the isolated surfactant
phase moves toward the oil corner. But taking advantage of the "re-
treat" of the anisotropic phase, the surfactant phase region may be
reconnected to the water corner. Indeed, at about 48°C, we noticed
the presence of a tiny domain of a new isotropic phase, which could
be in equilibrium with the former microemulsion phase (Fig. 3a).

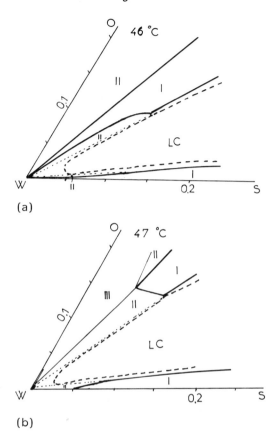

FIG. 2 Water/decane/$C_{12}EO_6$ phase diagram. a) t = 46°C. b) t = 47°C (see Fig. 1.)

With an increase in temperature, it rapidly grows taller, and the reconnection of these two surfactant phases is completed at about 51.5°C. This particular part of the diagram is of very limited stability both in composition and temperature, and moves toward the water-surfactant interface, closely following the upper limit of the anisotropic phase (Fig. 4b).

Therefore, except for the lowest temperatures, the phase behavior of $C_{12}EO_6$/decane/water closely resembles that of the ternary nonionic $C_{12}EO_5$/tetradecane/water: Both systems exhibit almost the same detailed features at about the same temperatures [4]. This full parallelism confirms one of our previous studies concerning the hydrophilic lipophilic balance/alkane carbon number

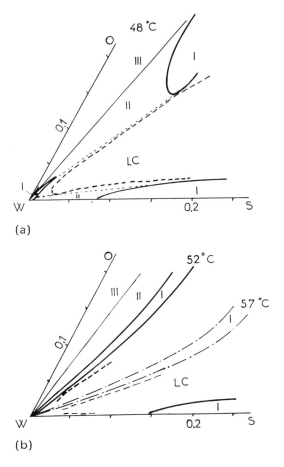

(a)

(b)

FIG. 3   Water/decane/$C_{12}EO_6$ phase diagram.   a) t = 48°C.   b) t = 52 and 57°C.

(ACN) equivalences [8], which showed that phase behavior should be kept constant if a decrease of 1 unit of HLB is compensated by an increase of 8 units of ACN.

The above qualitative description of the phase diagrams clearly suggests that the structure of the micellar aggregates should depend not only on the temperature but also on *both* the oil/surfactant and water/surfactant ratios. These structures express the way the oil/water solubilization is achieved by surfactant molecules with a strong tendency to associate. But in many cases, water dilution at constant structure appears highly unlikely if not impossible:

Interparticle and entropic effects should markedly influence the stability of these systems, mainly in the vicinity of critical points or anisotropic phases.

IV. SMALL-ANGLE NEUTRON SCATTERING AND STRUCTURAL RESULTS

Small-angle neutron scattering measurements were carried out at the I.L.L. (Grenoble, France) on a D17 spectrometer and at the Laboratoire Léon Brillouin (Saclay, France) on PACE apparatus. The wavelength and sample-detector distance were managed in such a way that the final range of transfer momentum q ($\sim$ scattering angle) was $0.007-0.2$ Å^{-1}. The scattered intensities were evaluated on an absolute scale after classical corrections for turbidity, incoherent scattering, and cell wall effects were performed. The water component was generally a mixture of H_2O and D_2O, and the alkane solvent was decane of various grades of deuteration [17-19].

For the samples investigated by small-angle neutron scattering, we chose two series of compositions. One series consisted of dilute systems, with about 3.5% w/w of disperse phase (oil + surfactant) in water, anticipating the oil-in-water nature of the solution. They are identified as A_0, B_0, C_0, Z_0, D_0, E_0, F_0 according to their molar oil/surfactant ratio, which were 0, 0.35, 0.70, 1, 1.5, 1.8, and 2.2, respectively. Thus, sample Z is not far from the demixing line at 25°C; sample D from the line at 30°C, and F at 35°C. The particular denomination of the Z point calls to mind that this sample corresponds to the critical oil content.

The other series consisted of five times more concentrated samples but with the same oil/surfactant ratio as the preceding ones (they are identified as A_1, B_1, . . .). They were chosen in such a way that they just belong to the most stable part of the one-isotropic-phase domain (Fig. 1A).

A few examples of experimental neutron scattering data are shown in Figs. 4–7. Figure 4 shows systems with very low oil content (sample B). The spectra comprise various maxima and minima, which depend on the temperature and the composition: Generally, for concentrated systems, the first maximum (i.e., at the smaller q) is due to interparticle (crowding) effects. But these extrema may also depend markedly on the isotopic composition of both the decane and water components. In the present case (low oil content), these "structural" extrema exist for larger scattering angles and are not practically influenced by interparticle effects. We noted that these large-q extrema are not at all noticeably washed out by an increase in temperature or concentration. This is quite evident for systems with higher oil/surfactant ratios (Fig. 5). Here, the curves have

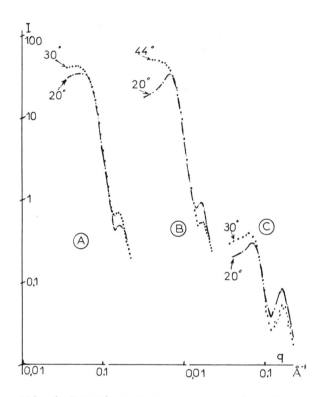

FIG. 4 Experimental I(q) spectra of small-angle neutron scattering for sample B ($\alpha = 0.35$) at various temperatures, various volume fractions ϕ, and different deuteration rates of the solvents (δ_W and δ_O). A: $\phi = 0.16$, $\delta_W = \delta_O = 1$. B: $\phi = 0.28$, $\delta_W = \delta_O = 1$. C: $\phi = 0.28$, $\delta_W = 0.3$, $\delta_O = 0.72$.

been shifted by a factor of 5, the ratio of the disperse phase concentrations. We are convinced, therefore, that these extrema are due to relatively precise structures, and not to ill-defined mixtures of ill-defined aggregates, at least for samples of this nonionic system. Hence, we have tried to characterize each sample by a representative particle of mean surfactant aggregation number (N), volume (V), and anisometry (p).

As a matter of fact, calculations show that these extrema of interest characterize the morphology of the cross section of the hydrophobic part of the isolated oil/water globule [17]: i.e., α_f, the penetration rate of the decane into the interfacial film, and the conformation of the hydrophobic chain, from which the mean

area per polar head (σ) can be calculated. Both the angular position and the intensity of these extrema are sensitive to the choice
of isotopic mixture for the water and oil solvents, as shown in Fig.
8; indeed, they are related to the excess of mean scattering length
of the surfactant layer relative to both the core and the exterior of
the globule, which clearly depends on α_f. They also depend on the
thickness of the film, i.e., on the conformation of the hydrocarbon
surfactant chain. Hence, by focusing our fitting computations on
these structural extrema obtained for two or three deuteration rates
of solvent, we are able to evaluate the ratio N/p independently of
the actual volume of the globule—i.e., the aggregation number (N)
or the ellipticity (p) when assuming a spheroidal shape. We also
get the most probable values of α_f, σ, and the chain conformation.
Let us emphasize that all these parameters are not independent but
are interrelated by geometric constraints and that these determinations are made possible by using the absolute values of the scattered
intensities.

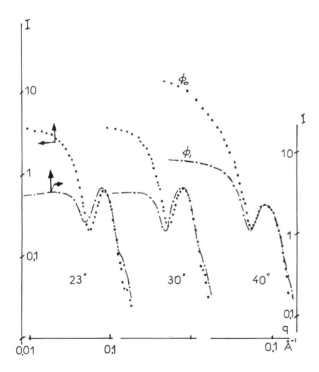

FIG. 5 I(q) spectra (see Fig. 4). Samples D (α = 1.5), ϕ_0 = 0.035,
ϕ_1 = 0.18, δ_0 = 1, δ_w = 0.7.

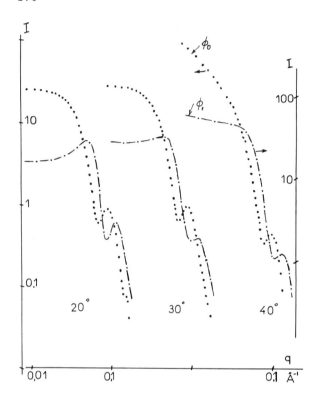

FIG. 6 I(q) spectra (see Fig. 4). Samples Z ($\alpha = 1$), $\phi_0 = 0.035$, $\phi_1 = 0.18$, $\delta_w = \delta_o = 1$.

On the other hand, the structure of the (external) hydrophilic part of the film mostly influences the part of the spectra at the largest q value. Although there the intensities are generally weak, making these determinations not very precise, we can get some idea of the conformation of the EO chain; and, knowing σ, we can deduce the rate of penetration of water into the film.

As a first conclusion, all of our results for all the samples of the present investigation show that:

The thickness of the hydrophobic part of the interfacial film is nearly equal to the length of the extended C_{12} chain. Such results have already been found from NMR studies on fluorinated oil/water microemulsions [20].

The thickness of the hydrophilic part is about 0.5 to 0.7 times the length of the extended EO_6 chain.

Now, if we want to determine the actual volume of the aggregate (and hence its anisometry when assumed to be spheroid shaped), we have to study the intensity scattered at the smallest q values, taking into account the interparticle effects.

The above discussion shows why a good morphological description may be obtained even when there are four independent parameters. As a matter of fact, all the determinations were performed in the same computation, taking into account the data of the whole q range. The technique can be summarized as follows. First we imagine model aggregates, taking into consideration the geometric constraints due to the properties of the molecules and the shape and volume of the particles. Then we calculate the theoretical spectra that would be scattered by such model particles, and by a fitting method we determine the best set of values of the representative parameters. The ultimate confidence in these values rests on obtaining whole sets of coherent results. In particular, for the same overall oil/water/surfactant composition, the same morphological results must be derived for all the different isotopic compositions of the solvents used. We must emphasize that the fitting is carried out on the *logarithm* of the intensities. In that way we ensure a nearly constant quality of the fitting throughout the whole q range investigated. [The mean typical agreement between experimental and theoretical curves I(q) is 5%.] Thus, there is no risk that one parameter will carry more weight than the others. Finally, we believe

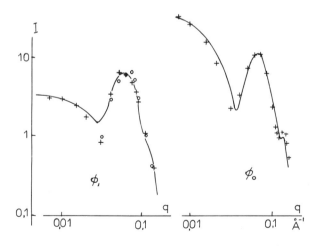

FIG. 7 I(q) spectra. Samples F (α = 2.2), δ_W = 0.7, δ_o = 1, ϕ_0 = 0.035, ϕ_1 = 0.18, t = 35°C. For ϕ = ϕ_1: (+) best fit for oblate particle (p = 0.5), (o) best fit for prolate particle (p = 2). Values of other parameters are in Table 2.

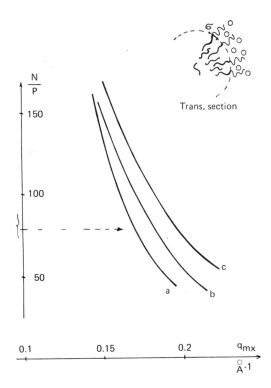

FIG. 8 Angular position q_{mx} of the maximum of I(q) of I(q) spectra calculated for B samples ($\alpha = 0.35$), which determines the structure of the transverse section of the oil/water glubule, e.g., the ratio N/p, irrespective of the proper values of N (aggregation number) and p (ellipticity of the globule). These particular calculations assume an extended conformation of the hydrocarbon part of the surfactant molecule. a: $\delta_O = \delta_W = 1$; b: $\delta_O = 0.72$, $\delta_W = 0.3$; c: $\delta_O = 0.52$, $\delta_W = 0.3$.

that the polydispersity effects cannot be essential when a fitting proves itself quite successful.

Although a presentation of our equations is beyong the scope of the present chapter and will be published elsewhere, we must stress here the *original* treatment we have derived to describe the interparticle effects, which must take into account both the crowding and critical phenomena. Since we are essentially concerned with hard-core, uncharged particles, which eventually interact only via very short-range attractive forces, we have found it particularly useful to adapt the Percus-Yevick equation for hard spheres with surface adhesion [21]. From the exact analytical solution derived by Baxter for a potential consisting of a hard core together with a rectangular thin attractive well (U_a), we have analytically calculated the Fourier transform of the direct correlation function, which enters the scattering structure factor directly. Indeed, this Percus-Yevick approximation for the potential assumed here indicates a transition occurring between disordered fluid states of the system (such as two micellar solutions) and the existence of a critical point that is in accordance with the van der Waals equation of state [22].

In our calculations, the well of the potential was assumed to be 2 Å in width. A change of 100% of this value would change only slightly the apparent value of the attractive potential for which the phase transition would occur. Therefore, our expression of the structure factor allows us to account for the scattered intensities in systems of any concentration, whether they are far from a transition point ($U_a < 1$ kT) or not ($U_a \approx 4$kT) (where k is Boltzmann's constant and T is the temperature). For example, in our computations on micellar systems of $C_{12}EO_m$ (m = 4, 5, 6, 8), we have found that a typical value of U_a/kT in the vicinity of the critical point is 4, quite a reasonable value indeed.

As a last point, it should be mentioned that this Percus-Yevick solution was originally derived for a spherical shape, and we have tentatively extended the equations to the case of anisometric particles. We then had to evaluate an equivalent volume fraction of the equivalent hard sphere. For that purpose we made use of the invariance property of the equation of state of fluids of hard-core particles when expressed in terms of the parameter $\pi = (\overline{R}.S/V)$ (where S is the surface of the particle, $\overline{R}$ is its mean radius of curvature, and V is its volume) [23]. The validity of this approximation results from the very short-range character of the potential, making the attractive and repulsive (hard-core) parts strongly correlated, whatever the mutual orientation of the particles.

Examples of curve fitting are shown in Fig. 7. Our final results are as follows. The curves of Fig. 5 clearly demonstrate that for some compositions the structures should not be strongly affected by changes in temperature or water concentration. But the structure of a few other systems certainly undergoes important modifications when conditions are varied, as indicated by the shifts of the extrema, both in position and in scattering level (Fig. 6). This is particularly striking for Z samples, which correspond to the critical oil/surfactant ratio.

As a result, we found that the changes of the structures precisely correlate to the evolution of the phase diagrams. In particular, we can make some distinction between oil-swollen micelles and oil/water microemulsions. Typical values of the structural parameters are given in Tables 1 and 2 as a function of temperature and composition. We comment on them as follows:

TABLE 1 Morphological Parameters of the Oil-Swollen Micelles (ϕ_0 = dilute samples; ϕ_1 = concentrate samples; α_f = oil penetration rate)

Sample	t (°C)	ϕ_0			ϕ_1		
		$<N>$	$<p>$	α_f	$<N>$	$<p>$	α_f
A	20	350			150		
	30	600			200		
	40	800			250		
B	20	400	4–5	0.2	150	1–2	0.2
	30	700	8	0.2	200	2	0.2
	45				250	3–5	0.25
C	20	350	2	0.3			
	30	600	5	0.35			
Z	24	400	1	0.4	350	1	0.4
	30	550	2	0.45	400	2	0.4
D	27	500	1	0.6			
	30	650	2	0.6	500	1	0.6

TABLE 2 Morphological Parameters of the Microemulsions [ϕ_1 = concentrate samples; α_f = oil penetration rate; σ = area per polar head (A^2); U_a = attractive potential (kT)]

Sample	t (°C)		$<N>$	$<p>$	α_f	σ	U_a
Z	40	0	500	0.4	0.45	47	4.2
		1	450	3	0.45	52	2.5
D	40	0	800	2–3	0.6	52	4
		1	600	1	0.7	54	2.5
E	40	0	1000	2–3	0.65	53	3.5
F	35	0	1400	2–3	0.7	56	3
		1	1200	2–0.5	0.7	51	1
	42	0	1500	3–0.3	0.7	54	4
		1	1300	3–0.4	0.7	54	2

1. The swollen micelles exist for temperatures less than about 30–32°C, that is, for molar oil/surfactant ratio (α) less than about 1.6–2. However, they may still exist until 45°C, when the water content is in the range 10–40%, although in that case the maximum oil solubilization is quite low. In the swelling process, the mean area per polar head seems constant everywhere ($\sigma \approx 50$ Å^2). The maximum of solubilization is monitored by the temperature, which at constant σ allows a variable oil penetration into the interface of the globules as the probable result of a controlled release of the constraints imposed on the exterior of the hydrophilic part of the micelles. (The values of σ correspond to the surface that separates the hydrocarbon part from the EO part of the surfactant.) Indeed, the penetration rate (α_f) of the oil into the surfactant layer markedly increases with the overall (α) oil/surfactant ratio, the limiting value of α_f being 0.6. On the other hand, the mean probable shape of the globule is definitely a prolate spheroid. If, in addition, the samples are dilute systems, the oil incorporation seems to be performed at nearly constant aggregation number with the demixing occuring when the spherical shape is attained—the higher the aggregation number, the higher the solubilizing power.

For the more concentrated systems, the particles must be more globular and the aggregation number increases with temperature, behavior that mimics microemulsion systems, as will become apparent below. The stability of these samples must then be related to the crowding (entropic) effect on globular particles. This analogy to microemulsions is revealed by the appearance of the phase diagram itself, where the water-swollen micelle domain is connected to the microemulsion domain.

It is important to emphasize that, in this scheme, the purely aqueous micelles appear as the continuous limiting case $\alpha \to 0$, whatever the concentration, as expected. (Our results concerning aqueous micellar structures of $C_{12}EO_6$ will be presented and discussed elsewhere, together with those of $C_{12}EO_m$, where m = 4, 5, 8.)

2. The microemulsions exist for larger oil/surfactant ratios and for temperatures above 35°C. In this case the oil incorporation induces a much less important further change in the oil penetration into the palisade layer: α_f seems to attain a saturation value. The aggregation number N noticeably increases with α, together with the volume of the oil/water globule. Thus we are no longer able to discriminate between prolate and oblate shapes; for all these samples, the aggregates seem to have the same mean anisotropy, but we probably have a mixture of deformable globules characterized by a given area per polar head. The value of σ is larger than the constant value for swollen micelles; its increase with α mainly accounts for the larger volume of the oil/water globules. And unlike the samples with low oil/surfactant ratios, the microemulsion aggregates are not

very dependent on their concentration: as a general rule, the larger the α ratio, the less dependent on the water/surfactant ratio is their size. The (apparent) value of the attractive potential U_a is relatively high ($\sim 4kT$), calling to mind that some of these isotopic systems are not far from a critical phase transition.

3. Finally, as seen in Figs. 1 and 6, the Z samples actually constitute a particular pair. While Z_1 is a "normal" sample within its series, the fitting of the data of the Z_0 sample was difficult and of lesser quality. In this case the "best" particle should be described as an oblate aggregate in which the area per polar head of the surfactant is only $47-48$ Å^2, that is, the smallest value ever found in the whole present investigation. This result certainly has to be related to the nearing emergence of the isolated liquid crystal domain for that composition, as we noted in our discussion of the phase diagram description. At that temperature, this anisotropic phase is most probably a lamellar phase, although our X-ray measurements have failed to determine it (due to a lack of contrast).

V. CONCLUSION

The complexity of the phase behavior of the aqueous system with "long" nonionic surfactant results from the conjunction of critical phenomena and of the presence of liquid crystal phases. Both these events are the direct expression of the strong tendency of the surfactant to associate with more or less deformed "palisade" layer, given the right balance between the hydrophilic and hydrophobic parts of that molecule. According to the oil/surfactant and water/surfactant ratios, the attractive forces, or rather the reduction of the inter- and intramolecular repulsive forces between hydrated EO groups, are responsible for the multiplicity of the micellar structures, which correlate the varying appearance of the phase diagram. This could be related to lateral tensions in the surfactant film, mainly exerted from the exterior of the particle, which monitors the mean radius of curvature at the hydrocarbon-EO chain interface (the R theory of Winsor [24]).

Therefore, our results suggest the following process of the solubilization. At low temperatures the decrease in tension may be performed at constant area per polar head at the level of this interface, allowing an increase of the oil penetration into the hydrophobic part of the film. If the concentration is low, this swelling can be carried out at nearly constant aggregation number, while crowding effects destabilize anisometric (prolate) particles at the expense of more globular aggregates. The swelling is completed when the spherical shape is attained.

Further oil incorporation is possible if the temperature allows a slight increase of the area per polar head, inducing some disorder into the surfactant layer. These microemulsion globules then may be large and deformable, and are less dependent on their concentration. When the tensions are sufficiently reduced (independently of the interparticle effects), lamellar liquid crystal phase may be formed even in systems with more than 80–90% water. These anisotropic-phase regions considerably distort the one-isotropic-phase domain, inducing its splitting for "optimal" oil/surfactant ratios. We have no result for the isotropic phase at higher temperatures, but we can suppose that it is a disordered suspension of lamellar aggregates.

VI. REFERENCES

1. Shinoda, K., Kunieda, H., Arai, T., and Saijo, H., *J. Phys. Chem. 88*, 5126 (1984) (and references).
2. Friberg, S., and Lapczynska, I., *Prog. Colloid Polymer Sci. 56*, 16 (1975).
3. Friberg, S., Buraczewska, I., and Ravey, J. C., in *Micelliza-tion, Solubilization and Microemulsions*, Mittal, K. L. (ed.), Plenum Press, New York, 1977, vol. 2, p. 901.
4. Kunieda, H., and Shinoda, K., *J. Dispersion Sci. Technol. 3*, 233 (1982).
5. Harusawa, F., Nakamuea, S., and Mitsui, T., *Colloid Polymer Sci. 252*, 613 (1974).
6. Florence, A. F., Treigier, J. P., Seiller, M., and Puisieux, F., *J. Colloid Interface Sci. 39*, 319 (1977).
7. Lang, J. C., in *Surfactants in Solution*, Mittal, K. L., and Lindman, B. (eds.), Plenum Press, New York, 1984, vol. 1, p. 35.
8. Buzier, M., and Ravey, J. C., *J. Colloid Interface Sci. 91*, 20 (1983).
9. Mathis, G., Leempoel, P., Ravey, J. C., Selve, C., and Del-Peuch, J. J., *J. Am. Chem. Soc. 106*, 6162 (1984).
10. Kuneida, H., and Friberg, S., *Bull. Chem. Soc. Jpn. 54*, 1010 (1981).
11. Bostock, T. A., McDonald, M. P., and Tidd, G. J. T., in *Surfactants in Solution*, Mittal, K. L., and Lindman, B. (eds.), Plenum Press, New York, 1984, vol. 3, p. 1805.
12. Smith, D. H., *J. Colloid Interface Sci. 102*, 435 (1984).
13. Kahlweit, M., Strey, R., Firman, P., and Haase, D., *Langmuir 1*, 281 (1985).
14. Nakagawa, T., in *Nonionic Surfactants*, Schick, M. (ed.), Marcel Dekker, New York, 1966, p. 558.

15. Maclay, W. N., *J. Colloid Interface Sci. 11*, 272 (1956).
16. Sugano, T., *Nippon Kagaku Zasshi 84*, 520 (1963).
17. Ravey, J. C., and Buzier, M., in *Surfactants in Solution*, Mittal, K. L., and Lindman, B. (eds.), Plenum Press, New York, 1984, vol. 3, p. 1759.
18. Ravey, J. C., Buzier, M., and Picot, C., *J. Colloid Interface Sci. 97*, 9 (1984).
19. Ravey, J. C., Stebe, M. J., and Oberthur, R., *Proceedings of the 5th Symposium on Surfactants in Solution*, Bordeaux, 1984, Mittal, K. L., and Bothorel, P. (eds.).
20. Stebe, M. J., Serratrice, G., Ravey, J. C., and Delpuech, J. J., *Proceedings of the 5th Symposium on Surfactants in Solution*, Bordeaux, 1984, Mittal, K. L., and Bothorel, P. (eds).
21. Baxter, R. J., *J. Chem. Phys. 49*, 2770 (1969).
22. Barboy, B., *J. Chem. Phys. 61*, 3194 (1974).
23. Nezbeda, I., and Boublik, T., *Molecular Phys. 51*, 1443 (1984).
24. Winsor, P. A., *Chem. Rev. 68*, 1 (1968).

10

Dynamic Processes in Water/Oil Microemulsions

A. M. CAZABAT, D. CHATENAY, P. GUERING, W. URBACH,
D. LANGEVIN, and J. MEUNIER   *Laboratoire de Physique, ENS,*
Paris, France

I. INTRODUCTION

Microemulsions [1] are transparent, isotropic, thermodynamically
stable dispersions of oil and water, stabilized by surfactant molecules.
    We shall consider here microemulsions containing ionic surfactants,
such as sodium dodecylsulfate (SDS), which is an anionic surfactant,
or benzyl hexadecyl dimethyl ammonium chloride (BHDC), which is a
cationic surfactant.
    In the case of SDS, low amounts of short-chain alcohol (such as
pentanol or butanol) are needed to obtain spontaneous emulsification.
    In these systems, at least if the alcohol content is not too large,
the oil and water domains are separated by well-defined interfaces.
The surfactant molecules are located at these interfaces, whereas the
alcohol molecules can also be found in the bulk phases.
    At low concentration of the dispersed phase, it is composed of
identical spherical isolated droplets. (Water droplets in a water-in-
oil microemulsion).
    At higher concentrations, the structure of the system depends
on the interactions between droplets. If they are repulsive, the
collisions are very short and no overlapping between interfaces of
colliding droplets occurs. If the interactions are attractive, the
duration of collisions increases, and transient clusters of droplets
are formed. Interfaces overlapping occurs during collisions, allowing
exchanges between touching droplets. These exchanges are achieved
by hopping of ions or molecules through the interfaces, or by transient opening of these interfaces with communication between the water
cores of the droplets. The probability [2] of such a transient merging is low in ternary systems ($\sim 10^{-3}$ per collision), where the hopping

processes are probably dominant. It is large (~ 1) in quaternary systems, where communications between water cores play an important role.

In any case, a good understanding of these aggregation and merging processes is needed to obtain information on structural and transport properties of microemulsions.

To this end, we have studied several ternary and quaternary systems. We have first determined the size of the droplets and the interactions between them by using quasi-elastic light scattering. Then the aggregation and merging processes have been probed using three different techniques, with various characteristic time and length scales: electrical conductivity, electrically induced birefringence, and self-diffusion measurements on fluorescent probes.

II. MATERIALS AND SAMPLES

Several ternary and quaternary systems have been studied at various concentrations of the dispersed phase, keeping constant the droplet size, which is achieved by adding to a given microemulsion increasing amounts of its continuous phase [1].

TABLE 1a   Ternary Systems

Droplet composition = water + BHDC

$$w_0 = \frac{water}{BHDC}$$

No. of molecules	Weight
7	0.33
11	0.5
15	0.66
22	1
24	1.1
27	1.25
33	1.5

Continuous phase composition = benzene

TABLE 1b Quaternary Systems

Name[a] W/O	Droplet composition		Alcohol (wt)		Continuous phase composition (cm^3)			
	w_0 (wt)		SDS		Alcohol	Water	Oil	
αTB	1	Water	0.42	Butanol	12.3	0.34	100	Toluene
ATB	1.25	Water	0.36	Butanol	13.5	0.4	100	Toluene
ATP	1.25	Water	0.7	Pentanol	17	0.4	100	Toluene

[a]The first letter in the microemulsion name is relative to w_0, the water to SDS ratio in W/O systems, the second to the oil (W/O), the third to the alcohol.

A. Ternary Systems

We prepared water/oil (W/O) microemulsions of water in benzene with benzyl hexadecyldimethylammonium chloride (BHDC) as surfactant. The water-to-soap molecules ratio was varied between 7 and 33, and the volume fraction ϕ of droplets between 0.01 and 0.15.

B. Quaternary Systems

Three series of quaternary microemulsions with sodium dodecylsulfate as surfactant were prepared. The oil was toluene (T), the aqueous phase was pure water, the cosurfactant was butanol (B) or pentanol (P). The composition of the droplets and of the continuous phase for each series can be found in Table 1. The volume fraction ϕ of droplets was varied between 0.01 and 0.20.

III. CHARACTERIZATION OF THE SAMPLES BY LIGHT SCATTERING

At low volume fraction ϕ of the dispersed phase, the droplets behave as permanent interacting Brownian particles [3,4].

A. Basic Formulas at Low ϕ

As the droplets' size is small compared to the optical wavelength λ, the scattered intensity is independent of the scattering angle θ and is inversely proportional to the osmotic compressibility $\partial \Pi / \partial \phi$, which can be written as

$$\frac{\partial \Pi}{\partial \phi} = \frac{kT}{\gamma}(1 + \beta\phi) \qquad \text{at low } \phi$$

where k is the Boltzmann constant and T is the absolute temperature.
The volume γ of the droplets and the virial coefficient β are then deduced from the intensity measurements in the low-ϕ range.
The autocorrelation function of the scattered light is simultaneously recorded on the samples. It is a single exponential,

$$g^{(2)}(\tau) = 1 + e^{-2D_c q^2 \tau}$$

where

$$q = \frac{4\pi n}{\lambda} \sin\frac{\theta}{2}$$

is the scattering wave vector, n is the sample index, and D_c is the collective diffusion coefficient of droplets. At low ϕ,

$$D_c = D_0(1 + \alpha\phi) \qquad D_0 = \frac{kT}{6\pi\eta_0 R_H}$$

where η_0 is the viscosity of the continuous phase and R_H is the hydrodynamic radius of droplets. Hence, R_H and the virial coefficient α (including both direct and hydrodynamic interactions) are deduced.

B. Experimental Results

Both static and dynamic measurements were performed on quaternary systems [4]. The corresponding values of $R = (3\gamma/4\pi)^{1/3}$, R_H, β, and α can be found in Table 2. There is no significant difference between R and R_H in these W/O microemulsions.

Only dynamic measurements were performed on ternary systems. The results for R_H and α are given in the same table.

TABLE 2 Droplets' Radius

	R (Å)	β	R_H (Å)	α
Ternary systems				
$w_0 = 7$			36	−11
11			41	−12
15			47	−13
22			55	−14
24			61	−19
27			66	−25
33			82	−30
Quaternary systems				
αTB	34	−12	36	−30
ATB	42	−20	41	< −30
ATP	44	+7	45	− 2

Positive values of β correspond to repulsive interactions; for example, the ATP system shows hard-spherelike behavior.

Negative values of β and α reflect attractive interactions between droplets. They are quantitatively explained if the interfaces overlapping is taken into account [3-5].

These attractive interactions are observed in all the systems studied here, except the hard-sphere one. We will now report on the influence of these interactions on the aggregation processes, as studied by other techniques.

IV. ELECTRICAL CONDUCTIVITY IN W/O MICROEMULSIONS

The electrical conductivity of the various W/O microemulsions was studied versus ϕ at constant droplet size.

As the continuous phase of W/O systems is not conducting, electrical conduction needs contact of droplets to allow charge

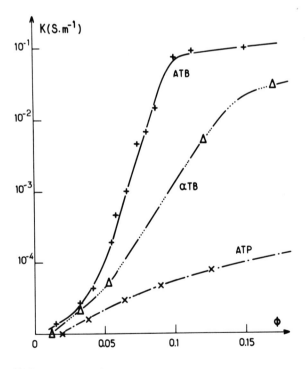

FIG. 1 Electrical conductivity versus ϕ for quaternary systems. Logarithmic scale.

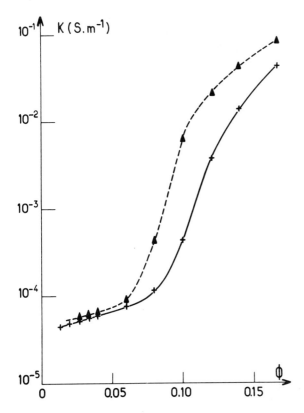

FIG. 2 Electrical conductivity versus ϕ for ternary systems.
Logarithmic scale: full line $w_O = 22$; dashed line $w_O = 33$.

transfer between them. This transfer can be achieved by charge
hopping, or transient merging of connected droplets with communication between the water cores.

Hence the electrical conductivity probes the connectivity of the
dispersed phase over macroscopic length scales and during short
time scales (probably below 10^{-7} s). When this connectivity is
achieved, a steep increase of the conductivity is observed, which
has been analyzed as a percolation process [6–8], with a percolation threshold ϕ_p.

The existence and position of this threshold depend on the interactions between droplets, which control the duration of the collisions and the degree of interface overlapping, hence the probability
of hopping or merging. This is illustrated in Fig. 1 for quaternary

systems and Fig. 2 for ternary systems: Building up of conductivity needs attractive interactions and ϕ_p decreases when the
strength of these interactions increases.

We shall now consider other techniques that allow one to probe
the clustering of droplets at shorter length scales and longer time
scales.

V. TRANSIENT ELECTRICALLY INDUCED BIREFRINGENCE

One way to evidence droplet association is to look for some anisotropy in the dispersed phase. Isolated spherical droplets do not
contribute, and such a method is sensitive only to droplet clusters:
dimers at very low ϕ, then larger aggregates.

A. Basic Features of the Kerr Effect

Anisotropic particules tend to orientate in an applied electrical field,
leading to an induced birefringence Δn, which is the difference between values of the refractive indices parallel and perpendicular to
the field.

The steady-state birefringence can be written as [9]

$$\Delta n = B \lambda E^2$$

where λ is the light wavelength, E is the electric field, and B is
the Kerr constant of the sample.

After switching off the field, the time decay of the birefringence
can be observed. It is related to the characteristic rotational times
τ_R of the particles.

Theoretical expressions for B and τ_R can be found in the literature [9], but only in very few cases: simply shaped particles (rods,
ellipsoids, etc.) without interactions.

Both the steady-state value Δn and the time decay of the bire-
frigence were studied. In all cases the proportionality between Δn
and E^2 was checked. The field has to be low enough to orientate
the anisotropic clusters existing in the microemulsion without inducing any structural change in it. This was checked carefully.

B. Predictions and Results at Low ϕ

At low ϕ, only isolated droplets and dimers are present. As the
isolated droplets do not contribute to the signal, B can be written as

$$B = f \cdot \gamma \cdot \phi_d$$

where F depends on the shape of the dimer (ellipsoid, biconcave shape, dumbbell shape) and its electrical and optical properties. γ is the droplet volume, ϕ_d the volume fraction of dimers:

$$\phi_d = A\phi^2$$

where A is expected to increase with attractive interactions. All factors are positive, then B has to be positive. In the dimer range, the time decay must be a single exponential. The decay time τ_R is the rotational time of a dimer, which can be calculated for a given dimer shape.

These predictions were compared with experiments in two ternary systems ($w_0 = 22$ and $w_0 = 33$) and in the three quaternary ones.

The agreement is quite satisfactory: B actually scales as ϕ^2 at low ϕ (note that this ϕ^2 range is very narrow in the ternary system with $w_0 = 33$; larger clusters appear for $\phi \gtrsim 1.2\%$). In this range, A is constant and increases with attractive interactions, as shown in Table 3.

To calculate A, we need the dimer shape. Several possibilities were considered, keeping in mind the specific behavior of ternary and quaternary systems. In quaternary systems, where the merging probability [2] is large, prolate ellipsoids or biconcave shapes

TABLE 3a   Quaternary Systems

Microemulsion Name	$\phi\%$	Biconcave shape $A = \phi_d/\phi^2$	Prolate ellipsoid $A = \phi_d/\phi^2$
ATP	2	0.25	0.08
	4	0.18	0.06
	6	0.17	0.06
αTB	1.65	2.2	0.7
	2	2.3	0.7
	2.7	2.3	0.7
ATB	0.65	4.5	1.5
	0.95	5	1.7
	1.7	5	1.7

TABLE 3b   Ternary Systems   *10Å overlapping   † no overlapping

Microemulsion			Dumbbell shape	
wt%	w_0	ϕ%	$A^* = \phi_d/\phi^2$	$A^\dagger = \phi_d/\phi^2$
1	22	1.2	4.9	4.2
		1.9	4.8	4.2
		2.5	5	4.5
1.5	33	1.2	25	22
		2	60 Not only dimers	50
		2.7	100	70

are expected. In ternary systems this probability is low and a dumbbell shape has to be considered.

The characteristic rotational times were calculated in these last systems and found to be in rather good agreement with measured

TABLE 4

Microemulsion	τ_R calculated (μs)	τ_R measured (μs)	$\bar{\tau}_{max}$ (μs)
ATP	0.23	–	~ 2
αTB	0.11	–	~ 20
ATB	0.17	–	~ 200
	Prolate ellipsoid		
$w_0 = 22$	0.33	0.4	~ 4
24	0.42	0.5	–
27	0.56	0.65	–
33	1.1	1.2	~ 35

Dumbbell shape

5 Å overlapping

ones. (These times are shorter in quaternary systems, and the experimental accuracy is too bad). They can be found in Table 4.

C. Results at Larger ϕ

At higher ϕ, larger clusters are formed. B first scales faster than ϕ^2, then passes through a maximum and decreases (see Figs. 3 [10] and 4). The time decay is no longer exponential, because slower contributions become visible. Roughly speaking, however, the mean decay time increases up to the point where B is a maximum, then stays more or less constant. The corresponding $\bar{\tau}_{max}$ values increase with attractive interactions (see Table 4).

All these features are qualitatively well understood by pointing out that the main contribution to the signal comes from clusters whose lifetime is larger than the rotational time: Large clusters with short lifetimes do not contribute. This explains both the maximum of B and the constant value of the relaxation time. This value $\bar{\tau}_{max}$ corresponds to a cluster lifetime and actually increases with attractive interactions.

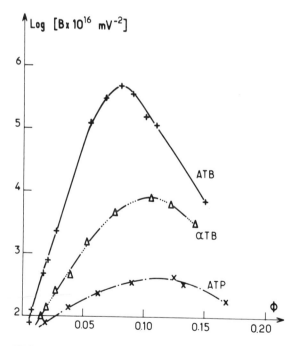

FIG. 3 Kerr constant B of quaternary systems. Logarithmic scale.

D. Negative Kerr Signals

So far we have discussed only how electrical birefringence can
evidence droplets aggregation, and what leads to positive expected
contribution. Let us note very briefly that the birefringence sig-
nals usually also contain a negative contribution, with shorter time
decay.

This negative term increases with ϕ (it scales as ϕ^2 is ternary
systems) and shows no maximum (see Fig. 5). This suggests that
some distortion of interfaces during collisions could be at the origin of this term, which should yield interesting information on the
interface dynamics during collisions.

We shall now consider another experimental technique, which
allows us to measure the self-diffusion coefficient of fluorescent

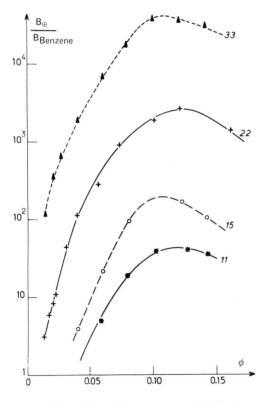

FIG. 4 Positive Kerr constant B^+ of ternary systems. Logarithmic
scale.

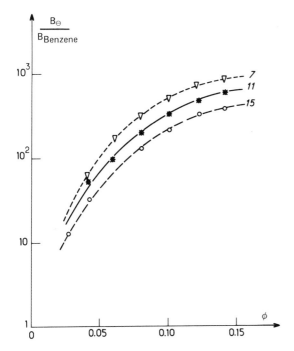

FIG. 5 Negative Kerr constant B⁻ of ternary systems.   Logarithmic scale.

probes, and to use it to get further information on the droplet clustering process.

VI. FLUORESCENCE RECOVERY AFTER PHOTOBLEACHING

Let us recall briefly the principle behind the experiment: When strongly lighted, fluorescent probes loose their fluorescent properties.
    A given volume of sample is illuminated by a strong laser beam, then the fluorescence intensity is measured in the same volume by using a low-intensity beam. The signal increases when new probes enter the studied volume, and their self-diffusion coefficient D_S can be deduced from the recovery time.
    This method [11] was improved by using a fringe pattern and by modulating the fluorescence signal [12].

The recovery curve is a single exponential with characteristic time

$$\tau = \frac{i^2}{4\pi^2 D_S}$$

where i is the interfringe spacing.

The study was performed on the ternary systems, which BHDC as surfactant [13]. The fluorescent probe is 3,3'-diethyloxadicarbocyanine iodide (DODCI). This probe is almost insoluble in benzene. Previous studies using picosecond spectroscopy have shown that the probe is embedded in the interfacial layer of droplets [14].

The measured values of D_S have been plotted in Fig. 6 versus the volume fraction ϕ of droplets. For each w_0, the infinite dilution value of D_S is identical to the one obtained by dynamic light scattering, that is, D_0. This supports the idea that the probe is linked to the droplet and does not perturb the system.

Let us consider first the case $w_0 = 22$ or 24. In Fig. 6 we have a plateau up to a certain value ϕ_c and then a rather sharp decrease by a fractor of about 2. This decrease of the probe mobility can be attributed to the onset of aggregation processes in the

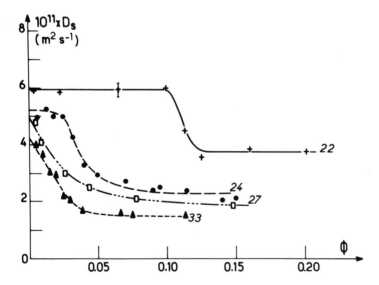

FIG. 6 Self diffusion coefficient of photochromic probes in ternary microemulsions versus volume fraction of droplets.

medium. As was observed for the percolation threshold ϕ_p of the electric conductivity, the threshold value ϕ_c decreases with increasing attractive interactions.

For the more attractive droplets ($w_0 = 27$ or 33), ϕ_c is almost zero, what means that even at low ϕ, most of the droplets belong to a cluster, in good agreement with electrical birefringence results in the case $w_0 = 33$.

Characteristic recovery times are in the range $0.1-1$ s, very large compared to the time between collisions and also to the lifetime of clusters. This explains why, after a decrease by a factor of about 2 for $\phi \sim \phi_c$, D_S stays constant for $\phi > \phi_c$. Over the time scale of the experiment, the average motion of the probe is not sensitive to clusters containing more than about three droplets.

VII. GENERAL CONCLUSION

We have studied droplets clustering in W/O microemulsion in relation to the strength of attractive interaction between droplets.

Several techniques were used, with different characteristic time and length scales:

Electrical conductivity probes macroscopic clusters over a very short time; transient induced birefringence probes orientation of clusters during their lifetime. This finite lifetime causes the measured decay times to stay fairly low, even above the electrical percolation threshold. Finally, self-diffusion probes connectivity over long times ($\sim$ s). Only very small clusters, with two or three droplets, are then observed.

The dynamic character of the clustering processes in microemulsions is clearly evidenced by these studies. This controls structural and transport properties of these systems, which play a major role in practical applications. So it is worth trying to get more quantitative information on the dynamic processes that continuously renew the structure of microemulsion phases.

VIII. REFERENCES

1. Schulman, J. H., Stoeckenius, W., and Prince, L., *J. Phys. Chem. 63*, 1677 (1959).
2. Zana, R., and Lang, J., *to be published*.
3. Calje, A. A., Agtero, W. G. M., and Vrij, A., in *Micellization, Solubilization and Microemulsions*, New York, 1977, vol. 2, p. 779.
4. Cazabat, A. M., and Langevin, D., *J. Chem. Phys. 74*, 3148 (1981).

5. Lemaire, B., Bothorel, P., and Roux, D., *J. Phys. Chem.* *87*, 1023 (1983).
6. Lagües, M., Ober, R., and Taupin, C., *J. Phys. Lett.* *39*, L-487 (1978).
7. Cazabat, A. M., Chatenay, D., Langevin, D., and Meunier, J., *Faraday Disc. Chem. Soc.* *76*, 291 (1982).
8. Safran, A., Webman, I., Grest, G. S., *to be published*; Webman, I., *private communication*.
9. Benoit, H., *Ann. Phys.* *6*, 561 (1961); *Molecular Electrooptics*, C. T. O. Konski ed. (M. Dekker Inc. N.Y. Basel) 1976.
10. Guering, P., Cazabat, A. M., *J. Phys. Lett.* *44*, L-601 (1983).
11. Axelrod, G., Koppel, D., Schlessinger, J., Elson, E., and Webb, W. W., *Biophys. J. 16*, 1055 (1976); Lanni, F., and Ware, B. R., *Rev. Sci. Inst. 53*, 905 (1982).
12. Davoust, J., Devaux, P. F., and Léger, L., *EMBO J. 1*, 1233 (1982).
13. Chatenay, D., Urbach, W., Cazabat, A. M., and Langevin, D., *Phys. Rev. Lett.* May 20, 1985, p. 2253.
14. B. Pouligny, Thesis, Bordeaux, France, 1983.

11

Determination of the Rigidity Constant of the Amphiphilic Film in Birefringent Microemulsions; Role of the Cosurfactant

JEAN-MARC DI MEGLIO, MAYA DVOLAITZKY, and
CHRISTIANE TAUPIN *Laboratoire de Physique de la Matière
Condensée, Collège de France, Paris, France*

I. INTRODUCTION

The amphiphilic lamellae of birefringent microemulsions are known to exhibit undulations. The quantitative study of the corresponding angular disorientation is well explained by a competition between the high flexibility and the interaction between layers. We have determined, for the first time, the rigidity constant of the interfacial film and evidenced that one role of the cosurfactant is to lower the interfacial rigidity.

Lyotropic systems containing oil and water have been widely studied for some years [1]. Two main situations occur: Either oil and water are organized in a periodic array, or they form isotropic dispersions called microemulsions [2]. Recently the flexibility of the interfacial film has been invoked to explain the variety of these structures [3], and much interest has been devoted to special lamellar phases that appear in the vicinity of microemulsions in the phase diagram for a smaller amount of alcohol (cosurfactant) [4]. In a previous paper [5], it was shown by the spin labeling method that these lamellae are undulated, and that the angular disorientation amplitude of the undulations increases with the degree of swelling (i.e., the oil/water ratio) of the phase. In this chapter, these undulations are studied quantitatively in the frame of de Gennes' calculations [3], which presents them as resulting from the competition between the flexibility of the film and the interlamellar interactions. This allows us to measure the constant of rigidity K and to verify the predictions of the model with regard to swelling. We have shown for the first time the effect of the cosurfactant on the film flexibility.

II. EXPERIMENTAL

The compositions of the samples are given in Table 1. The amount of alcohol corresponds to the minimum amount required to obtain the transparent birefringent phase. Samples with a slightly higher quantity of alcohol have also been investigated (see Sec. VII). All these phases show classical lamellar textures when observed with a microscope between crossed polarizers; their homogeneity has been checked with a phase-contrast setup.
    The labeled surfactant of formula

$$CH_3-(CH_2)_{11}-\underset{\underset{O\quad N-O}{\diagup\diagdown}}{C}-(CH_2)_3-{}^+N\bigcirc O,\ CH_3SO_4^-$$

resides exclusively in the interfacial film [6] and was used at a molar fraction of around 0.1% with respect to sodium dodecylsulfate (SDS). The spin label measurements have been performed on a

TABLE 1   Composition of the Samples

Swelling Rate (O/W ratio)				
Sodium dodecylsulfate (SDS, surfactant)	0.4 g	0.4 g	0.4 g	0.4 g
Tridistilled water	1 ml	1 ml	1 ml	1 ml
1-Pentanol (cosurfactant)	0.23 ml	0.32 ml	0.43 ml	0.72 ml
1-Pentanol required to obtain an isotropic microemulsion	0.54 ml	0.69 ml	1 ml	1.6 ml

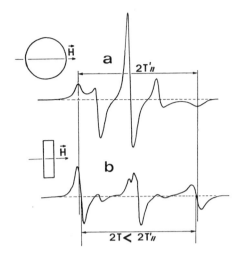

FIG. 1 (a) "Powder spectrum" obtained with a spherical sample container (isotropic distribution of the lamellae). (b) "Oriented spectrum" obtained with a parallel glass wall container. Note the discrepancy between the two extreme splittings.

Varian E-9 EPR spectrometer. Two types of sample containers were used [5], which give rise either to an homeotropic orientation (checked by observation with a microscope between crossed polarizers) or to an isotropic distribution of the lamellar domains. The spectra of these samples differ from the ones of usual lamellar phases (for instance, smectic thermotropic liquid crystals); the main effect is a discrepancy between the extreme splittings measured either on an oriented spectrum or on the corresponding powder spectrum (Fig. 1). This indicates that in spite of the homeotropic aspect, a signficant fraction of the surfactant molecules are not parallel to the magnetic field; this is due to a continuous angular spread around a mean orientation corresponding to smooth undulations of lamellae [5]. This angular spread can be determined by the synthesis of the oriented spectrum using the linewidths and splittings deduced from the synthesis of the powder spectrum according to classical procedures [7].

III. EFFECT OF TEMPERATURE ON ANGULAR DISTRIBUTION

The spectra of a sample with a 4 oil-to-water ratio have been recorded as a function of a temperature between 20 and 70°C. The

temperature was controlled within ±0.2°C in the cavity of the EPR
spectrometer by a Varian E-257 variable accessory. Figure 2 represents the extreme splittings measured on oriented (O) and on
isotropic [or powder (P)] spectra. According to the procedure
discussed above, first described in Ref. 5, we have determined
the square of the spread angle θ as a function of the temperature.
We found that $\phi^2 \sim 10^{-2}T$ (rad) (Fig. 3).

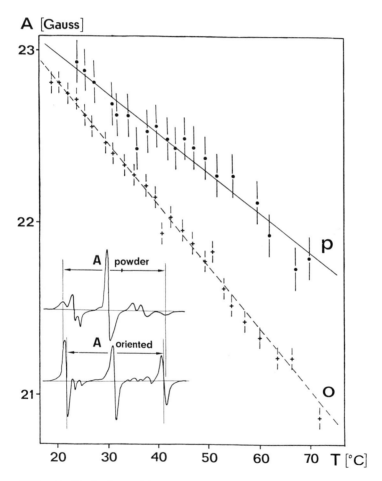

FIG. 2 Variation with temperature of the extreme splittings for
p (powder spectrum) and o (oriented spectrum).

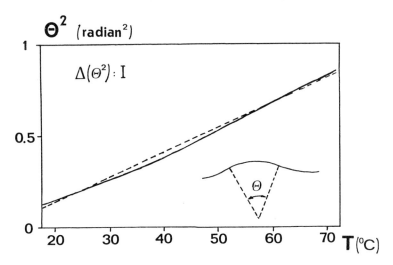

FIG. 3 Variation with temperature of the square angular spread of the normal to the lamellae (dashed line: $\theta \sim 10^{-2}T$).

IV. INTERPRETATION: ROLE OF FLEXIBILITY

In order to understand this behavior, we use the model described in Ref. 3 for multilamellar systems. We assumed the lamellae to be water lamellae separated by cyclohexane as shown in Fig. 4.

The free energy per unit area is written as

$$F = \frac{1}{2}K(\nabla_{\perp}^{2}z)^{2} + \frac{1}{2}U''z^{2} \tag{1}$$

where K is the rigidity constant of the interface [8], z is the local displacement of the water layer with respect to the reference flat plane (Fig. 4), and U" is the curvature of the interaction potential between layers. The geometry of Fig. 4 is the simplest one we can imagine that allows us to vary the distance between layers and thus to destroy the lamellar order.

The first term of Eq. (1) represents the energy needed to curve the interface, and the second term is determined by the interactions between layers.

We have evaluated U" from the following competitive interactions: (see the appendix to this chapter for the detailed computation U" = $U''_{vdW} + U''_{rep}$)

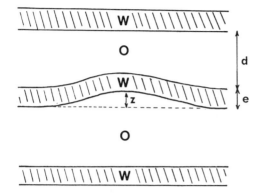

FIG. 4 Geometry used in the model.

Van der Waals attractive interactions, which lead to a collapse of the lamellar structure:

$$U''_{vdW} = -\frac{A}{\pi}\left[\frac{1}{d^4} + \frac{1}{(d + 2e)^4} - \frac{2}{(d + e)^4}\right]$$

with A being the Hamaker constant for water embedded in cyclohexane [$= A_{131}(H_2O, C_6H_{12})$], d the distance that separates two lamellae, and e the thickness of a lamella (Fig. 4). Steric repulsive interactions, which were first introduced by Helfrich [9]:

$$U''_{rep} = +5.04\left[\frac{(kT)^2}{Kd^4}\right]$$

We think that these interactions are stronger than the van der Waals ones because of the low expected value for K. This hypothesis will be checked further.

The Hamaker constant is equal to 1.5×10^{-13} erg [10]. d and e were deduced from geometric calculations, taking an area per polar head of around 60 Å^2 [11]; this leads to d = 160 Å and e = 40 Å for the studied sample. This theoretical approach allows us to estimate the average θ^2:

$$\theta^2 = \frac{kT}{\pi K} \log\left(\frac{\xi_u}{a}\right) \qquad (2)$$

where ξ_u is a characteristic length equal to $(K/U")^{1/4}$ and a is a molecular size of around 10 A [3]. Thus the experimental behavior ($\theta^2 \sim T$) is in good agreement with the model. This allows us to estimate the rigidity constant K from the following formula by an iteration procedure [recall that $\xi_u = (K/U")^{1/4}$]:

$$K = \frac{kT}{\pi\theta^2} \log \left(\frac{\xi_u}{a}\right)$$

[We assume that K is fairly independent of the temperature [12].] We find $K = 6.4 \times 10^{-15}$ erg by taking a = 10 A. The assumption that repulsive interactions are greater than van der Waals interactions is thus well checked: $U"_{rep}/U"_{vdW} = 70$. This value is much lower than the value obtained for lecithin lamellar systems ($K_{lec} = 2 \times 10^{-12}$ erg [13]); we may understand this by the fact that in lecithin systems interactions inside the interfacial film are much stronger because electrostatic interactions between polar heads are not screened by the cosurfactant and steric interactions in the hydrophobic part of the film may be strong because of the two alkyl chains per surfactant molecule.

V. EFFECT OF SWELLING

It has been shown [5] that local order parameters of surfactant molecules are independent of the degree of swelling (S = 0.40);

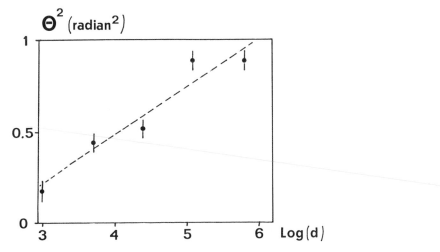

FIG. 5 Plot of θ^2 versus the logarithm of the distance between lamellae.

furthermore, amounts of cosurfactant needed to obtain the samples correspond to a straight line in the phase diagram. These two experimental facts prove that the chemical composition in the film does not change [14] with the degree of swelling, hence the rigidity must remain the same. It is thus of interest to study the influence of the swelling of the phases on the angular spread, since it constitutes a fine test of the model: U'' is to a very good approximation proportional to $1/d^4$, so θ^2 must be linear with log d:

$$\theta^2 = \frac{kT}{\pi K} \log d + \text{constant}$$

This is shown in Fig. 5. The agreement between theory and experiment is good and leads to another determination of K that is independent of the previous one: We find $K = 4 \times 10^{-14}$ erg.

VI. DYNAMIC EFFECTS ON THE SPECTRA

Figure 6 represents the linewidth used in the synthesis of the high-field part of the spectra versus the swelling degree; it is clear that a dynamic effect is correlated with the oil-to-water ratio. Because of the flexibility, the interfacial film is locally curved, and when the surfactant molecules diffuse laterally on this curved surface,

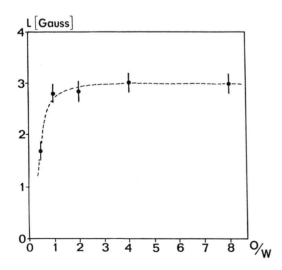

FIG. 6 Linewidth used in the synthesis of the high-field peak in a powder spectrum as a function of the swelling rate.

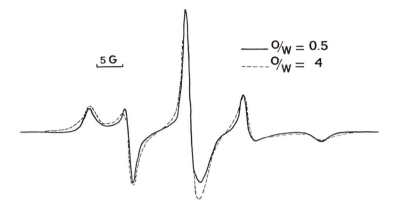

FIG. 7 Powder spectrum of a 4 O/W sample (dashed line) compared to a powder spectrum of a 0.5 O/W sample (full line). Note the discrepancy between the linewidths.

they undergo a disorientation of their mean axis [15]. If the disorientation time is of the same order as the characteristic time of ESR experiments (10^{-8} s), this will promote a broadening of the spectral lines [7].

The curve of Fig. 6 can thus be easily understood: For small swelling rates (O/W < 4), the dispersion angle of the lamellae is very small and does not lead to a strong disorientation of the labeled surfactant; whereas for strong swelling rates (O/W > 4), the dispersion angle becomes important and the mean axis of the label molecule may disorient itself by more than 1 rad during the time scale of the experiment. This phenomenon allows us to evaluate the radius of curvature of the interfacial film according to the procedure developed in Ref. 16 for the sample with a 4 O/W ratio. Taking as a reference the spectrum where the interfacial film is almost flat (O/W = 0.5) (Fig. 7), we estimate from a simulation method [17] an angular correlation time of around 5×10^{-7} s; the lateral diffusion constant D_{lat} is also estimated from ESR experiments according to the procedure developed in Ref. 18 and found equal to 3×10^{-7} cm^2 s^{-1}. This gives a radius of curvature of around 80 Å ($R^2 = 4 D_{lat} \tau_R$ [16]), which is comparable to the correlation length deduced from Eq. (2) (with a mean value for K of 2.3×10^{-14} erg, $\xi = 60$ Å).

VII. EFFECT OF THE COSURFACTANT (ALCOHOL)

We have performed experiments on samples (O/W = 4) that contained up to 80% more alcohol than is required to obtain the birefringent

Di Meglio et al.

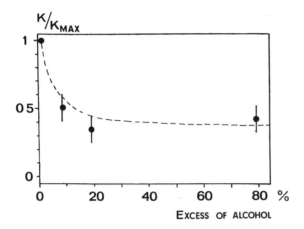

FIG. 8 Plot of K/K_{max} versus excess of alcohol (excess with respect to the amount needed to obtain a birefringent microemulsion).

phase. The samples are still homogeneous and present lamellar
structures (oily streaks) when observed with a microscope.

The angle of dispersion increases with the excess of cosurfactant. The added quantitiés do not change the mean distances between lamellae: This indicates a decrease of the rigidity due to a
modification of the chemical composition in the film. Figure 8 represents the ratio K/K_{max} versus the excess of alcohol.

VIII. CONCLUSION

Smooth undulations observed in birefringent microemulsions are shown
to be a consequence of the competition between the high flexibility
of the interfacial film and interlamellar interactions. The validity of
the model [3] is verified by studying both the influence of temperature and oil-to-water ratio. It has been shown that the role of the
cosurfactant is indeed to lower the rigidity of the film; this explains
the transition from organized phases toward isotropic microemulsions.

IX. APPENDIX: COMPUTATION OF U"

We have taken into account both van der Waals and steric interactions between one layer and its two nearest neighbors.

We have for the energy W per unit area ($W = W^+ + W^-$):

$$W^+ = - \left(\frac{A}{12\pi}\right) \left[\frac{1}{(d-z)^2} + \frac{1}{(d-z+2e)^2} - \frac{2}{(d-z+e)^2}\right]$$

$$+ 0.42 \frac{(kT)^2}{K(d-z)^2} \qquad \text{(upper layer)}$$

$$W^- = - \left(\frac{A}{12\pi}\right) \left[\frac{1}{(d+z)^2} + \frac{1}{(d+z+2e)^2} - \frac{2}{(d+z+e)^2}\right]$$

$$+ 0.42 \frac{(kT)^2}{K(d+z)^2} \qquad \text{(lower layer)}$$

assuming that $z \ll d$, we have for W:

$$W = - \left(\frac{A}{6\pi d^2}\right) \left(1 + \frac{3z^2}{d^2}\right) - \left[\frac{A}{6\pi(d+2e)^2}\right] \left[1 + \frac{3z^2}{(d+2e)^2}\right]$$

$$+ \left[\frac{2A}{6\pi(d+e)^2}\right] \left[1 + \frac{3z^2}{(d+e)^2}\right] + 0.84 \frac{(kT)^2}{K[1 + 3z^2/(d+e)^2]}$$

The curvature of the potential is defined by

$$U'' = \frac{\partial^2 W(z)}{\partial z^2} \qquad (z = 0)$$

and we finally get for U'',

$$U'' = - \left(\frac{A}{\pi}\right) \left[\frac{1}{d^4} + \frac{1}{(d+2e)^4} - \frac{2}{(d+e)^4}\right] + 5.04 \frac{(kT)^2}{Kd^4}$$

X. ACKNOWLEDGMENTS

We would like to thank Jean-Francois Joanny and Loic Auvray for many helpful discussions, and also one of the referees for a remark about the physics of the interaction potential. This research has received partial finanacial support from PIRSEM (C.N.R.S.) under AIP no. 2004.

XI. REFERENCES

1. Ekwall, P., in *Advances in Liquid Crystals*, Brown, G. H., (ed.), Academic Press, New York, 1975, vol. 1, p. 1.
2. Hoar, T. P. and Schulman, J. H., *Nature 152*, 102 (1943).
3. De Gennes, P. G., and Taupin, C., *J. Phys. Chem. 86*, 2294 (1982).
4. Dvolaitzky, M., Ober, R., Billard, J., Taupin, C., Charvolin, J., and Hendricks, Y., *Compt. Rend. Acad. Sci. II 295*, 45 (1981).
5. Di Meglio, J. M., Dvolaitzky, M., Ober, R., and Taupin, C., *J. Phys. Lett. 44*, L-229 (1983).
6. Dvolaitzky, M., and Taupin, C., *Nouveau J. Chim. 1*, 355 (1977).
7. Berliner, L. J. (ed.), *Spin Labelling*, Academic Press, New York, 1976.
8. Helfrich, W., *Z. Naturforsch. 28C*, 693 (1973).
9. Helfrich, W., *Z. Naturforsch. 33a*, 305 (1978).
10. Visser, J., *Adv. Colloid and Interface Sci. 3*, 331 (1972).
11. Dvolaitzky, M., Guyot, M., Lagues, M., Le Present, J. P., Ober, R., Sauterey, C., and Taupin, C., *J. Chem. Phys. 69*, 3279 (1978).
12. Cantor, R., *Macromolecules 14*, 1186 (1981).
13. Servuss, R. M., Harbich, W., and Helfrich, W., *Biochim. Biophys. Acta 436*, 900 (1976).
14. Graciaa, A., Lachaise, J., Martinez, A., Bourrel, M., and Chambu, C., *Compt. Rend. Acad. Sci. B 282*, 547 (1976).
15. Seelig, J., and Limacher, H., *Mol. Cryst. Liq. Cryst. 25*, 1051 (1974).
16. Di Meglio, J. M., Paz, L., Dvolaitzky, M., and Taupin, C., *J. Phys. Chem., in press*.
17. McCalley, R. C., Shimshick, E. S., McConnell, H. M., *Chem. Phys. Lett. 13*, 115 (1972).
18. Devaux, P., and McConnell, H. M., *J. Am. Chem. Soc. 94*, 4475 (1972); Sackmann, E., and Träuble, H., *J. Am. Chem. Soc. 94*, 4492 (1972).

12

The Correlation Between Phase Behavior and Structure of a Four-Component System: Poly-Ethylene Glycol/ Sodium Dodecyl Sulfate/Pentanol/H$_2$O

D. Y. CHAO, M. L. CHEN, and Y. G. SHEU *Industrial Technology Research Institute, Hsinchu, Taiwan, Republic of China*

I. INTRODUCTION

The change in the concentration of sodium dodecylsulfate (SDS) can directly affect the correlation between phase behavior and structure of a four-component system, polyethylene glycol/sodium dodecylsulfate/pentanol/H$_2$O. Experimental results indicate that the phenomena of microemulsion and lamellar liquid crystals can take place at a concentration of SDS greater than 5% and 17% by weight, respectively. In our experiments, these liquid crystals have been shown to be hydrophilic and their rheological properties to be non-Newtonian.

Unlike a regular surfactant [1,2] with a long hydrocarbon chain length, the low-molecular-weight polymer of polyethylene glycol [2, 3], which has both hydrophilic and hydrophobic (a short ethylene chain) properties, can be treated as a special kind of polymerlike surfactant. Use of this polymerlike surfactant in the system sodium dodecylsulfate/pentanol/H$_2$O may contribute to increasing the solubility of pentanol in water solution and interact with SDS to form polyelectrolytelike conformations. The formation of these conformations, depending on composition, temperature, polymer and surfactant concentration, and chain length, are responsible for the variation of the surface characteristics, structure, and phase behavior of a polyethylene glycol/sodium dodecylsulfate/pentanol/H$_2$O system. The chief aim of this work was to investigate the properties and performance of low-molecular-weight polyethylene glycol employed in the system and to relate the structure and phase behavior to the composition of this four-component system at room temperature in terms of the surface and interfacial tension, rheological properties, and morphological studies.

II. EXPERIMENTS

A. Materials and Method

Reagent-grade polyethylene glycol (PEG) with a molecular weight of 400 was purchased from Merck & Co. Sodium dodecylsulfate and pentanol were supplied by Merck & Co., and reported to be 99% pure. All of these chemicals were used in the experiments without further purification.

Samples of solutions were prepared by dissolving sodium dode-cylsulfate in a fixed 50 g of distilled, deionized water, followed by addition of polyethylene glycol and pentanol, respectively. The total amount of each sample was 100 g by weight. These samples, containing different amounts of sodium dodecylsulfate, polyethylene glycol, and pentanol, and with a constant concentration of water (50 g), were used for the structure and phase behavior analysis.

B. Apparatus

A Du Noüy Ring tensiometer was calibrated using water and ethanol, and used to measure the surface and interfacial tension for samples of solutions including liquid crystals at room temperature. The experimental error of surface tension measurements for these samples of solutions with no liquid crystals was estimated to be approximately ±0.2%, and of solutions with liquid crystals to be greater than ±0.4%. Rheological measurements for samples after one day of preparation were carried out at $25 \pm 0.05°C$ with a Brookfield viscometer. Microscopic examination of all the samples using polarized light was also conducted at $25 \pm 0.05°C$. Details of the Du Noüy Ring tensiometer, Brookfield viscometer, and polarized light microscope are given in Refs. 4–6.

III. RESULTS AND DISCUSSION

The phase diagram of a four-component system, PEG/SDS/pentanol/ H_2O, is presented in Fig. 1. Notations given in this phase diagram represent different phases, in which $L_1 + L_2$ means a mixture of two liquid phases; M the microemulsionlike phase; LC the liquid crystalline phase; L the liquid phase or solution; M-1 the aqueous micellar phase 1; M-2 the aqueous micellar phase 2 in equilibrium with liquids or solutions; and G a mixture of solid crystals in equilibrium with liquid crystals and solutions. In the region of $L_1 + L_2$, since the solubility of pentanol in water solutions is very low, the addition of excess pentanol to PEG solution continning SDS results in phase separation. The oil-water interfacial tensions for samples labeled a, b, c, d and e, given in this region, with a constant ratio of pentanol and PEG equal to 9:1 and with various concentration

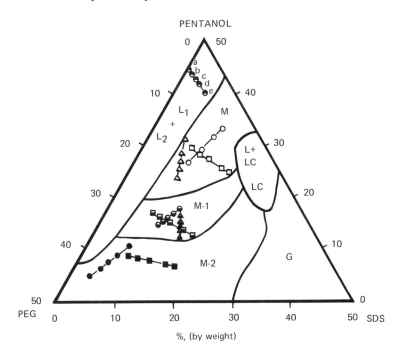

FIG. 1 Phase diagram for a four-component system, PEG/SDS/
Pentanol/H$_2$O. (The amount of H$_2$O used in the system was fixed
at 50% by weight, ○ and △, □, ○, △, □, ●, and ■ represent samples
taken from different regions.)

of SDS were measured at 25 ± 0.05°C, and are shown in Fig. 2.
We noted that interfacial tensions between oil and water for these
five samples decrease drastically as the SDS concentration is gradually increased. We also noted that as long as the concentration of
SDS reaches 5% by weight, phase separation behavior in this region
tends to disappear. This is due mainly to increased adsorption of
SDS molecules at the oil-water interface, causing large amounts of
pentanol diffusing into PEG aqueous solutions to form the oil-in-
water transparent microemulsionlike phase. In Fig. 3, it appears
that, for samples given in the region of M, containing SDS and
pentanol in a weight ratio of 1:3 and with various concentrations of
PEG, the surface tension at 25 ± 0.05°C is seen to be constant, almost independent of the concentrations of PEG employed in the composition. Under the same experimental conditions, the surface tensions for these samples with ratios of SDS/PEG and PEG/pentanol
equal to 1:2 and 2:5 are observed to be independent of the amounts

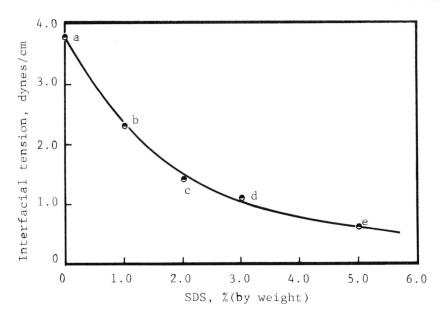

FIG. 2 Interfacial tension between pentanol and PEG aqueous solution versus SDS concentration at 25 ± 0.05°C.

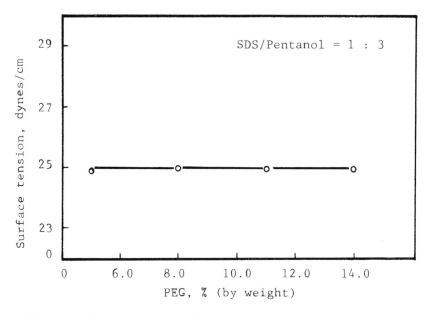

FIG. 3 Surface tension of microemulsionlike phase versus concentration of PEG at 25 ± 0.05°C.

of SDS and pentanol used, respectively, as shown in Figs. 4 and 5. Obviously, our results, given in Figs. 3 through 5, illustrate that the change in composition for these samples has almost no effect on their surface characteristics. We then studied the rheological properties of these samples in the same region given above. It was found that a linear relationship between apparent viscosity and shear rate at $25 \pm 0.05°C$ for a sample comprising 10% PEG, 15% SDS, 25% pentanol, and 50% H_2O, shown in Fig. 6, represented simple Newtonian flow behavior. Since the average distance between micellar and PEG molecules and between swollen micellar molecules themselves (oil-in-water microemulsion) is relatively weak, molecules do not interfere appreciably with each other's freedom of movement. By combining the evidence described above, it appears that the oil/water (O/W) transparent microemulsionlike structures are likely to be formed.

In Figs. 5 and 7, it seems that the surface tensions at $25 \pm 0.05°C$ for samples given in the region of M-1 with ratios of PEG/pentanol and SDS/pentanol equal to 3:2 and 2:3 and with various amounts of SDS and PEG increase slightly with increasing SDS and PEG concentrations, respectively. This is because the interaction of SDS with PEG may lead to polyelectrolytelike structures, which reduce the adsorption of SDS molecules at the interface, and therefore the surface tension increases slightly. In the same region described above, the surface tension for samples with a ratio of

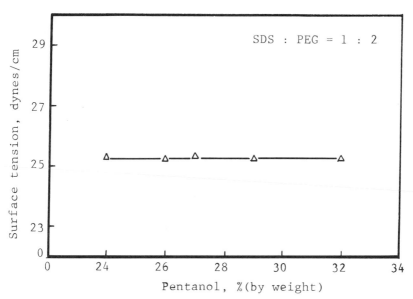

FIG. 4 Surface tension of microemulsionlike phase versus amount of pentanol at $25 \pm 0.05°C$.

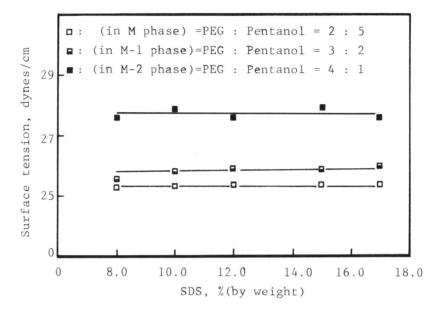

FIG. 5 Surface tension of M, M-1, and M-2 phases versus concentration of SDS at 25 ± 0.05°C.

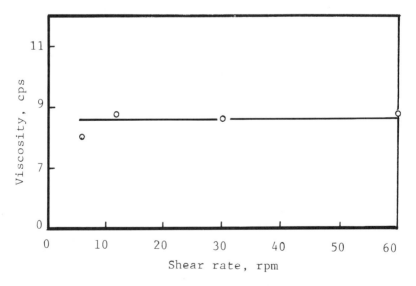

FIG. 6 Viscosity versus shear rate at 25 ± 0.05°C for a sample containing 10% PEG, 15% SDS, 25% pentanol, and 50% H_2O given in the region of M.

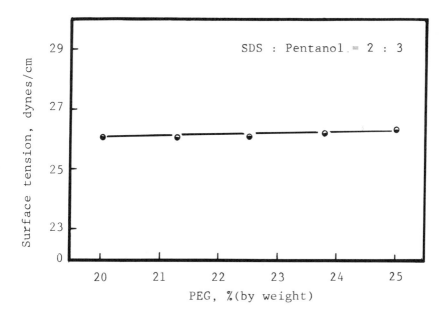

FIG. 7 Surface tension versus concentration of PEG at 25 ± 0.05°C for samples given in the region of M-1.

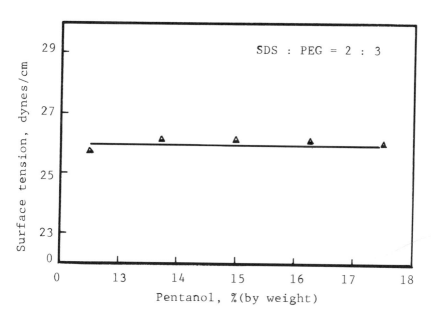

FIG. 8 Surface tension versus amount of pentanol at 25 ± 0.05°C for samples given in the region of M-1.

SDS /PEG equal to 2:3 is seen to be independent of the amounts of
pentanol used, as shown in Fig. 8. It may be explained that since
the saturation adsorption of SDS and pentanol and, possibly, PEG
molecules at the interface has been reached, the surface concentration associated with the surface tension of solutions does not change
significantly upon further addition of pentanol. Figure 9 shows that
the surface tensions for samples selected from the region of M-2
containing a ratio of SDS /pentanol equal to 2:3 largely increase as
the concentration of PEG is gradually increased. This is possibly
due to the formation of polyelectrolytelike structures resulting from
the interaction between SDS and PEG molecules, strongly decreasing
the adsorption of SDS molecules at the oil-water interface. In comparing Figs. 5, 7, and 9, the values of surface tensions for samples
given in the region of M-2 are larger than those of samples in the
region of M-1, indicating that more polyelectrolytelike structures
formed in the M-2 region are responsible for increased surface tension of solutions. In addition, Fig. 5 also shows that the surface
tension for samples of the M-2 region with a ratio of PEG /pentanol
equal to 4:1 increases with increasing concentrations of SDS and is
seen to be higher than that of samples with a ratio of PEG /pentanol
equal to 3:2 given in the region of M-1. The explanation of this

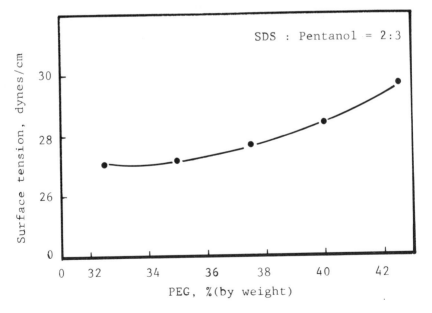

FIG. 9 Surface tension versus concentration of PEG at 25 ± 0.05%C
for samples given in the region of M-2.

behavior is the same as described above. However, the formation of
large amounts of polyelectrolytelike structures in the region of M-2
turns out to be true.

Both liquid phase (L) and liquid crystalline phases (LC) were
separated from samples of solutions taken from the region of L + LC
by a centrifuge at a rate of 500 rpm for 30 min. Surface tension,
interfacial tension, and viscosity measurements at 25 ± 0.05°C for
L, LC, and L + LC phases given in the phase diagram of Fig. 1
are shown in Table 2. It seems that the surface tension of the L
phase for samples with a fixed concentration of PEG (1% by weight),
which is almost equivalent to that of pentanol (see Table 1) and
smaller than that of LC phase, does not depend on the concentration
of SDS and pentanol used in the compositions, whereas the surface
tension of the LC phase changes with the composition, increases
with increasing amounts of SDS and decreasing with decreasing
amounts of pentanol. It can be considered that lamellar liquid crystals, due to their parallel-oriented hydrocarbon chains of SDS molecules adsorbed at the interface, are more ordered than pure pentanol
in the surface layer. Furthermore, under the influence of the dispersion force, the interaction between parallel-ordered molecules in
the LC phase is much stronger than that of the disordered molecules present in the L phase. This interaction makes molecules even
more ordered, and therefore the surface tension of the LC phase
increases. Table 2 illustrates how the L-LC interfacial tension for
samples containing a ratio of pentanol/PEG approximately equal to
29:1 increases with increasing concentrations of SDS, and for samples with a ratio of pentanol/SDS nearly equal to 26:20, increases
as the concentration of PEG is increased. This can be explained
by the fact that the interaction between SDS and PEG molecules
tends to form polyelectrolytelike structures that greatly reduce the
adsorption of SDS molecules at the interface, resulting in the increasing oil-water interfacial tension.

TABLE 1 Surface Tension of
Substances at 25 ± 0.05°C

Substance	Surface tension (dynes/cm)
Pentanol	27.6
SDS	40.1
PEG	47.7
H_2O	72.3

TABLE 2 Surface Tension, Interfacial Tension, and Viscosity for L, LC, and L + LC Phases in a PEG/SDS/Pentanol/H$_2$O System at 25 ± 0.05°C

Composition, %(by weight) (PEG/SDS/pentanol/H/O)	Surface tension for LC phase (dynes/cm)	Surface tension for L phase (dynes/cm)	Interfacial tension (dynes/cm)	Viscosity		
				LC (cps)	L + LC (cps)	L (cps)
1/18.5/30.5/50	30.0	27.1	1.1	287	18.3	10.7
1/19/30/50	29.5	27.2	1.6	296	18.9	11.3
1/20/29/50	32.0	27.1	1.8	442	77.4	12.2
1/21/28/50	32.6	27.6	2.3	513	175.5	13.0
0/20/30/50			1.0			
1/20/29/50			1.8			
3/20/27/50			2.1			
5/20/25/50			3.3			

Table 2 clearly indicates that the apparent viscosities for all the
phases tend to increase when the concentration of SDS is increased.
This is because the strong interaction between planar micelles themselves and between planar micelles and PEG molecules, yielding large
effective phase volume, coupled with the steric hindrance experienced, causes the viscosity to be higher. It is also noteworthy that
the apparent viscosities of the LC phase, having large amounts of
lamellar liquid crystals, are much higher than those of the L and
L + LC phases, as a result of a much stronger interaction between
ordered molecules. To explore further the structures of liquid crystals, the rheological properties of these liquid crystals were examined
and presented in Fig. 10. The apparent viscosity decreases with an
increase in the shear rate for a liquid crystal consisting of 0% PEG,
20% SDS, 30% pentanol, and 50% H_2O, as shown in Fig. 10. Since
the planar micelles are large in two dimensions and restrict each
other's rotational degree of freedom to a very marked extent, ordering of these planar micelles creates lamellar liquid crystals. In addition, the interactions between lamellar liquid crystals are strong, and
a networklike structure created in the solutions is readily broken
down by shear stress. This results in reduced viscosity.

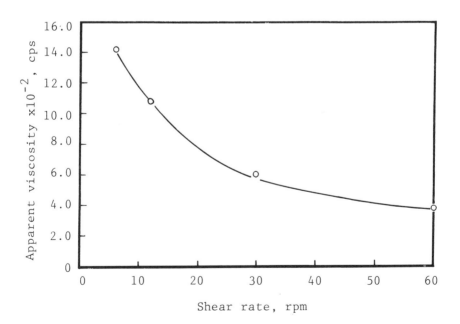

FIG. 10 Apparent viscosity versus shear rate at 25 ± 0.05°C for a
liquid crystal containing 0% PEG, 20% SDS, 30% pentanol, and 50%
H_2O chosen from the region of L + LC.

In order to examine in detail structures present in this four-component system, microscopic examination was made of samples of solutions chosen from both M-1 and M-2 pjases under polarized light at 25 ± 0.05°C. They exhibited a dark area between crossed polarizers, this being indicative of isotropic aqueous micellar phases. Samples given in the region of L + LC also were examined, and they showed the presence of liquid crystalline phases with anisotropic structures. These structured solutions with unique structural direction exhibited liquid crystals with circular shapes known as lamellar liquid crystals [7,8]. Lamellar liquid crystals appeared initially at a concentration of SDS equal to and greater than 17% by weight, and increased as the concentration of SDS was further increased. Similarly, microscopic analysis of samples in G region was done at 25 ± 0.05°C. It appears that the sizes of the lamellar liquid crystals become much smaller in comparison to that of the liquid crystals, that the area composed of solutions is reduced substantially, and that large amounts of solid crystals appear as the SDS concentration is increased appreciably.

To test the surface characteristics of liquid crystals, a drop of these liquid crystals was placed on the surface of water solutions. They spread rapidly and then mixed in water. Thus, these liquid crystals can be considered to be hydrophilic.

IV. CONCLUSIONS

Experimental results on the surface tension and viscosity studied at 25 ± 0.05°C suggest that sufficient amounts of SDS molecules (5% by weight) adsorbed at the oil-water interface may lead to the O/W transparent microemulsionlike phase.

Microscopic examination of samples using polarized light at 25 ± 0.05°C exhibits liquid crystals with circular shapes known as lamellar liquid crystals. Lamellar liquid crystals appear initially at a concentration of SDS equal to and greater than 17% by weight, and increase as the concentration of SDS is further increased. These liquid crystals have tested hydrophilic, and their rheological properties have been shown to be non-Newtonian.

Under polarized light, the phenomena of solid crystals in equilibrium with liquid crystals and/or solutions can take place at a concentration of SDS equal to and greater than 25% by weight. These mixed phases tend to form gellike structures with high viscosities.

V. ACKNOWLEDGMENT

We are grateful to Union Chemical Research Laboratories, ITRI, Hsinchu, Taiwan, for their support of this research.

VI. REFERENCES

1. Mukerjee, F., *Adv. Colloid Interface Sci. 1*, 241 (1967).
2. Attwood, D., and Florence, A. T. (eds.), *Surfactant Systems*, Chapman and Hall, New York, 1983.
3. Gallaugher, A. F., and Hibbert, H., *J. Am. Chem. Soc. 59*, 2514 (1973).
4. Harkins, in *Physical Methods of Organic Chemistry*, Weissburger, A. (ed.), Interscience, New York, 1959, vol. 1.
5. Van Wazer, J. R., Lyon, J. W., Kim, K. Y., and Colwell, R. E., *Viscosity and Flow Measurement, A. Laboratory Handbook of Rheology*, Interscience, New York, 1963.
6. Hartshorne, N. H., and Stuart, A., *Crystals and the Polarizing Microscope*, Edward Arnold, London, 1960.
7. Rosevear, F. B., *J. Am. Oil Society 31*, 628 (1954).
8. Benton, W. J., Fort, T., Jr., and Miller, C. A., SPE (Society of Petroleum Engineers of Aime) 7579 (1978).

13

Random Bicontinuous Winsor Microemulsions: Curvature, and Interfacial Tensions

LOÏC AUVRAY *Laboratoire de Physique de la Matière Condensée;*
Collège de France, Paris.

J. P. COTTON *Laboratoire Léon Brillouin, CEA-CEN, Saclay,*
Gif sur Yvette, France

RAYMOND OBER *Laboratoire de Physique de la Matière Condensée,*
Collège de France, Paris, France

CHRISTIANE TAUPIN *Laboratoire de Physique de la Matière Con-*
densée, Collège de France, Paris, France

I. INTRODUCTION

In this chapter we demonstrate experimentally, by using the neutron scattering technique, that Winsor-type microemulsions containing comparable volumes of oil and water have a random bicontinuous structure as predicted by recent theories. We relate the water-surfactant correlations to the average curvature of the surfactant layer in the microemulsions and observe the progressive inversion of this curvature as the water-in-oil ratio is inverted. We also propose a description of the structure of the interface between a Winsor middle phase and a phase in excess—oil, for example. Depending on the surfactant film spontaneous curvature (or water salinity), the interface is either closed by a flat Langmuir film, in which case a wetting layer of the third (water) phase separates the film and the microemulsion, or the interface is open, i.e., traversed by oil pores. This description is in agreement with recent ellipsometry experiments and interprets the variations of the middle-phase ultralow interfacial tensions with salinity.

II. MAIN FEATURES OF WINSOR MICROEMULSIONS

The "Winsor III microemulsions" [1] appear when an oil-brine-ionic surfactant-cosurfactant mixture separates into three phases. The middle-phase microemulsions coexisting with an upper oil phase and a lower water phase are associated with extremely low interfacial tensions [2] ($\sim 10^{-3}$ dyne/cm) and have potential applications in enhanced oil recovery [2]. Several particularities are noteworthy: These phases are intermediate between oil-in-water microemulsions obtained at low salinity and water-in-oil microemulsions obtained at high salinity; they cosolubilize maximal quantities of oil and water; and they exist only in a well-defined "optimal" [2] range of brine salinity.

According to recent theories [3-5]. these microemulsions could not be described classically as systems of water-in-oil or oil-in-water droplets and realize the interesting possibility of being random dispersions of oil and water, stabilized at the microscopic scale by a surfactant interfacial film. The film would be in a state of almost zero interfacial tension [6], so that the total film interfacial energy is comparable to the entropy of mixing of the oil and the water [3]. The microemulsion structure (and phase behavior) may then depend on the curvature elasticity of the film [7]. The formation of a highly disordered and labyrinthic structure would be possible if the film were very flexible (film rigidity K smaller than kT [7,20] and without spontaneous curvature. This last parameter, C_0 (in Å^{-1}), is related to the hydrophilic-lipophile balance or to Winsor's R ratio and depends in particular on the electrostatic repulsions between the surfactant polar heads, and is thus tunable by varying the water ionic strength.

By contrast to the well-known equation giving the radius R of, say, water-in-oil spheres,

$$R = \frac{3\phi_w}{C_s \Sigma} \tag{1}$$

the mean size ξ of the oil and water elementary volumes of the random structure ($\xi \sim 100 \text{ Å}$) is related to the microemulsion composition by the geometric constraint

$$\xi = \frac{6\phi_o \phi_w}{C_s \Sigma} \tag{2}$$

where ϕ_o and ϕ_w are the oil and water volume fraction ($\phi_o + \phi_w = 1$), C_s is the surfactant concentration, and Σ is the area per surfactant molecule in the film ($\Sigma \sim 60 \text{ Å}^2$).

Equation (2) interpolates continuously between the case of water-in-oil droplets ($\phi_w \ll 1$) and the case of oil-in-water droplets ($\phi_0 \ll 1$). In the models of random microemulsions, the average mean curvature of the film $<C>$ (positive by convention for W/O droplets) varies continuously with ϕ_0 from negative values ($\phi_0 < 0.5$) to positive values ($\phi_0 > 0.5$). In particular, it vanishes by symmetry when the microemulsions contain as much oil as water (inversion point, $\phi_0 = \phi_w$).

We have performed a systematic study [8–10] of these microemulsions in a very representative and well-known system [11] by using the small-angle scattering of X-rays and neutrons. These techniques probe the microemulsion structures at the scales 10–1000 Å, which means that both the local ($r < \xi$) and large-scale ($r > \xi$) structures are explored. Associated with the method of contrast variation with deuterated molecules, neutron scattering has enabled us to separate the partial structure factors of the water, oil, and surfactant; this is particularly interesting because the organization and the correlations of the surfactant molecules are very different in the bicontinuous microemulsions and in the droplet microemulsions. The main results are the following:

1. A well-defined surfactant interfacial film evidenced by the asymptotic behavior of the scattered intensities exists even in "critical" Winsor microemulsions [8].

2. The characteristic length of the microemulsions structure, ξ, defined as a mean radius of curvature of the film and drawn from the spectra in the intermediate range of scattering vector q (semilocal scale), follows experimentally the prediction of Eq. (2) [8].

3. When $\phi_0 = \phi_w$, the fluctuations of the water and surfactant concentrations, measured by the partial film-water structure factor at q = 0, are not correlated [9].

4. The intensity scattered only by the oil and water exhibits a pronounced peak at a scattering vector q* that is proportional to the surfactant concentration C_S [8–10].

Point 2 is the agreement with experiments on other systems [12,13], suggesting that the Winsor middle-phase microemulsions have indeed a random structure. From point 3 and an argument of Widom [5], we have deduced that the mean curvature of the surfactant layer is zero on average when the Winsor microemulsions contain as much oil as water ($\phi_0 = \phi_w$) and proved for the first time unambiguously that random microemulsions indeed exist. Observation 4 shows that the correlation between elementary volumes of oil (or water) are of the excluded volume type. This is contrary to the Kaler-Prager [14] calculation based on the Talmon-Prager model [3], but occurs if the size (of order $q*^{-1} \sim \xi$) of the elementary volumes of oil and water is well defined and does not fluctuate very much as is assumed in the de Gennes model [4]. This provides an indirect confirmation that the effects of the film rigidity are important in microemulsions.

This chapter is divided into two parts. In the first, we extend the contrast variation experiments of Ref. 9, relate them to the film curvature $<C>$, and study the change in $<C>$ when ϕ_0 is varied. In the second part, we present a structure model of the interfaces between a Winsor middle phase and the phases in excess and relate the variations of the corresponding interfacial tensions with the water ionic strength to the variations of the spontaneous curvature of the surfactant film.

III. CURVATURE OF THE SURFACT ANT FILM AND SCATTERING EXPERIMEMENTS

A. Samples

The microemulsions are monophasic microemulsions whose compositions are close to those of the corresponding middle phases [11]. They are made of brine (H_2O/D_2O, NaCl) at the optimal salinity (S = 6.5%), toluene (C_7H_8/C_7D_8), sodium dodecyl sulfate (SDS), and 1-butanol (with a deuterated hydroxyl group). We vary the oil volume fraction ϕ_0 between 0.4 and 0.7 at constant surfactant massic concentration (C_{ms}, in g/ml).

B. Scattering Experiments

The neutron scattering experiments were done at Laboratoire Léon Brillouin ("Orphée" reactor, Saclay, France) (Spectrometer P.A.C.E., isotropic radial multidetector, λ = 7 Å, $\Delta\lambda/\lambda$ = 10%). The scattering vector range is $5.7 \ 10^{-3} < q < 6 \times 10^{-2} \ \text{Å}^{-1}$. For each value of ϕ_0, the contrast variation experiments were done on series of six to nine samples made with different mixtures of H_2O and D_2O.

C. Theory

Assuming that a microemulsion is globally incompressible and constituted of three distinct geometric parts, the oil, the water, and the surfactant film (respective volume fractions ϕ_0, ϕ_w, and ϕ_f, scattering length density n_0, n_w, and n_f), the intensity scattered by a unit volume of the sample is the sum of only three contributions:

$$i(q) = (n_w - n_o)^2 \chi_{ww}(q) + (n_f - n_o)^2 \chi_{ff}(q) \qquad (3)$$
$$+ 2(n_w - n_f)(n_f - n_o)\chi_{wf}(q)$$

The partial structure factors are defined by the relation

$$\chi_{ij}(q) = \int d^3r < \delta\phi_i(0)\delta\phi_j(r) > e^{iq\cdot r}$$

where $\delta\phi_i(r)$ is the local fluctuation of the volume fraction of the ith component ($i = w, f$). The brackets mean average value.

The cross-term $\chi_{wf}(q)$ has been studied [10], and its value at $q = 0$ is particularly interesting. At zero angle, $\chi_{fw}(0)$ measures the cross-correlation between the thermodynamic fluctuations of the water volume fraction ($\delta\phi_w$) and the film volume fraction ($\delta\phi_f$) in a macroscopic volume V: $\chi_{wf}(0) = V<\delta\phi_w\,\delta\phi_f>$. As first noticed by Widom [5] and developed previously [9,10], this quantity, calculable from the thermodynamic models of microemulsions [3–5], has a simple geometric interpretation. As the water volume fraction in the volume V fluctuates by the quantity $\delta\phi_w$, the interfacial film moves; the variation δS of the interfacial area in V is proportional to the volume swept by the film (i.e., to $\delta\phi_w$), and the proportionality coefficient is the film curvature $<C>$. As S is proportional to ϕ_f, $\chi_{wf}(0)$ is on average proportional to $<C>$ and to $<\delta\phi_w^2>$, that is, to $\chi_{ww}(0)$. Up to a numerical factor and to the first order in d (where d is the thickness of the film),

$$\frac{\chi_{wf}(0)}{\chi_{ww}(0)} \propto <C>d \tag{4}$$

One recovers the result, already used to show the microemulsions inversion [15], that $\chi_{wf}(0)$ is positive for W/O droplets and negative for O/W droplets.

In the mean field approximation, the equilibrium relation (2) is still applicable to describe the macroscopic fluctuations in the volume V. As $C_s\Sigma = 6\phi(1 - \phi_w)/\xi$, one has (assuming ξ fixed as in Ref. [4], $\delta\phi_f \propto \delta C_s\Sigma \propto [(1 - 2\phi_w)/\xi]\,\delta\phi_w$. The average curvatures of the film $<C>$ and $\chi_{wf}(0)$ are both proportional to $(1 - 2\phi_w)$ and vanish when $\phi_0 = \phi_w = 0.5$ [5,9]. We notice that the same relation (2) can be used to discuss in the mean field approximation the quadratic average of the film fluctuations, i.e., the intensity scattered at zero angle by the film, $\chi_{ff}(0)$. This last quantity is analyzed in Ref. [10].

D. Results

Knowing n_0, n_w, and n_f, $\chi_{wf}(q)$ was determined by fitting the three unknown functions $\chi_{ww}(q)$, $\chi_{wf}(q)$, and $\chi_{ff}(q)$ to the observed intensities according to Eq. (3). Figure 1 represents the

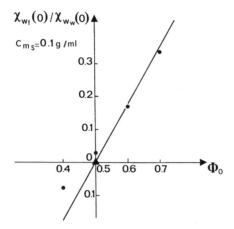

FIG. 1 Variations of the ratio $\chi_{wf}(q)/\chi_{ww}(q)$ extrapolated at zero angle as a function of the oil volume fraction at constant surfactant concentration. This ratio is proportional to the average curvature of the surfactant layer in the microemulsions (the triangle is a datum from Ref. 9.)

variations of the experimental ratio $\chi_{wf}(0)/\chi_{ww}(0)$ as a function of ϕ_0 at constant C_s. This ratio increases continuously with ϕ_0, changing its sign for $\phi_0 = 0.05 \pm 0.03$, in agreement with the prediction of the models of random bicontinuous microemulsions, Figure 1 shows clearly that the Winsor microemulsions cannot be described by a droplet model and that the film mean curvature is progressively inverted as the water and oil proportions in the samples are progressively inverted.

IV. SPONTANEOUS CURVATURE EFFECT ON THE MICROEMULSION INTERFACIAL PROPERTIES

The ultralow interfacial tensions $\gamma_{m/o}$ and $\gamma_{m/w}$ between a middle-phase microemulsion (m), the upper oil phase (o), and the lower brine phase (w) have been much studied experimentally [2,11] and theoretically [16–18]. However, their relations with the middle-phase structure are yet unclear. In the frame of the models of random bicontinuous microemulsions, we present here a description of the structure of the m/o and m/w interfaces, fully developed in Ref. 19, which interprets the variations of the middle-phase interfacial tensions which the brine salinity.

We consider for definiteness the middle-phase-oil phase interface. Because the structure of the microemulsion is random and bicontinuous, it seems at first sight that the microscopic interface between the oil and the water (occupied by the surfactant film) in the oil-microemulsion interfacial region cannot be flat at the scale of a few hundred angstroms (a few ξ). There should be continuous paths connecting the oil channels in the bulk of the microemulsion and the oil in the upper phase. In this case, we say that the o/m interface is open, and we notice [Fig. (2a)] that the surfactant film in the o/m interfacial zone is on average bent toward water.

If the film spontaneous curvature C_0 is positive (favoring water-in-oil microemulsion, high-salinity case), this structure is energetically favorable, but this is not the case if C_0 is negative (favoring the film bending toward oil, i.e., oil-in-water microemulsion, low-salinity case).

If $C_0 < 0$, an alternative structure of the o/m interface may be that of a closed interface (Fig. 2c): The closed o/m interface is the superposition of a microemulsion-water open interface, where the surfactant layer is on average bent toward oil, a water-wetting layer (whose thickness is of order ξ) and a flat Langmuir surfactant film separating the water-wetting layer and the oil upper phase.

Limiting the discussion to these two extreme structures of the microemulsion interfaces, the main results of the model, which are supported by recent ellipsometry experiments [21,22], are as follows:

1. The spontaneous curvature, C_0, has a characteristic value, C_0^* which is of the order of ξ^{-1}, ξ being the microemulsion characteristic length.
2. Depending on the relative value of C_0 with respect to C_0^*, there is three different situations:

a   b   c

FIG. 2 Microscopic structures of the interface between a Winsor middle phase and the oil upper phase in excess at the scale of a few hundred angstroms. The water salinity (spontaneous curvature) decreases from (a) to (c). In (a), the spontaneous curvature of the surfactant film, C_0, is positive or very small, and the film is bent toward water. In (c), C_0 is smaller than $-C_0^*$ ($\sim -\xi^{-1}$), the microemulsion surfactant film is preferentially bent toward oil, and the interface is closed by a flat Langmuir film and a water-wetting layer.

(a) If $C_0 < -C_0^*$ (low salinity), the middle phase minimizes its
     surface free energy with an open w/m interface and a
     closed m/o interface.
(b) If $-C_0^* < C_0 < C_0^*$, the two interfaces are open.
(c) If $C_0 > C_0^*$ (high salinity), the w/m interface is closed
     whereas the m/o interface is open.

Because of the free energy excess of the wetting layer and the
Langmuir surfactant film, $\gamma_{m/o}$ is larger than $\gamma_{m/w}$ in case (a) and
smaller in case (c). The intermediate case (b), which corresponds
to a small spontaneous curvature, is particularly interesting, be-
cause the difference $\Delta\gamma = \gamma_{m/o} - \gamma_{m/w}$ depends only on the curvature energy and is related directly to C_0:

$$\Delta\gamma \propto -KC_0C_s\Sigma \tag{5}$$

Taking $K \sim 10^{-14}$ erg [7,20], $1/C_0 \sim 100$ Å, and $1/C_s\Sigma \sim 100$ Å,
one estimates $\Delta\gamma \sim 10^{-2}$ dyne/cm, which is the right order of magnitude. Note that if $C_0 = 0$, one expects the middle phases to contain as much oil as water [5] and that $\Delta\gamma = 0$. This is indeed what
is observed in many Winsor systems at the optimal salinity [2].

We conclude from the results of the above simple model and from
the scattering experiments that the main features of the phase behavior, structure, and interfacial properties of the Winsor middle
phase are now being distinguished and are well described by recent
theories, presenting them as random bicontinuous dispersions resulting from a competition between the oil-water dispersion entropy
and the surfactant film surface and curvature energies.

V. REFERENCES

1. Winsor, P. A., *Solvent Properties of Amphiphilic Compounds*,
    Butterworths, London, 1954.
2. Reed, R. L., and Healy, R. N., in *Improved Oil Recovery by
    Surfactant and Polymer Flooding*, Shah, D. O., and Schechter,
    R. S. (eds.), Academic Press, New York, 1977.
3. Talmon, Y., and Prager, S., *J. Chem. Phys. 69*, 2984 (1978).
4. De Gennes, P. G., Jouffroy, J., and Levinson, P., *J. Physique
    43*, 1241 (1982).
5. Widom, B., *J. Chem. Phys. 81*, 1030 (1984).
6. Schulman, J. H., and Montagne, J. B., *Ann. N.Y. Acad. Sci.
    92*, 366 (1961).
7. de Gennes, P. G., and Taupin, C., *J. Phys. Chem. 86*, 2294
    (1982).

8. Auvray, L., Cotton, J. P., Ober, R., and Taupin, C., *J. Physique 45*, 913 (1984).

9. Auvray, L., Cotton, J. P., Ober, R., and Taupin, C., *J. Phys. Chem. 88*, 4586 (1984).

10. Auvray, L., Ph.D. Thesis, Université de Paris-Sud, France (1985). See also Auvray, L., Cotton, J. P., Ober, R., and Taupin, C., *Proceedings of the Symposium on Complex and Supramolecular Fluids*, Annandale, New Jersey, June 17–21, 1985, *to appear*.

11. Cazabat, A. M., Langevin, D., Meunier, J., and Pouchelon, A., *Adv. Colloid and Interface Sci. 16*, 175 (1982).

12. Kaler, E. W., Bennett, K. E., Davis, H. T., and Scriven, L. E., *J. Chem. Phys. 79*, 5685 (1983).

13. De Geyer, A., and Tabony, J., *Chem. Phys. Lett.*, *to appear*.

14. Kaler, E. W., and Prager, S., *J. Colloid and Interface Sci. 86*, 359 (1982).

15. Lagües, M., Ober, R., and Taupin, C., *J. Physique Lett. 39*, L487 (1978).

16. Huh, C., *J. Colloid and Interface Sci. 71*, 408 (1979).

17. Talmon, Y., and Prager, S., *J. Chem. Phys. 76*, 1535 (1982).

18. Borzi, C., and Widom, B., *this volume*.

19. Auvray, L., *J. Physique Lett, 46*, L163 (1985).

20. Di Meglio, J. M., Dvolaitzky, M., and Taupin, C., *J. Phys. Chem. 89*, 871 (1985).

21. Beaglehole, D., Clarkson, M. T., and Upton, A., *J. Colloid Interface Sci. 101*, 330 (1984).

22. Meunier, J., *J. Physique Lett.*, *submitted for publication*.

14

Percolation in Interacting Microemulsions

S. A. SAFRAN, GARY S. GREST, and AMY L. R. BUG *Exxon Research and Engineering, Corporate Research Science Laboratories, Annandale, New Jersey*

ITZHAK WEBMAN *Rutgers University, New Brunswick, New Jersey*

I. INTRODUCTION

In this chapter, a phenomenological theory for the effects of interactions on the conductivity of water-in-oil microemulsions in the dilute limit of spherical droplets is reviewed. In the single-phase region, clustering due to attractive interactions tends to dramatically lower the volume fraction of droplets needed for conduction (percolation threshold). The dynamic rearrangements of the equilibrium clusters of globules modify the critical exponents for the onset of the percolation transition. The results are in good agreement with recent experiments on water-in-oil microemulsions.

Microemulsions [1] are multicomponent fluids characterized by equilibrium globular, domain, or networklike structures on length scales ~ 100 Å. There has been some controversy [2] over whether a new term is needed to describe these systems—exemplified by water, oil, and surfactant mixtures—or whether they are best classified as particular examples of multicomponent or micellar solutions. However, these systems are novel and of unique interest precisely because two of the components (e.g., the oil and the water) tend to form domains with the third component (e.g., the surfactant) at the interfaces between these domains—and *not* because they form random, multicomponent mixtures. In some regions of concentration and temperature, microemulsion systems form spherical globules of water-in-oil or oil-in-water [3,4] (for a treatment of shape transitions in microemulsions, see Ref. 5). These globules maintain their spherical form as many parameters of the system are varied, as is the case of Aerosol OT-water-oil systems [3]. In these phases, the

microemulsions are model systems for studying the properties of
interacting colloidal particles.

The organization of the globules in solution is important in
determining macroscopic properties such as phase behavior and
transport. This organization is a function of the interactions between globules and has been most extensively studied in systems
of spherical globules [3,4,6,7] using light and neutron scattering.
In addition to scattering probes of the interparticle correlations,
one can relate the conductivity (in a water-in-oil system) to the
connectedness of the clusters of microemulsion droplets. For a
completely random system composed of conducting globules in a
non- or poorly conducting background fluid, the conductivity shows
a large increase when a macroscopically connected set of globules
first occurs as a function of the volume fraction. The volume fraction of globules, ϕ_p, at which this happens is termed the percolation
threshold.

In this chapter we summarize some current theoretical work on
the effects of interactions and dynamics on percolation transitions
in microemulsions. Recent conductivity measurements on water-in-
oil systems have shown that interacting, spherical microemulsions
undergo a nonconducting to conducting transition at very small
($\sim 10\%$) volume fractions of dispersed phase [8-10]. Monte Carlo
simulations show how clustering due to attractive interactions plays
an important role in dramatically lowering the percolation threshold
of colloidal systems. The microemulsions are particularly interesting, since they are a *prime* example of an interacting, colloidal system that shows a percolation threshold; most other materials that
show percolation transitions consist of *quenched* mixtures of conducting and nonconducting materials. In addition, the dynamic
nature of the microemulsion droplets is shown to change the functional form of the conductivity near the percolation transition. In
Sec. II, we summarize theory and experiment on the effects of attractive interactions on the static properties of microemulsions. The
role of these interactions in lowering the percolation threshold and
on the conductivity is discussed in Sec. III.

II.   INTERACTIONS AND PHASE SEPARATION

The macroscopic behavior of microemulsion systems—e.g., phase
equilibrium—is determined by the global structure of the globules
in solution. The correlations between globule positions that lead to
this organization are determined by the interactions between globules.
The effects of interactions are most simply treated—both theoretically
and experimentally—for spherical droplets. To treat the interactions,
the internal degrees of freedom of the individual globules are

assumed to be such that the globules always remain spherical, although they may change their size as the conservation laws dictate. The globule radius ρ and the density of globules n (number per unit volume) are determined for monodispersed spheres from the constraints of surfactant and internal phase conservation:

$$\phi = \frac{4\pi\rho^3 n}{3} \quad \text{and} \quad v_s = 4\pi\rho^2 \delta n$$

where δ is a typical surfactant length. The radius ρ is then proportional to ϕ/v_s, where ϕ is the volume fraction of dispersed phase and v_s is the surfactant volume fraction.

In Refs. 11 and 12 it is shown how the attractions lead to a "liquid"-"gas" type of equilibrium between two microemulsion phases. The two phases consist of globules of the same type and size. The number density of globules is different, however, in the two phases. One phase is typically dense in globules ("liquid") and the other dilute ("gas"). The transition from a single phase to these two coexisting phases is driven by changing the temperature [3] of the globule size [6].

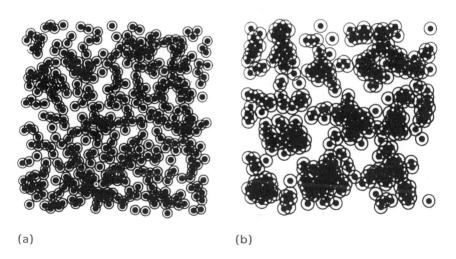

(a)   (b)

FIG. 1 Typical clusters obtained in simulations [13] of interacting globules in two dimensions. The dark area shows the hard core, while the circle around each globule indicates the conductivity shell δ. The δ shell is chosen so that system is at the percolation threshold; overlap of circles signifies connectedness. (a) No attraction. (b) Strong attraction leads to clustering.

The attractive interactions between globules are also responsible for more microscopic effects than the phase equilibrium alone. Even in the single phase, attractions lead to clustering, which can be measured by scattering probes. Figure 1 shows the effects of interactions in a two-dimensional system where the clustering can be seen. This figure is from the simulations described in Ref. 13. Similar simulations for a three-dimensional system, with hard-sphere particles that interact with a square-well potential $V(r) = \infty$ if $r < 1$, $V(r) = -\varepsilon$ if $1 < r \leqslant (1 + \lambda)$, and $V(r) = 0$ for $r \geqslant (1 + \lambda)$, show the effects of clustering in the interparticle structure factor $S(Q)$ [6]. Here r is the separation between the sphere centers measured in units of the sphere diameter, ε is the depth of the attractive well, and λ is the range of the attraction. Comparison of these calculations with experimental measurements of $S(Q)$ show that for Aerosol OT-water-oil systems, the attraction is extremely short-ranged, with a range of about 3 Å ($\lambda \stackrel{\sim}{\sim} 0.03$ for 100-Å-diameter globules). In addition, it was found that the interaction strength ε was approximately linear in the globule radius [6].

III. PERCOLATION IN MICROEMULSIONS

The previous section discussed the consequences of attractive interactions on the static properties of microemulsions, such as the phase diagram and equilibrium density fluctuations. In this section we examine the effects of interactions on the conductivity of microemulsions with an analysis of percolation in interacting, colloidal systems.

The concept of a percolation transition is often used in interpreting the conductivity of disordered systems. In systems with completely random distributions of particles, the percolation transition signifies the first emergence of an infinite cluster at some critical value of the volume fraction ϕ_p. If the particles are conducting and the matrix (or background fluid) is insulating, the system will show no conductivity for $\phi < \phi_p$. There will be a continuous transition at ϕ_p to a conducting state. For microemulsion systems, this transition is measurable through the ionic conductivity. However, a detailed treatment of the microscopic mechanism of the charge transfer is beyond the scope of this work.

Recent measurements of the conductivity of microemulsions have suggested the existence of a percolation threshold as a function of volume fraction, temperature, and globule size [8–10,14]. The transition, which occurs as either temperature or size is varied, shows percolation for very small volume fractions ($\sim 10\%$). Furthermore, the transition occurs near the temperature or size at which "liquid"-"gas" phase separation takes place. Some studies have attributed this transition to the onset of a bicontinuous phase [15].

However, recent scattering experiments in Aerosol OT-water-oil microemulsions have shown that these systems maintain the integrity of the spherical globules up to rather high volume fractions [16]. In addition, there is *no* change in the particle size and shape as the critical point for phase separtion is approached (or, for that matter, even in the two-phase regime) as a function of temperature [3]. Thus, instead of indicating a transition from a spherical to a bicontinuous state, the conductivity transition is interpreted to occur at the percolation threshold of rigid globules. The low value of the threshold is due to the effects of interglobular interaction [9,10,14].

Our recent Monte Carlo simulations [13,17,18], have shown that the percolation transition for spherical particles with a short-range attraction is a strong function of their interactions. As a critical point (or coexistence curve) is approached, the clustering of the particles that occurs tends to increase the connectivity of the system; the percolation threshold is decreased. Although this behavior is *not* universal—there are situations where the interactions can *increase* the percolation threshold [13]—the lowering of ϕ_p by attractive interactions is a general feature of systems with short-range hopping of the conducting species. In addition to the interactions, the dynamic nature of the Brownian motion of the microemulsion globules has subtle effects on the dependence of the conductivity of the system for $\phi < \phi_p$, as discussed below.

The first misconception that was elucidated by these simulations was the notion of a universal value of ϕ_p for spheres. Although there are systems with universal values of ϕ_p (e.g., a binary mixture of conducting and insulating closely packed spheres [19], where $\phi_p = 0.17$), microemulsions are a unique experimental realization of the percolation [20,21] of *interacting spheres with a hard core in a continuum background*. The threshold depends on the distance over which two adjacent globules can transfer charge. To define connectedness in this continuum system, it is necessary to introduce shell of width $\delta/2$ (in units of the hard-core diameter) around each hard core. When two of these δ shells overlap, the corresponding spheres are said to be connected and hence in the same cluster. The physical origin of δ is the finite range of hopping of charge between two globules or a finite range of deformation of two adjacent spheres, which may transfer charge via a *local* exchange of water and/or surfactant.

The Monte Carlo simulation is described in detail in Refs. 13, 17, and 18. The interparticle potential is the square-well interaction described in Sec. II. Standard Metropolis [22] algorithms for systems of 108, 500, and 2048 particles were studied in both two and three dimensions. In the noninteracting case ($\varepsilon = 0$), the percolation probability is a function of both the shell size δ and the volume fraction ϕ, as shown in Fig. 2, where the shell dependence of the

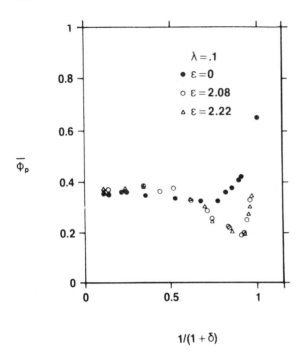

FIG. 2 Monte Carlo results for the critical volume fraction for percolation ϕ_p as a function of the conductivity shell size (δ) for both the noninteracting ($\varepsilon = 0$) and attractively interacting systems in three dimensions. The effective volume fraction $\bar{\phi}_p = \phi_p (1 + \delta)^3$ is plotted vertically, while $1/(1 + \delta)$ is plotted horizontally.

effective volume fraction for percolation, $\bar{\phi}_p = \phi_p (1 + \delta)^3$, is shown. If δ is small, the system percolates only at high volume fractions in the absence of interactions. In the limit $\delta \to 0$ (and $\varepsilon = 0$), $\phi_p \sim 0.65$, the value for random close packing. At large values of δ, the results approach the known value of $\phi_p \sim 0.35$ for overlapping spheres [19]. Away from the coexistence curve (i.e., when the attractive interaction is small), the transition is observed [10] at high values of the volume fraction. This is consistent with a *small* value of δ for these systems. (For a discussion of the equivalent problem in two dimensions, see Ref. 13.)

As the attractive interaction strength is increased ($\varepsilon > 0$), $\bar{\phi}p$ drops dramatically (see Fig. 2). The lowering of the percolation threshold is associated with the formation of correlated clusters in the interacting system. These clusters are both anisotropic and fractallike, and their percolation threshold is expected to be lower than that of random spheres. For small values of the interaction range λ and the conductivity shell δ, ϕ_p typically drops from values near random close packing ($\phi_p \approx 0.65$) to values of ϕ_p of the order of 10% as ε is increased. The largest effects occur near the critical point; the minimum in Fig. 2 occurs very close [11] to the critical volume fraction where concentration fluctuations are the largest. The observed percolation transitions in microemulsions are therefore sensitive indicators of the effects of concentration fluctuations on the transport behavior. Again, the lowering of ϕ_p with increasing interaction is nonuniversal; it occurs only for small values of the shell δ. For large values of δ, the interactions can actually *increase* the percolation threshold; moderate attractions cause the clusters to become more compact, and ϕ_p is increased. The competition between these two effects of attractive interactions on percolation is discussed in more detail in Ref. 13.

In addition to being the prime physical example of percolation in an interacting, equilibrium system, microemulsions are also a model system for the study of dynamic effects on percolation [14,23,24]. In the usual static models of systems of conducting regions embedded in an insulating matrix, the conductivity vanishes below ϕ_p. However, for microemulsions, the clusters continuously rearrange due to Brownian motion, resulting in a finite conductivity for $\phi < \phi_p$. Unlike the static case, the charge carriers are not trapped in the finite clusters. A charge on a water globule can propagate by either hopping to a neighboring globule or via the diffusion of the host globule. If the typical hopping time of the carriers between globules is much shorter than some characteristic time related to the motion of the globules, a steep increase in the overall carrier diffusion is expected as $\phi \to \phi_p$. This is due to transition from a regime of transport dominated by globular motion and cluster rearrangement to a regime of transport dominated by the motion of charge carriers on a large, connected cluster of globules.

The effects of globule dynamics on the conductivity in microemulsions were discussed by Lagues [23], who suggested that for $\phi < \phi_p$, the conductivity σ increases as $(\phi_p - \phi)^{-\tilde{s}}$, where $\tilde{s}$ differs from its static value. (The static percolation system with finite σ below ϕ_p is one where the charge carriers have finite conductivity in the matrix—i.e., the system is a mixture of good and bad conductor.) In our Monte Carlo simulations, a scaling argument— different from that of Lagues—is presented which suggests that the exponent $\tilde{s} = 2\nu - \beta = 1.3$ in three dimensions. These results are applicable to percolation in a system where the charge carriers hop on slowly moving colloidal particles. The value of the exponent $\tilde{s}$ differs from the exponent $s \approx 0.7$ for the static case. Our value of $\tilde{s}$ is in agreement with the Monte Carlo simulations [18] as well as with the experimental results on microemulsions [14,23]. In all these cases, the conductivity transition is smeared around the value of ϕ_p appropriate to the *static* percolation problem. Thus, the previous discussion of the shift of ϕ_p with interaction strength (as modified by either temperature or globule size) is completely applicable as far as the value of the threshold, while the dynamical effects of globule motion and cluster rearrangement are responsible for the change in the exponent.

The scaling argument for the dynamical case has also been extended to predict the frequency dependence of the conductivity at $\phi = \phi_p$. For small values of the frequency $\omega < 1/T_R$ (where T_R is the cluster rearrangement time), the conductivity is frequency independent. This behavior is followed by a power law regime with $\sigma(\phi = \phi_p) \underset{\sim}{\sim} w^\chi$ with $\chi = \mu/(\tilde{s} + \mu)$, where μ is the conductivity exponent for $\phi > \phi_p$. Again, this relation is in agreement with recent measurements [14].

IV. ACKNOWLEDGMENT

The authors are grateful for stimulating interactions with S. Bhattacharaya, W. Dozier, J. S. Huang, M. W. Kim, and J. Stokes.

V. REFERENCES

1. For a general survey, see *Surfactants in Solution*, Mittal, K., and Lindman, B. (eds.), Plenum Press, New York, 1984.
2. B. Lindman and P. Stiles, in *Surfactants in Solution*, Mittal, K., and B. Lindman (eds.), Plenum Press, New York, 1984, p. 1651.
3. Kotlarchyk, M., Chen, S. H., Huang, J. S., Kim, M. W., *Phys. Rev. A29*, 2054 (1984); Huang, J. S., and Kim. M. W., *Phys. Rev. Lett. 47*, 1462 (1981).
4. Brunetti, S., Roux, D., Bellocq, A. M., Forche, G., and Bothorel, P., *J. Chem. Phys. 83*, 1028 (1983); *J. Colloid and Interface Sci. 88*, 302 (1982); Ober, R., and Taupin, C., *J. Phys. Chem. 84*, 2418 (1980).
5. Safran, S. A., Turkevich, L. A., and Pincus, P. A., *J. Physique Lett. 45*, L69 (1984).
6. Huang, J. S., Safran, S. A., Kim, M. W., Grest, G. S., Kotlarchyk, M., and Quirke, N., *Phys. Rev. Lett. 53*, 592 (1984).
7. Kim, M. W., Dozier, W. D., and Klein, R., *J. Chem. Phys. 84*, 5919 (1986).
8. Eicke, H. F., Kubick, R., Hasse, R., and Zschokke, I., in *Surfactants in Solution*, Mittal, K., and Lindman, B. (eds.), Plenum Press, New York, 1984, 1533.

9. Cazabat, A. M., Chatenay, D., Guering, F., Langevin, D., Meunier, J., Sorba, D., Lang, J., Zana, R., and Pailette, M., in *Surfactants in Solution*, Mittal, M., and Lindman, B. (eds.), Plenum Press, New York, 1984, p. 1737; Cazabat, A. M., Langevin, D., Meunier, J., and Pouchelon, A., *J. Phys. Lett.* *43* L89 (1982).

10. Kim, M. W., and Huang, J. S., *Phys. Rev.* *A34*, 719 (1986).

11. Safran, S. A., and Turkevich, L. A., *Phys. Rev. Lett.* *50*, 1930 (1983).

12. Safran, S. A., Turkevich, L. A., and Huang, J. S., in *Proceedings of Symposium on Surfactants in Solution*, Mittal, K., (ed.), Plenum Press, New York, *in press*.

13. Bug, A. L. R., Safran, S. A., Grest, G. S., and Webman, I., *Phys. Rev. Lett.* *55*, 1896 (1985).

14. Bhattacharaya, S., Stokes, J., Kim, M. W., and Huang, J. S., *Phys. Rev. Lett.* *55*, 1884 (1985); H. A. Van Dijk, *Phys. Rev. Lett.* *55*, 1003 (1985).

15. Chatenay, D., Urbach, W., Cazabat, A. M., and Langevin, D., *Phys. Rev. Lett.* *20*, 2253 (1985).

16. Kotlarchyk, M., Chen, S. H., Huang, J. S., and Kim, M. W., *Phys. Rev. Lett.* *53*, 941 (1984).

17. Safran, S. A., Webman, I., and Grest, G. S., *Phys. Rev.* *A32*, 506 (1985).

18. Grest, G. S., Webman, I., Safran, S. A., and Bug, A. L. R., *Phys. Rev.* *A33*, 2842 (1986).

19. Scher, H., and Zallen, R., *J. Chem. Phys.* *53*, 3759 (1970); Blanc, R., and Guyon, E., *Percolation Structures and Processes—Annals of the Israel Physical Society, Vol. 5*, Deutcher, G., Zallen, R., and Adler, J. (eds.), Israel Physical Society, Jerusalem, 1983, p. 229.

20. Coniglio, A., De Angelis, U., and Forlani, A., *J. Phys. A 10*, 1123 (1977).

21. Chiew, Y., and Glandt, E., *J. Phys. A 16*, 2599 (1983).

22. Binder, K. (ed.), *Monte Carlo Methods in Statistical Physics*, Springer, Berlin, 1979.

23. Lagues, M., *J. Phys. Lett.* *40*, L331 (1979).

24. Kutner, R., and Kehr, K. W., *Phil. Mag. A48*, 199 (1983).

15

Structure and Tension of the Interfaces in a Model Microemulsion

CARLOS BORZI,* REINHARD LIPOWSKY,† and BENJAMIN WIDOM
Cornell University, Ithaca, New York

I. INTRODUCTION

This chapter presents a technique for finding the main features of the structure and tension of the interfaces of an ideal system that models Winsor III microemulsions. The model is that of Talmon and Prager [1], further simplified by de Gennes and Taupin [2], and thermodynamically investigated by Widom [3], who found its mean field free energy after introducing a cutoff of possible cell sizes. The final results, along with the discussion of the wetting problem, were presented at the twentieth Faraday Symposium [4].

We shall also cite here some new results on the parameter sensitivity of the model, whose complete properties are going to be published elsewhere [5].

II. OUTLINE OF THE MODEL

The system consists of two main components (A and B), immiscible in each other because of polarity (e.g., brine and oil) but solubilized after adding an appropriate blend of amphiphiles.

When in equilibrium, such a system shows two or three phases depending on the value of the H/L balance of the amphiphile blend as it was shown in Fig. 1 of Ref. [7].

Present affiliations:
*Instituto de Física de Liquicos y Sistemas Biológicos, La Plata, Argentina.
†Sektion Physik der Univerität München, München, FRG.

(A single-phase situation can also be reached, both experimentally and theoretically, within the framework of the model, but we shall not consider such a possibility here.)

A cubic partition of the volume of the system is considered, each cubelet being filled randomly with pure A or B and the amphiphile being placed in a bidimensional film between the resulting domains. The free energy (Φ) for the model, in mean-field approximation, depends on two densities: ϕ (the volume fraction of either of the main components, say, A) and τ (the dimensionless concentration of amphiphile), and on a geometric variable x, the cell size of the cubic partition scaled by a molecular length (a). This molecular length imposes the above-mentioned cutoff of cell size, which is required because A and B cannot be mixed beyond their molecular sizes. The system is then specified by three dimensionless parameters: λ the mean curvature of the amphiphile film), q (its stiffness), and p (the spontaneous mutual solubility of the main components, which is assumed to be very low.) The last two are scaled with temperature (θ) and typical values are p = 7 and q = 10 [3]. Lowering θ causes p and q to increase. For higher values of p, the second-order phase transition, which gives tricritical character to the critical end points [3], disappears without further consequences. There are no qualitative changes due to higher values of q, which imply just a stiffer film. Increasing θ causes p and q to decrease. There is a critical value of q below which the α and γ phases become identical. This is a critical end point [5]. We do not consider lower values of p because they would imply high spontaneous solubility of the main components, and this is not the physical case we are modeling.

The system choses its cell size for a given composition as the one that minimizes the overall free energy, which has two sheets (due to the imposed cutoff) joined at what has been called the "crossover locus" [3]:

$$\Phi = \min \begin{array}{l} \Phi(\phi, \tau; x(\phi, \tau) \\ \Phi(\phi, \tau; x = 1) \end{array}$$

We have generated phase diagrams for the model, in the densities space, by means of the tangent plane construction for different values of the parameters in the range of interest. In Fig. 1 we sketch the influence of λ on the three-phase tie-triangle. The vertices of those tie-triangles are the boundary conditions required to solve the Euler-Lagrange equations derived from the variational principle of the surface tension in the density functional theory of the interface [6]. In order to write such a variational principle explicitly, we shall consider an extension of the widows version of the model in which the cell size is allowed to adjust by itself through the

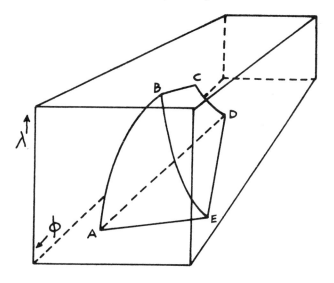

FIG. 1 General topology of the three-phase region of the model. The AED triangle is the isoceles tie-triangle corresponding to the symmetric case $\lambda = 0$. As λ increases, the triangle shrinks up to the line BC (locating the α-β critical end point). The lines AB, EB, DC are the loci of the coexisting α, β, and γ phases, respectively. There exists a complementary symmetric part of the figure corresponding to negative values of λ.

interface (that is, in the convex-up part of the free-energy surface hidden in the tangent plane construction in order to minimize the surface tension functional. This extension was recently discussed when analyzing fluctuations in the cell size of the model [7] and justified in Ref. 5. The resulting expression for the surface tension is then:

$$\sigma = \min {\int_{-\infty}}^{\infty} [U(\dot{\underset{\sim}{y}}) + \frac{1}{2} \dot{\underset{\sim}{y}}' \, G(y) \dot{\underset{\sim}{y}}] \; dt \tag{2}$$

where U is the extended version of the free energy minus its tangent plane, $\underset{\sim}{y}$ is a vector of components ϕ, τ, x (the dot indicates a derivative with respect to t, which is the distance through the interface, and the prime indicates transposition), and where G is a symmetric tensor whose components could, in principle, depend on the densities. We shall discuss and fix those coefficients later.

   The equilibrium profiles of the three variables will give the value of the model surface tension scaled by the one corresponding to the system without surfactant (that is, the binary mixture A-B to which the model converges in the limit $\tau = 0$.)

III. DYNAMICAL ANALOGY AND PARABOLIC APPROXIMATION

Equation (2) suggests a dynamical analogy to the surface tension problem [6] in which σ represents the action, the distance through the interface: t (which is the integration variable) represents the time; and, consequently, the integrand represents the Lagrangian of a frictionless particle that takes infinite time to slide along the $-U$ surface between any two of its three peaks (which correspond to the coexisting bulk phases.)

Let us look first at the "potential energy" U. By plotting its sections directly as functions of one and two of its variables, we have concluded that such a surface is smooth enough to be approximated by the truncations of its expansions around the minima; that is, we expand U around the peaks (which are located at the vertices of any of the tie-triangles in Fig. 3). Then we truncate those expansions at second order in the densities. (Note that, around the α and γ phases, the expansion in the third coordinate, the dimensionless cell size x, should be truncated at first order because of the cutoff [4].) Finally, we think of the surface resulting from taking the minimum among the three truncations for each set of values of ϕ, τ, and x as an approximation to the original U function:

$$U(\underset{\sim}{y}) = \min \left\{ \frac{1}{2} \underset{\sim}{z}'w^{\alpha}\underset{\sim}{z} + v^{\alpha}x; \ \frac{1}{2} \underset{\sim}{y}'w^{\beta}\underset{\sim}{y}; \ \frac{1}{2} \underset{\sim}{z}'w^{\gamma}\underset{\sim}{z} + r^{\gamma}x \right\} \qquad (3)$$

where

$$z = \begin{pmatrix} \phi \\ \tau \end{pmatrix}; \quad \underset{\sim}{y} = \begin{pmatrix} \phi \\ \tau \\ x \end{pmatrix}; \quad W_{ij}^{\alpha,\gamma} = \frac{\delta^2 U}{\delta z_i \, \partial z_j} \Big|_{\underset{\sim}{y} = \underset{\sim}{y}^{\alpha,\gamma}}$$

$$v^{\alpha,\gamma} = \frac{\delta U}{\delta x} \Big|_{\underset{\sim}{y} = \underset{\sim}{y}^{\alpha,\gamma}}; \quad W_{ij}^{\beta} = \frac{\delta^2 U}{\delta y_i \, \partial y_j} \Big|_{\underset{\sim}{y} = \underset{\sim}{y}^{\beta}}$$

The surface in Eq. (3) has three sheets joined at the seams where they intersect each other. The projection of that seam onto the (ϕ, τ) plane proves to be very close to the original crossover locus, and is one of the reasons for the success of the "parabolic approximation" [8] in the particular case of this model. The way the approximated surface looks, compared to the original one, is shown at the tops of Figs. 2–6 in the next section.

We still have to fix the coefficients of the second-order term in the gradient expansion of the free energy (the components of the tensor G in Eq. 2). This term, in the extended van der Waals theory [6], is supposed to take into account the inhomogeneities in the densities as expanded up to second order in their gradients. The coefficients are assumed to be proportional to the correlation of fluctuations among densities. From previous analysis [3,7], we know that the cross-terms are much smaller than the direct ones, so we consider just a diagonal tensor. (The inclusion of nondiagonal terms does not significantly affect the shape of the profiles nor the relative values of the surface tensions, as verified by direct computation. We shall come back to this point at the end of the section.)

The $\dot\phi^2$ term takes into account interactions (or, rather, correlations) between the main components, occurring through the cells walls. It should then be weighted by x^2. The $\dot\tau^2$ term considers amphiphile-amphiphile interactions on the dividing film where they are located and thus has the same weighting factor x^2. We rescale the overall "kinetic energy" by x^2, and the resulting expression is

$$K = M_1 \, \dot\phi^2 + M_2 \, \dot\tau^2 + M_3 \, (\ln \dot x)^2$$

where $M_1 \sim 10$ come from the solution of the A-B mixture model. $M_2 \sim M_1/100$ results from analysis of the correlation of fluctuations [3,7], and $M_3 \sim 0.35 \, M_1$ comes from the solution of the limiting case of the present model in which there is no surfactant [9].

In order to solve Eq. (2), we now rescale the coordinates with the square root of the corresponding "masses," then diagonalize the approximated potential energy (the "parabolic" one) on each one of its sheets. We get, analytically, families of exponential solutions on each sheet of the U surface. The solution of the surface tension problem comes from matching the exponential forms and the first derivatives somewhere along the path [8]. We take advantage of the arbitrary position of the dividing surface and fix the matching point at $t = 0$ ($t = -T/2$ and $t' = T/2$ when double-matching is required, as we shall discuss below, T being determined by the conservation of energy). The match occurs, of course, at some point on the surface seam.

There are three features that should be noted:

1. The cutoff in the cell size variable x leads to parabolic solutions for that coordinate, rather than exponentials, on the α and γ sheets.

2. The β-sheet intrudes itself between the other two, as pictured in Figs. 3 and 6, and it requires that we match twice when looking for the α-γ direct path.

   These two facts together make it impossible to reach complete analytical solution of the problem. In a model system without cutoff (that is, with rounded peaks in all the coordinates) or with all the paths lying on only two sheets (which means all single-matching problems), the solution for the structure and tension of the interfaces would be fully analytical within the parabolic approximation.

3. In order to consider nondiagonal terms in the "kinetic energy" tensor (assuming that they were bigger than what they are in our case), the procedure to follow would be the standard one for dealing with small-oscillation problems in classical mechanics: simultaneous diagonalization of both potential and kinetic energies. We did that for purposes of checking and, as has been mentioned, results were not affected significantly.

IV. STRUCTURE AND TENSION OF THE INTERFACES

A. Symmetric case ($\lambda = 0$)

The structure is symmetric at a point equally far from both critical endpoints. The system shows, naturally, three phases, and the tangent plane construction yields an isoceles tie-triangle (Fig. 1).

The α-β and β-γ profiles are identical because of the symmetry (the ones corresponding to ϕ are inverted with respect to the β bulk value). The ϕ profiles approach their limiting values exponentially, doing so much more slowly on their way to the β-value (because of the bigger osmotic compressibility of that phase). The τ profiles have similar asymptotic behavior and, in addition, show a bump in the middle reflecting the expected excess concentration of amphiphile at the interface. (Such bumps are also typical in mean field theory when there is more than one independent density [6].) The x profile goes exponentially to its β-value but it is sharp on the α (or γ) side because of the cutoff. All the features discussed above are shown in Fig. 2, where we also picture (at the top) how the parabolic approximation looks in the three independent variables.

The α-γ direct path, which requires that we match twice and then obtained the solution numerically, is shown in Fig. 3. The full symmetry of the $\lambda = 0$ case can be appreciated in that figure.

B. Typical Asymmetric Case ($0 < \lambda < \lambda_c$)

In the typical asymmetric case we move toward the α-β critical end point by increasing the value of the curvature parameter λ. Alternatively, we could approach the β-γ critical end point by going

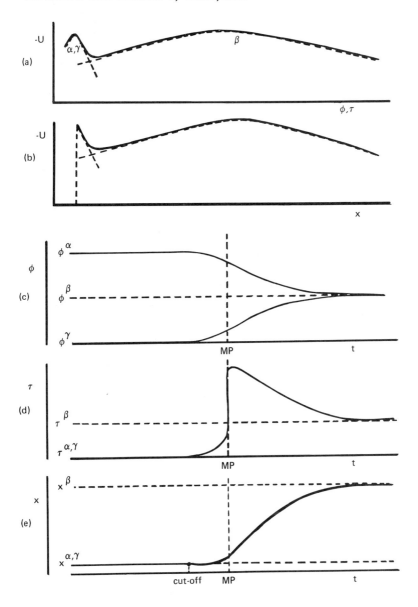

FIG. 2 α-β and β-γ interfaces for λ = 0 (symmetric case). (a) and (b) show how the parabolic approximation looks (dotted lines) in comparison with the original surface (heavy lines). (c), (d), and (e) show the resulting profiles.

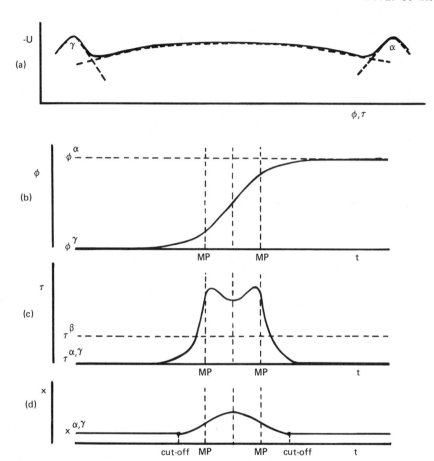

FIG. 3 α-γ interface (direct path) for $\lambda = 0$. In (a) we show the parabolic approximation for the "rounded-peak" variables ϕ and τ (for x there are sharp peaks on both sides.) (b), (c), and (d) show the respective profiles.

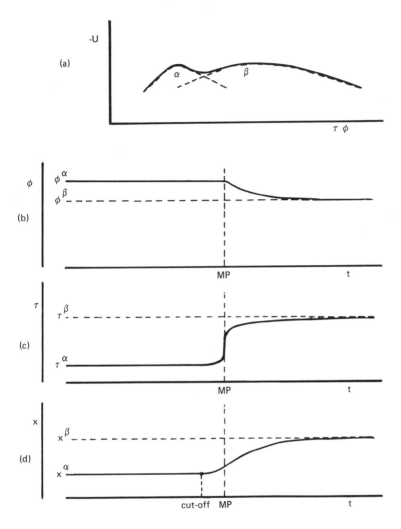

FIG. 4 α-β interface for $0 < \lambda < \lambda_c$, when the bump in τ has completely disappeared. Again (a) shows the parabolic approximation and (b), (c), and (d) the corresponding profiles.

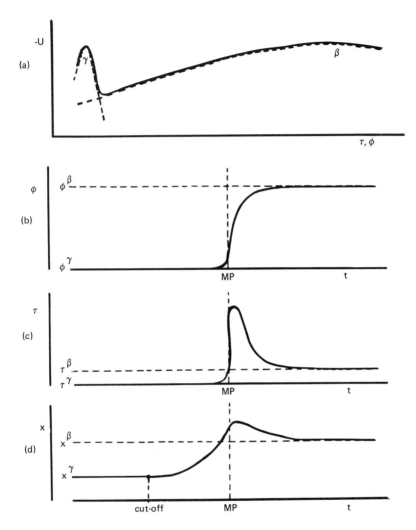

FIG. 5   β-γ interface for $0 < \lambda < \lambda_c$.

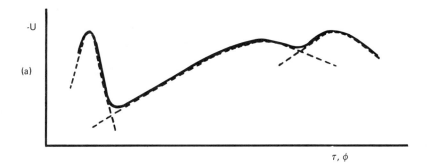

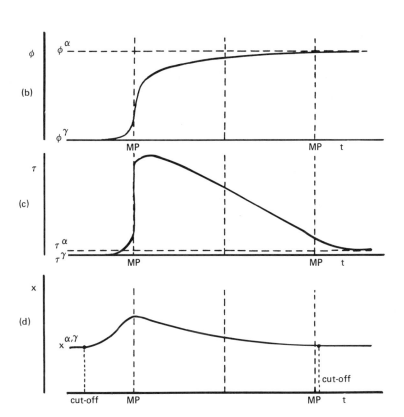

FIG. 6 α-γ direct path for $0 < \lambda < \lambda_c$.

to negative values of λ, but because of the intrinsic symmetry of
the model it is sufficient to analyze the $\lambda > 0$ case.)

The α-β interface, which disappears for $\lambda = \lambda_c$, shows a
smoother profile, and the bump in τ disappears completely, in
accord with expectations, close to λ_c. That is the case shown in
Fig. 4.

The noncritical interface β-γ reflects the expected mean field
behavior, approaching its β-value (which is the near-critical phase)
much more slowly. This is shown in Fig. 5. We note the appear-
ence of a bump in the cell size profile, which is an indirect manifestation of the increased adsorption of amphiphile at that noncritical
interface when approaching the critical end point.

The α-γ interface becomes asymmetric, as shown in Fig. 6.

The values of the surface tension as determined by Eq. (2)
with the profiles above are three orders of magnitude lower than
the oil-water tension (τ_0) taken as reference. It is in good agreement with experimental results [10,11]. Further results following from
the detailed analysis of the relative values of those surface tensions
corresponding to different interfaces, as functions of the parameters
of the model, are discussed in Ref. (5).

V. SUMMARY AND COMMENTS

We have modelled the free-energy surface of a model system by
means of paraboloids, which are asymptotic to the peaks of the real
free energy. We could then obtain the main features of the structure and tension of the interfaces of that model in a quasi-analytical
way. This is remarkable because the standard numerical solution
of this type of problem turns out to be intolerably cumbersome when
the number of independent densities is bigger than 2 (it is 3 in the
present case.)

The results for the profiles of the independent variables are
the ones expected in the mean field approximation, and elucidate
the consequences of the cutoff imposed on the cell size. It should
be noted that the profiles reported above are preliminary results
and show only qualitative shapes and tendencies. Quantitative plots
may be found in Ref. 5.

The numerical values of the surface tension are of the order of
the experimental ones, reaffirming the reliability of the model.

VI. REFERENCES

1. Talmon, Y., and Prager, S., *Nature* (London) 267, 333 (1977);
    J. Chem. Phys. 69, 2984 (1978); *J. Chem. Phys.* 76, 1535 (1982).

2. De Gennes, P. G., and Taupin, R., *J. Phys. Chem.* **86**, 2294 (1982).
3. Widom, B., *J. Chem. Phys.* **81**, 1030 (1984).
4. Borzi, C., Lipowsky, R., and Widom, B., Phase Transitions in Adsorbed Layers, Oxford, December 17–18, 1985, Faraday Symposium No. 20.
5. Balbuena, P., Borzi, C., and Widom, B., Physica A (in honor to Prof. P. W. Kasteleyn), *in press.*
6. Rowlinson, J. S., and Widom, B., *Molecular Theory of Capillarity*, Clarendon Press, Oxford, 1982.
7. Borzi, C., *J. Chem. Phys.* 82, 3817 (1985).
8. This scheme of approximation by paraboloids is due to Lipowsky, and Nilges, M., *(unpublished)*. Nilges, M., Diploma Thesis, University of Munich, 1984.
9. Borzi, C., and Widom, *unpublished results.*
10. Pouchelon, A., Meunier, J., Chatenay, D., and Cazabat, A. M., *Chem. Phys. Lett.* 76(2), 277 (1980).
11. Klaus, E., Jones, J., Nagarajan, R., Ertekin, T., Chung, Y., Arf, G., Yarzumbeck, A., and Dudenas, P., *Soc. Petroleum Eng. J.*, February 1983, p. 76.

16

Mechanism of Formation of Six Microemulsion Systems

HENRI L. ROSANO, JOHN L. CAVALLO, and GEORGE B. LYONS
City College of The City University of New York, New York

I. INTRODUCTION

Oil-in-water (O/W) emulsions of sodium long-chain sulfate/n-hydro-carbons/5% NaCl were titrated to clarity (85% transmittance or better) with long-chain dimethylamine oxide. These data confirm that micro-emulsion formation appears to be dependent on specific interactions among the constituent molecules at the O/W interface. Six emulsions (5 O/W; 1 W/O) were titrated to clarity with a cosurfactant. The volume of the continuous phase was varied from 10 to 60 ml. The mole fractions of cosurfactant/surfactant and cosurfactant/continuous phase were determined at various temperatures (15 to 55°C). At 30°C, free energy, enthalpy changes accompanying cosurfactant ad-sorption during microemulsion formation were found to vary from -19.6 to -6.4 kJ/mole and -25 to $+18.7$ KJ/mole, respectively. En-tropy change of formation was positive in five cases (from 3.5×10^{-2} to 8.3×10^{-2} kJ/K mole) and -1.8×10^{-2} kJ/K mole in one case. These small values help to explain why the manner of com-bining the various components may be important in the formation of these systems.

During the titration of an emulsion with a cosurfactant, the sys-tem often undergoes viscosity changes before clearing. After each addition the system was allowed to equilibrate. Phase separation was observed, and it was concluded that O/W microemulsions may be described as hydropathic oleomicellar solutions.

The most significant difference between emulsions (opaque) and microemulsions (transparent) lies in the fact that putting work into an emulsion or increasing the surfactant concentration usually im-proves the stability. This is not the case with microemulsions, which appear to be dependent for their formation on specific interactions

among the constituent molecules. If these interactions are not realized, neither putting in work nor increasing the surfactant concentration will produce a microemulsion. On the other hand, once the
conditions are right, spontaneous formation occurs and little mechanical work is required [1].

Terms such as microemulsion [2], swollen micellar solution [3],
micellar emulsion [4,5], middle phase [6], unstable microemulsion
[7], and spontaneous transparent emulsions [8] have been used to
describe these systems. The broadest definition may simply be a
transparent dispersion of oil, water, and surfactant that forms spontaneously upon addition of a cosurfactant. Microemulsions are either
O/W or W/O dispersions. Two criteria must be considered: formation and stability.

A. Formation

As originally suggested by Schulman et al. [9,10], microemulsions
form when the surfactant and cosurfactant, in just the right ratio,
produce a mixed adsorbed film that reduces the interfacial tension
γ_i below zero. They concluded that γ_i must have a metastable negative value, giving a negative free energy variation $-\gamma_i$ dA (where
dA is the change in interfacial area) responsible for spontaneous
dispersion.

The interfacial tension γ_i in the presence of a mixed film is
given by

$$\gamma_i = \gamma_{O/W} - \pi_i$$

where $\gamma_{O/W}$ is the O/W interfacial tension without the film present
and π_i is the interfacial surface pressure of the film. At equilibrium γ_i becomes zero. If this concept of zero interfacial tension is
accepted, stabilization of the microemulsion is implied.

However, this model does not seem to be a conceptually valid
one, since $\gamma_i = 0$ would not require the dispersed phase to be distributed in spherical droplets as is found in the systems under
discussion [11].

Rosano et al. [7] have considered the dynamic role of the cosurfactant by lowering the interfacial tension during the titration
of a coarse emulsion into a transparent dispersion. It has been
pointed out that during the addition of a cosurfactant to an emulsion of either O/W or W/O, excess cosurfactant accumulates at the
oil/water interface during transport, reducing the interfacial tension to well below the positive equilibrium value. The surfactant
retards the cosurfactant interfacial transport; a prolonged low interfacial tension helps in the formation of a large increase in the interfacial area. Evenutally, γ_i regains a positive value responsible for
the resolution of the system into microemulsion droplets.

B. Stability

Emulsion and microemulsion stability do not appear to be dependent only on the value of the γ_i, but mainly on the structure of the film surrounding each droplet [12,13]. For a given oil/surfactant pair, cosurfactant steric requirements determine the volume of the dispersed phase that can be stabilized. These systems are oil and cosurfactant dependent. Surfactant, cosurfactant, nature of the oil, and aqueous phase are four interacting variables that determine the size of the dispersed phase droplet when microemulsions are formed.

Only specific component combinations can produce transparent systems. In addition, the various components must be put together in just the right order to produce a microemulsion. Consequently, these experimental results lead to two basic questions:

1. Are these systems kinetically stable, since they may show path dependency in their formation?
2. Are they thermodynamically stable, even though their occasional path dependency properties may reflect activation energy barriers that they must overcome during their formation?

This chapter discusses the thermodynamic properties and preparation associated with the formation of six O/W microemulsion systems.

II. EXPERIMENTAL

A. Chemicals

N-Octane, n-decane, n-dodecane, n-tetradecane, and n-hexadecane were all 99%+ (gold label Aldrich Chemical Co., Milwaukee, Wis.). Sodium myristyl sulfate (Cyclo Chemicals Corporation, Miami, Fla.), sodium cetyl sulfate (Henkel Chemical Co., Fort Lee, N.J.), sodium dodecyl sulfate, potassium hydroxide, toluene, and 1-pentanol (Fisher Scientific Co., Fair Lawn, N.J.) were all reagent grade. Octyl and nonylphenolethylene oxide compounds were all 100% active (GAF Corp., New York, N.Y.). Dodecyl, tetradecyl, and hexadecyldimethylamine oxide (Onyx Chemical Co., Jersey City, N.J.) were all 30% active. All chemicals were used as received. Freshly distilled water was used in all preparations, and all glassware was first thoroughly cleaned with a fresh sulfuric acid/potassium dichromate solution.

B. Method of Preparation of Transparent Systems

Microemulsion systems were prepared in water-jacketed beakers maintained at constant temperature, using the titration technique [14]. An initial coarse macroemulsion was prepared from an oil/surfactant mixture added to an aqueous phase. The system was then titrated to clarity using a cosurfactant delivered from a microburet. Continuous stirring was maintained throughout the titration process to ensure homogeneous mixing. Percent transmittance was measured as the system started to clear with a Spectronic 20 spectrometer (Bausch & Lomb Co., Rochester, N.Y.) at 520 nm.

Another method employed in the preparation of microemulsions was to add the cosurfactant in various ratios between oil and aqueous phases before combining them. It was found that the manner of combining the components could affect the formation of the dispersion. From the results, it appears that the formation of these spontaneously forming transparent systems may be, in certain cases, path dependent [14]. In this study, five O/W emulsions and one W/O emulsion were titrated to clarity (transmittance > 90%) with a cosurfactant. The systems investigated are shown in Table 1. Other oil-in-water microemulsions were prepared with n-octane, n-decane, n-tetradecane, or n-hexadecane, 5% saline and sodium lauryl, sodium myristryl or sodium cetyl sulfate. The systems were titrated to clarity with dodecyl, tetradecyl, or hexadecyldimethylamine oxides; see Table 2.

Table 1

SYSTEMS	CONTINUOUS PHASE	DISPERSED PHASE	SURFACTANT	COSURFACTANT
1	water	2 ml n-octane	0.5 ml C_9phenol-1.5-EO + 0.5 ml C_9phenol-4-EO	C_8phenol-9-EO
2	water	2 ml n-decane	0.5 ml C_9phenol-1.5-EO + 0.5 ml C_9phenol-4-EO	C_9phenol-10-EO
3	water	2 ml n-hexadecane	0.5 ml C_9phenol-1.5-EO + 0.5 ml C_9phenol-4-EO	C_9phenol-9-EO
4	toluene	2 ml water	1.98×10^{-3} M SDS	1-pentanol
5	saline 0.375 N KOH	2.3 ml n-hexadecane	2.3×10^{-3} M stearic acid	1-pentanol
6	5% NaCl saline	1 ml n-decane	1 gm. SDS	DDAO

TABLE 2

Na cetyl sulfate				
LDAO	C8	C10	C12	C14	C16
C16	NC	NC	NC	NC	NC
C14	5.5(90%)*	NC	NC	NC	NC
C12	3.0(88%)	NC	NC	NC	NC

Na myristyl sulfate				
LDAO	C8	C10	C12	C14	C16
C16	NC	NC	NC	NC	NC
C14	6.6(100%)	6.9(95%)	7.8(92%)	7.3(96%)	NC
C12	3.6(96%)	5.4(97%)	NC	NC	NC

Na lauryl sulfate				
LDAO	C8	C10	C12	C14	C16
C16	9.0(100%)	NC	NC	NC	NC
C14	3.3(90%)*	2.1(98%)	3.0(90%)	4.0(88%)	NC
C12	1.8(99%)	3.3(100%)	NC	NC	NC

C. Viscosity and Percent Transmittance

In a water-jacketed beaker maintained at 30°C, an emulsion consisting of 2.0 ml n-decane/0.5 ml nonylphenol-1.5-EO + 0.5 ml nonylphenol-4-EO/35 ml water was titrated to clarity with nonylphenol-10-EO. Upon each addition of nonylphenol-10-EO, percent transmittance and viscosity (Brookfield Synchro-Lectric Model LVT, Stoughton, Mass.) were determined.

D. Phase Volume and Particle Size Determination

Volumes of upper and lower phases were determined for 15 individual systems prepared with 2.0 ml n-octane/0.5 ml nonylphenol-1.5-EO + 0.5 ml nonylphenol-4-EO/30 ml water. To each system increasing amounts of nonylphenol-9-EO were added. Each of the systems was shaken thoroughly and stored in graduated cylinders for 30 days at room temperature (22°C). The mean drop diameter of the lower phase was determined with a Coulter Sub-Micron Particle Analyzer, Model N4 (Coulter Electronics, Hialeah, Fla.).

III. RESULTS

A. Long-Chain Dimethylamine Oxide Microemulsions

These systems illustrate the specificity involved in microemulsion formation. It was observed that with sodium cetyl sulfate, n-octane produced a transparent gel. For sodium myristyl sulfate, transparent systems were more prevalent for all oils used except n-hexadecane. No microemulsions were formed when hexadecyldimethylamine oxide was used as a cosurfactant. With sodium lauryl sulfate, microemulsion systems were obtained, but again no transaprent systems were found when n-hexadecane was used as the oil phase.

B. Thermodynamic Properties

Figure 1 represents a typical plot of the minimum volume of octylphenol-9-EO, at various volumes of aqueous phase, required to titrate an emulsion made with 2.0 ml n-hexadecane/0.5 ml nonylphenol-1.5-EO + 0.5 ml nonylphenol-4-EO to clarity at 45°C and at 55°C. For example, at 45°C, if an initial emulsion containing 30 ml of water was titrated, the system remained milky until 1.5 ml of octylphenol-9-EO had been added. Beyond this volume of cosurfactant, the system remained clear. The same process was repeated with various volumes of water (continuous phase). A straight line was obtained, as shown in the plot. The value of the slope (milliliters of octylphenol-9-EO per milliliter of water) and intercept (milliliters of octylphenol-9-EO at zero milliliters water) was determined using a

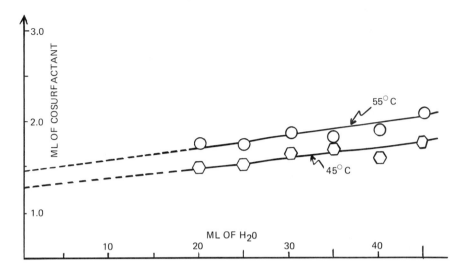

FIG. 1 2 ml n-hexadecane/0.5 ml nonylphenol-1.5-EO + 0.5 ml
nonylphenol-4-EO/a variable volume of water, titrated to clarity
with octylphenol-9-EO at 45°C and 55°C.

linear regression process. The mole fraction X_b of octylphenol-9-
EO in the continuous phase and at the interface X_i was then determined. Using the equation,

$$\Delta G = -RT \ln \frac{X_i}{X_b}$$

free energy values accompanying the adsorption of octylphenol-9-
EO at the interface during the transformation of a primary coarse
emulsion into a transparent system were determined. The same procedure was repeated at various temperatures. Plotting the change
in free energy versus temperatures, the entropy change accompanying cosurfactant adsorption was calculated, viz.,

$$\left(\frac{\partial \Delta G}{\partial T} \right)_p = -\Delta S$$

Table 3 shows the calculated free energy, enthalpy, and entropy
values for the six systems investigated at 30°C.

Table 3

SYSTEM	ΔG (kJ/mole)	ΔH (kJ/mole)	ΔS (kJ/K mole)
1	-17.4	-6.8	3.5×10^{-2}
2	-18.1	-6.0	4.0×10^{-2}
3	-19.6	-25.0	-1.8×10^{-2}
4	-6.4	18.7	8.2×10^{-2}
5	-14.8	8.7	7.8×10^{-2}
6	-13.9	11.3	8.3×10^{-2}

C. Viscosity and Percent Transmittance During Microemulsion Formation

Relative viscosity and percent transmittance for a system prepared with 2.0 ml n-decane/0.5 ml nonylphenol-1.5-EO + 0.5 ml nonylphenol-4-EO and 35 ml water were measured upon addition of nonylphenol-10-EO; see Fig. 2. Upon addition of 2.5 ml of cosurfactant, the viscosity increased to 40 CP (centipoise), and filaments were first observed while the system was being stirred. A maximum viscosity of 110 CP was reached when 3.10 ml of cosurfactant was added. Filaments were still observable up to 4.2 ml of nonylphenol-10-EO. The viscosity of the system decreased with further addition of nonylphenol-10-EO, and the system cleared at 4.85 ml. With still more nonylphenol-10-EO added, the system remained clear (> 85% transmittance) but the viscosity increased again.

D. Phase Volume Changes

Figure 3 shows the change in phase volume for 15 individual systems prepared with 2 ml n-octane/0.5 ml nonylphenol-1.5-EO + 0.5 ml nonylphenol-4-EO/30 ml water and varying amounts of octylphenol-9-EO. Initially, the systems started out as two phases, both the upper and lower phases being clear. By adding octylphenol-9-EO, the upper phase started to become milky and increase in volume. When 1 ml of this cosurfactant was added, a one-phase system was obtained. Uppon adding still more cosurfactant, the upper phase started to decrease in volume as the lower phase began to clear. The particle size of the lower phase region was determined. It was found that the particle size decreased as the concentration of octylphenol-9-EO was increased.

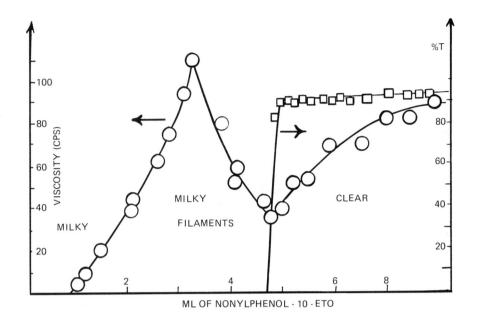

FIG. 2 Change in viscosity and percent transmittance for 2 ml
n-decane/0.5 ml nonylphenol-1.5-EO + 0.5 ml nonylphenol-4-EO/
35 ml water and titrated with nonylphenol-10-EO at 30°C.

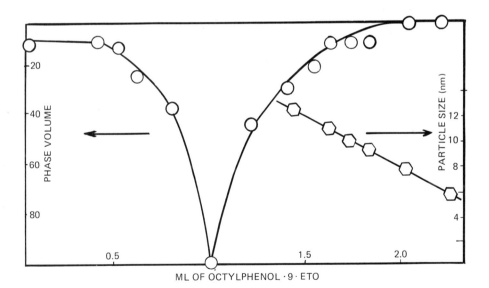

FIG. 3 Phase volume change and particle size determination for 2 ml
n-octane/0.5 ml nonylphenol-1.5-EO + 0.5 ml nonylphenol-4-EO/30 ml
water and titrated with nonylphenol-10-EO at 22°C.

267

IV. DISCUSSION

Rosano [14] has shown that the formation of transparent systems
are affected by many factors, including the type and composition
of the surfactant and cosurfactant, the dispersed phase volume,
and the structure of the oil. For a system prepared with 2.0×10^{-3} mole stearic acid and 2.0 ml of oil, 16 ml of 0.375 N KOH,
they reported that out of 52 alcohols, only five produced transparent
O/W systems (transmittance > 79%). These results illustrate that
only specific component combinations can produce microemulsions.
As seen in Table 2, for emulsions prepared with various long-chain
sodium sulfates, water, and oil, only specific combinations will produce transparent systems when titrated with a long-chain dimethylamine oxide as a cosurfactant.

For the six systems investigated (Table 3), it can be seen that
the free energy values are all negative, indicating that the transformation of an emulsion into a microemulsion is a spontaneous process.
However, the driving forces for these processes are small and may
explain why the method of preparation sometimes affects their formation. The correct order of preparation probably lowers activation
energy barriers and initiates their formation. Experimentally, the
first consideration in the preparation of these transparent systems
seems to be (a) that enough primary surfactant has to be present
to cover the interfacial area, and (2) that the primary emulsion be
as finely dispersed as possible. The method of preparation used
has to produce the best kinetic conditions favorable to the dispersion of the dispersed phase into a microemulsion [15].

Microemulsion formation involves: (1) a large increase in the
interfacial area (e.g., a droplet of radius 120 nm will disperse into
1728 microdroplets of radius 10 nm—a 12-fold increase in the interfacial area); and (2) the formation of a mixed surfactant/cosurfactant
film at the oil/water interface, which is responsible for a very low
transitory interfacial tension. It is reasonable to assume that these
two effects contribute both positively and negatively to the entropy
values. For the systems investigated, the entropy values were found
to be positive with one exception (system 3).

Under dynamic conditions, during the transformation of an initial
coarse emulsion into a microemulsion, changes in the systems viscosity often are a basic characteristic prior to microemulsion formation. This effect is observed, in particular, for systems prepared
with both nonionic surfactants and cosurfactants. Figure 2 illustrates the change in viscosity and percent transmittance for a system prepared according to system 3 and titrated to clarity with
nonylphenol-10-EO. Initially, a low-viscosity coarse emulsion is
formed. Upon adding cosurfactant, the viscosity increases and
reaches a maximum value [16]. Within this region the system remains

milky and filament structures are observed throughout. Upon adding more cosurfactant, the viscosity decreases to a minimum value, at which point the system spontaneously clears (95% transmittance). It is interesting to note that once the system becomes transparent, further addition of cosurfactant does not decrease the transmittance. Rosano et al. [14] have shown that for O/W microemulsions prepared with stearic acid, n-hexadecane, 0.375 N KOH, and 4-methylcyclohexanol, the transmittance decreases once the capacity for the cosurfactant in the aqueous phase and at the interface is exceeded. Since both sites are saturated, the excess cosurfactant penetrates the interfacial film and dissolves in the oil phase. For O/W systems this means an increase in droplet size and a concomitant decrease in transmittance. For nonionic systems (systems 1, 2, and 3), once maximum transparency is reached, excess cosurfactant increases the systems bulk viscosity, probably due to further cosurfactant adsorption on the droplet surfaces, and/or formation of micelles in the bulk.

Emulsions made of 1.95% sodium dodecylsulfate, 3.75% 1-butanol, 46.3% toluene, and varying weight percentages of a stock 48% sodium chloride solution were prepared [17]. Below 4.7% sodium chloride, a two-phase system was formed with an upper clear oil phase and a lower milky aqueous phase, displaying oil-in-water emulsion characteristics. Above 6.4% sodium chloride, a water-in-oil emulsion dispersion started to form, indicated by the milky upper oil phase and clear lower aqueous phase. In the region between 4.7 and 6.4% sodium chloride, a three-phase system with a clear upper oil phase, a clear lower aqueous phase, and a middle surfactant phase was obtained. Within the three-phase region, filament structures were observed if the system was shaken gently. The formation of the filament structures seemed to be due to the ultralow values of γ_i within this region. When the overall change in the interfacial tension of the upper and lower phases was plotted against sodium chloride concentration, the value of the interfacial tension was a minimum within the three-phase region [18,19]. It seemed that achieving low values in interfacial tension was a necessary step in microemulsion formation. Once the value was sufficiently low ($< 10^{-3}$ dyne/cm), spontaneous dispersion occurred with little or no mechanical work required. It also is worth mentioning that phase separation rates also varied as a function of salt concentration. Fastest separation was reported for the system in which the middle phase contained equal volumes of oil and water. It has been suggested by Robbins [20] that spontaneous microemulsification will occur in the middle of the three-phase region and that the middle phase will have to be transparent or translucent.

During the preparation of microemulsions under dynamic conditions (titration method), two questions should be addressed: (1) What is the role of the cosurfactant in forming microemulsions?

(2) Are these systems undergoing phase volume changes as the concentration of cosurfactant is increased? Recent spectroscopic evidence [21] indicates that the role of the cosurfactant is to reduce
the rigidity of the interfacial film, allowing the transition from a
well-organized phase toward an isotropic microemulsion. Figure 3
illustrates the phase volume behavior for system 2 as the concentration of cosurfactant (in this case, octylphenol-9-EO) is increased.
Upon the addition of octylphenol-9-EO, the upper phase starts to
become milky and increase in volume. Once a total of 1 ml has been
added, a one-phase system is obtained. Within this region the system is milky and birefrigent, and filament structures are prevalent.
As the cosurfactant concentration is further increased, the upper
phase volume starts to decrease simultaneously as the lower phase
begins to clear. When just the right amount of cosurfactant has
been added, a one-phase transparent system is obtained. It seems
that the resolution of an emulsion prepared with both nonionic surfactants and cosurfactants into a microemulsion follows a phase-
inversion process. As the cosurfactant concentration increased,
the particle size in the lower phase was found to decrease. Upon
the addition of cosurfactant to an initial coarse emulsion, the droplets start to elongate and reach a maximum (formation of filamentlike
structures). As the cosurfactant concentration is further increased,
these structures start to break into smaller fragments until finally
microdroplets are formed [22,23]. This effect was observed with
systems 1 and 3.

Shinoda [24] has proposed a general schematic illustration of
the change of solution behavior of surfactant with the hydrophilic-
lypophilic balance (HLB) in a surfactant system (see Fig. 4). For
a system prepared with equal volumes of oil and water plus a lypophilic surfactant (or surfactant mixture), at equilibrium a two-phase
system is obtained with a clear lower aqueous phase and a milky
upper oil phase. When the system becomes more hydrophilic, a
three-phase system is formed, with upper and lower phases clear
and a middle surfactant phase. Further increasing the hydrophilicity,
again a two-phase system is formed, with a milky lower aqueous
phase and clear upper oil phase. The same basic trend in solution
behavior is observed for nonionic surfactants if the temperature is
increased, resulting in a decrease in the HLB value [25]. For ionic
surfactants, if, instead of increasing the temperature, the salt concentration is increased (decreasing the HLB), the solution behavior
will also show two- to three- to two-phase behavior as previously
mentioned [17,19].

This schematic diagram can be related to Fig. 3. In the initial preparation of our O/W emulsion systems, only 2 ml of oil was
used with a larger volume of water. Therefore, half of the diagram
in Fig. 4 should be considered (represented by the dotted line).

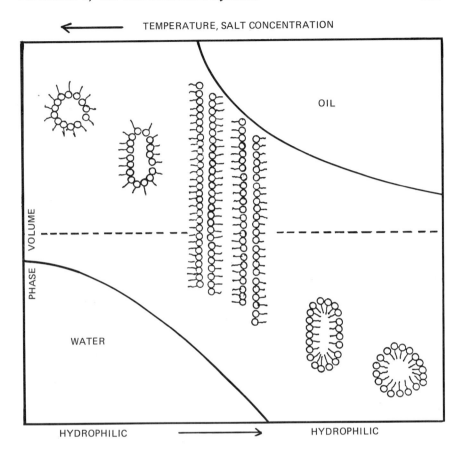

FIG. 4 Schematic illustration of the change in solution behavior of
surfactant with hydrophilic-lypophilic (HLB) in a water, oil, surfactant system. (Reproduced from Ref. 24, Fig. 2.)

For a primary emulsion prepared with a small O/W volume ratio and
a lypophilic surfactant or (surfactant mixture), the HLB value of
the system will increase upon titrating the system with a hydrophilic
cosurfactant. The upper phase increases, until a one-phase system
is formed (filament structures are observed). Upon further increasing the system HLB, a one-phase transparent O/W dispersion is obtained. Therefore, it is concluded that microemulsions of the O/W
type correspond to the lower right side of Fig. 4, whereas microemulsions of the W/O type correspond to the upper left region.
Kahlwett et al. [26] have studied the phase behavior of ternary
systems (H_2O-oil-nonionic surfactant) and quarternary systems

(H_2O-oil-nonionic surfactant-electrolyte). In the case of ternary systems the temperature was varied, while for quaternary systems the salt concentration was varied. Their results verify that maximum mutual solubility between oils and surfactants may be explained in terms of "simple" phase behavior with respect to the existence of tricritical points in ternery systems.

Using Fourier transform proton and carbon-13 pulsed-gradient spin-echo (PGSE) NMR techniques, Lindman et al. [27] have investigated seven microemulsion systems by determining the multicomponent self-diffusion coefficient for systems prepared with oil/water and surfactant and a cosurfactant of either short- or long-chain alcohols. They reported that for the ionic surfactant/short-chain alcohol/hydrocarbon/water type of system, (1) no distinct separation into hydrophobic and hydrophilic domains was observed, (2) these systems show no extended aggregates, and (3) the internal interfaces were determined to be flexible and highly disorganized, an idea already suggested by di Meglio et al. [28]. Microemulsions formed with nonionic surfactants of the polyethyleneoxide type, hydrocarbon, and water are probably similar in structure to those formed with short-chain alcohols. Microemulsions prepared with long-chain alcohols are somewhat different in structure and form distinct droplets of water in a hydrophobic medium. In a review article, de Gennes and Taupin [29] have also concluded that a flexible interface is an absolute requirement for maintaining some microemulsion type systems, an idea already advanced by Schulman [2]. However, in some cases a bicontinuous structure may also occur. Chatenay et al. [22] have found that for polyphasic systems away from the three-phase domain, the interaction between the droplets is hard-spherelike; whereas in the middle of the three-phase region, the structure is bicontinuous, as evidenced from conductivity measurements. These studies therefore indicate that not all interfaces fall into the same classical picture of well-organized oil or water droplets with distinct boundaries. What is seen is that internal interfaces have either a quite limited spacial extension or are very dynamic and flexible in nature and break up and re-form at a very high rate, or both.

Other structural models have been proposed to describe the structure and formation of microemulsions: Taupin and co-workers [30] have presented a model of hard oil and water globules with a relatively sharp transition between them; Scriven [31] one of a complex, periodic, three-dimensional network of both oil and water continuity; and Talman and Prager [32] one of hard, randomly arranged pholyhedra; whereas Friberg et al. [33] proposed a random structure with varying curvatures. Robbins [34] has shown that for a W/O microemulsion system, the activity of the water in the droplet core must be reduced for the water to flow into the droplets. This effect was readily verified by water vapor pressure measurements. Based on vapor pressure analysis of O/W microemulsions, two distinct regions of transparency exist [35]. For small volumes of oil, encapsulated noninteracting droplets are formed; whereas a dynamic equilibrium between the breaking and re-formation of droplet interfaces has been proposed for larger volumes. Similar behavior was also suggested by Weatherford [36] for W/O systems.

V. CONCLUSION

Transparent ternary, quaternary, and polyphasic systems have all been called microemulsions. There is disagreement on whether these systems are thermodynamically stable or not. Limiting ourselves to the six quaternary systems (oil/water/surfactant/cosurfactant) investigated in this study, it has been concluded that microemulsions are thermodynamically stable [34]. However, the order or preparation plays a major role in their formation. It is also assumed that once the right method of preparation is found, the activation energy barriers these systems must overcome during their formation are significantly lowered; the free energy values corresponding to the adsorption of cosurfactant at the interface are small negative numbers, indicating that the process of transforming an emulsion into a microemulsion is spontaneous, although the driving force is small; microemulsions of the O/W type correspond to the lower right region of the phase diagram shown in Fig. 4; and the word microemulsion as proposed by Schulman [2] may be an unfortunate term to describe these transparent systems. The word microemulsion, as opposed to macroemulsion, should designate only transparent emulsions. In the case of W/O microemulsions, Hoar and Schulman [8] in 1943 called these systems oleopathic hydromicellar solutions and hydropathic oleomicellar solutions for O/W microemulsions. In view of our results, these terms are more appropriate to describe our systems.

VI. ACKNOWLEDGMENTS

H. L. R. would like to thank Dr. H. Peper and gratefully acknowledge the financial assistance of the Gillette Company, Boston, Mass., and the Lever Bros. Co., Edgewater, N.J., in supporting this work.
    J. L. C. would like to thank Dr. C. J. Cante and the General Foods Corporation, Tarrytown, N.Y., for encouragement, financial, and technical assistance.

VII. REFERENCES

1. Tadros, Th. F., in *Structure/Performance Relationships in Surfactants*, Rosen, M. J. (ed.), 253rd ACS Symposium Series ACS, Washington, D.C., 1984, p. 154.
2. Stoeckenius, W., Schulman, J. H., and Prince, L. M., *Kolloid Z. 169*, 170 (1960).
3. Shinoda, K., and Kunieda, H., *Kolloid Z. 42* (a), 381 (1973).
4. Adamson, A. W., *J. Colloid and Interface Sci. 39*(2), 261 (1969).
5. Gerbacia, W. E., Rosano, H. L., and Zajac, M., *Am. Oil Chem. Soc. 53*, 101 (1976).
6. Robbins, M. L., in *Micellization, Solubilization, and Microemulsions*, Mittal, R. L. (ed.), Plenum Press, New York, 1977, vol. 2, pp. 713–753; Scriven, L. E., in *Micellization, Solubilization, and Microemulsions*, Mittal, K. L. (ed.), Plenum Press, New York, 1977, vol. 2, pp. 877–893.
7. Rosano, H. L., Lan, T., Weiss, A., Whittam, J. H., and Gerbacia, W. E., *J. Phys. Chem. 85*, 468 (1981).
8. Hoar, T. P., and Schulman, J. H., *Nature 152*, 102 (1943).
9. Bowcott, J. E., and Schulman, J. H., *Z. Electro-Chem. 59*(4), 283, (1955).
10. Schulman, J. H., and Montague, J. H., *Ann. N.Y. Acad. Sci. 92*, 366 (1961).
11. Gerbacia, W. E., and Rosano, H. L., *J. Colloid and Interface Sci. 44*, 242 (1973).
12. Gerbacia, W. E., Rosano, H. L., and Whittman, J. H., *J. Colloid and Interface Sci.*, Kerker, M., (ed.), Academic Press, New York, 1976, vol. II, pp. 245–256.
13. Rosano, H. L., Lan, T., Weiss, A., Gerbacia, W. E. F., and Whittam, J. H., *J. Colloid and Interface Sci. 72a*(2) (1979).
14. Rosano, H. L., *J. Soc. Cosmet. Chem. 25*, 609 (1974).
15. Rosano, H. L., U.S. Patent 4,146,499 (Mar. 27, 1979).
16. Rosano, H. L., Weiss, A., and Gerbacia, W. E., in *Proceedings of the VIIth International Congress on Surface Active Substances*, Moscow, September 12–16, 1976, vol. 1, p. 453 (1977).
17. Rosano, H. L., Jon, D., and Whittam, J. H., *J. Am. Oil Chem. Soc. 59*(8), 360, (1982).
18. Bellocq, A. M., Bourbon, D., and Lemmcean, B., *to be published*.
19. Healy, R. N., Reed, R. L., and Stenmark, D. G., *SPE J. 16*, 147, 1976.
20. Robbins, M., *private communication*.
21. Di Meglio, J. M., Dvolaitzky, M., and Taupin, C., *to be published*.

22. Chatenay, D., Guering, P., Urback, W., Cazabet, A. M., Langevin, P., Meuiner, J., Leger, L., and Lindman, B., *to be published.*

23. Di Meglio, J. M., Paz, L., Dvoalaitzky, M., and Taupin, C., *J. Phys. Chem. 88,* 6036 (1984).

24. Shinoda, K., *Progr. Colloid Polymer Sci. 68,* 1 (1983).

25. Shinoda, K., Kunieda, H., Arai, T., and Saijo, H., *J. Phys. Chem. 88,* 5126 (1984).

26. Kahlweit, M., Lessner, E., and Strey, R., *J. Phys. Chem. 87,* 5032 (1983).

27. Lindman, B., Stilbs, P., and Moseley, M., *J. Colloid and Interface Sci. 83,* (1981).

28. Di Meglio, J. M., Dvolaitzky, M., Ober, R., and Taupin, C., *J. Physique Lett. 44,* L229 (1983).

29. De Gennes, P. G., and Taupin, C., *J. Chem. Phys. 86,* 2294 (1982).

30. Lagues, M., Ober, R., and Taupin, C., *J. Phys. Lett. 39,* 487 (1978).

31. Scriven, L. E., in *Micellization, Solubilization and Microemulsions,* Mittal, K. L. (ed.), Plenum Press, New York, 1977, vol. 2, p. 877.

32. Talmon, Y., and Prager, S., *J. Chem. Phys. 69,* 517 (1978).

33. Friberg, S., Lapezynska, I., and Gillberg, G., *J. Colloid and Interface Sci. 56,* 19 (1976).

34. Robbins, M. L., Preprints, 48th National Colloid Symposium, 1974, p. 174.

35 Cavallo, J. L., and Rosano, H. L., *to be published in J. Phys.*

36. Weatherford, W. D., *J. Dispersion Sci. Technol. 6,* 467, 1985.

17

Static SANS Measurements on Concentrated Water-in-Oil Microemulsions

E. CAPONETTI* and L. J. MAGID   *University of Tennessee, Knoxville, Tennessee*

I. INTRODUCTION

Small-angle neutron scattering measurements were performed on concentrated water-in-oil microemulsions comprised of potassium oleate/water/n-hexadecane/and either 1-pentanol or 1-hexanol, with a volume fraction of dispersed phase ranging from 0.33 to 0.47. The scattering patterns establish, even without recourse to models, that structures characterized by dimensions of tens of angstroms are present in the solutions. Of models tested, monodisperse oblate ellipsoids with a water core and a permeable shell gave reasonably satisfactory fits; this geometry is not considered conclusively establlshed, however. Over the range of compositions covered, there appears to be no drastic change in structure, nor any profound difference between 1-pentanol and 1-hexanol as cosurfactant.

Recently there has been considerable interest in the effect of cosurfactant structure on the microstructure of microemulsions. Four-component systems comprised of sodium dodecylsulfate (SDS), water, a 1-alkanol (C_4 to C_7), and an aliphatic or aromatic hydrocarbon have been studied most extensively [1,2]. For the SDS-containing (nominally) water-in-oil (W/O) microemulsions, direct confirmation of the presence of discrete aggregates with water cores [3–5] and/or the existence of bicontinuous domains [6–8] has been obtained using small-angle scattering techniques (neutron or X-ray). In addition, a few water-in-oil microemulsions containing other

---

*Postdoctoral Fellow at the University of Tennessee, 1984. Permanent affiliation: Istituto di Chimica Fisica, University of Palermo, Palermo, Italy.

TABLE 1

Code	KOl	n-C_{16}(H or D) amount/mole	1-C_5OH ratio	1-C_6OH
A	1.0 g / 1.0	5.0 ml / 5.54	2.0 ml / 5.96	–
B	1.0 g / 1.0	5.0 ml / 5.52	2.0 ml / 5.97	–
C	1.0 g / 1.0	5.0 ml / 5.54	–	2.0 ml / 5.18
D	1.0 g / 1.0	5.0 ml / 5.52	2.0 ml / 5.97	–
E	1.0 g / 1.0	4.42 ml / 4.85	1.99 ml / 5.89	–
F[b]	1.0 g / 1.0	4.42 ml / 4.84	2.02 ml / 5.99	–
G[c]	1.0 g / 1.0	4.36 ml / 4.79	2.06 ml / 6.22	–

[a]Volume occupied by surfactant, water, and one-half of the alcohol present.
[b]With H_2O and n-$C_{16}D_{34}$.
[c]With D_2O and n-$C_{16}D_{34}$.

surfactants [9–15] have been investigated using small-angle neutron scattering (SANS).

Inferences concerning the microstructures present in microemulsions have also been drawn from techniques other than small-angle scattering. A representative system is one containing potassium oleate (KOl), water, n-hexadecane (nC_{16}), and either 1-pentanol (1-C_5OH) or 1-hexanol (1-C_6OH) and the surfactant. Shah et al. [6–18] made measurements of the electrical resistance, viscosity, and NMR properties (chemical shifts, T_2's of the water present) in the microemulsions formed by successive additions of water to a suspension of 5 ml of n-hexadecane and 2 ml of alcohol per gram of KOl. It is well known that in these systems there is a water-to-surfactant molar ratio (ca. 8 for KOl) below which proper reverse micelles or W/O microemulsion aggregates do not form: see, for

D_2O (or H_2O)	Water/oil ratio, v/v	Shah's designation	ϕ, dispersed phase[a]
1.0 ml / 17.8	0.2	Cosolubilized	0.33
2.5 ml / 44.7	0.5	Microemulsion	0.43
2.5 ml / 44.6	0.5	Microemulsion	0.43
3.27 ml / 57.0	0.65	Microemulsion	0.47
2.55 ml / 45.3	0.58	Microemulsion	0.46
2.49 ml / 44.2	0.56	Microemulsion	0.45
2.51 ml / 44.4	0.58	Microemulsion	0.46

example, the light scattering investigations of Sjöblom and Friberg [19] on $KOl/1$-$C_5OH/H_2O/n$-C_{10} and those of Bellocq and Fourche [20] on $KOl/1$-$C_6OH/H_2O/n$-C_{12}. However, Shah et al. [18] concluded that their system was molecularly dispersed (cosolubilized) at all water-to-KOl ratios when 1-C_5OH was the cosurfactant, while for the case of 1-C_6OH, true microemulsion aggregates were present; see Fig. 1.

Boussaha et al. [21], using positron annihilation, inferred that the transition from cosolubilization to formation of W/O aggregates occurred at 0.2 and 0.4 v/v water-to-oil for 1-C_6OH and 1-C_5OH, respectively. Atik and Thomas [22] employing flourescence quenching of $Ru^*(II)$ by $K_3Fe(CN)_6$, inferred that the 1-C_5OH system may have proper water pools (which exchange their contents rapidly) below 0.4 v/v water-to-oil.

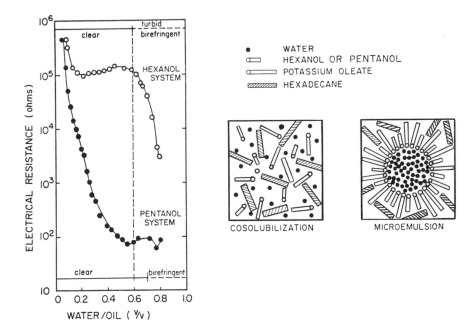

FIG. 1 Electrical resistance data and proposed states of organization for KOl microemulsions, from Ref. 18.

We therefore thought it interesting to perform small-angle neutron scattering measurements for the $KOl/1-C_nOH/water/n-C_{16}$ systems, with n equal to 5 or 6, using compositions studied previously by nonscattering methods. Table 1 presents the compositions of the solutions used. For many W/O microemulsions that contain aggregates, there are attractive interactions between the aggregates [23–27]; the strength of these attractions depends on the alcohol chain length and on the oil's chain length and chemical character (aliphatic versus aromatic; this is analogous to solvent quality in solutions having polymers as solutes). n-Hexadecane is an oil for which the interactions are expected to be hard-spherelike.

II. EXPERIMENTAL

A. Materials

Scattering measurements were carried out on three compositions, in which D_2O was the only deuterated component, with $1-C_5OH$ as cosurfactant and with different amounts of D_2O, and on one composition with $1-C_6OH$ as cosurfactant. Another set of three solutions were of approximately the same molar concentration, but were made up with D_2O or $n-C_{16}D_{34}$ or both, $1-C_5OH$ being the alcohol. Compositions are given in Table 1; the volume fractions listed are upper limits of the disperse phase, because some of the alcohol is probably distributed in the hexadecane rather than the particles. The deuterated hexadecane was obtained from Cambridge Isotopes Laboratories, and was stated to be of 96% isotopic purity. In the analysis of results, allowance was made for the 4% proton content.

B. SANS Measurements

Neutron scattering measurements were performed on the 30-m (source-to-detector) instrument of the National Center of Small-Angle Scattering Research located at the High Flux Isotope Reactor, Oak Ridge National Laboratory. Samples A-D of Table 1 were contained in cylindrical quartz spectrophotometric cells of 2 mm path length and E-G in 1-mm cells, which were thermostatted at $25 \pm 0.05°C$ by means of circulation from an external bath. Correction for detector background and sensitivity and for scattering by the empty cell and conversion of patterns to radial averages were by programs provided by the Center, as earlier [28]. Scattering cross sections, $d\Sigma/d\Omega$ $(cm^{-1}) = I(Q)$, were obtained from these averages by factors also provided by the Center, determined from scattering of substances of known cross section, or by calibrations with H_2O or Al-4, a porous sample provided by the Center as a secondary standard. In view of the high fraction of "solute" components, the usual subtraction of scattering by pure solvent seemed questionable. For samples A-D, we subtracted instead a pattern from an H_2O/D_2O solution of a ratio selected to have the same incoherent scattering length density as the solution. A similar difference pattern could be obtained by subtracting intensities from hexadecane multiplied by the fraction of hexadecane in the solution. Residual incoherent scattering was accounted for by a parameter (B) in the least-squares fit making a contribution to intensity invariant with Q. Results for sample-to-detector distances of 1.4, 2.4, and 8.1 m are reported, the range of Q being ~ 0.01 to 0.35 $Å^{-1}$. For samples E, F, and G, only corrections for detector background and empty cell were made, and incoherent scattering was accounted for by least-squares parameter, B. In these cases, samples-to-detector distances at the High Flux Isotope Reactor were 12, 4, and 1.4 m, the Q range being approximately the same. We combined with this set measurements at 4.6 m obtained at the Center's SANS system at the Oak Ridge Research Reactor. Agreement between the measurements with the two systems was good. Runs at the different distances were combined using programs provided by the Center.

III. RESULTS

The scattering curves obtained for compositions A-D are presented
in Fig. 2, and for E-G, in which scattering contrast was varied by
use of different deuterated components for essentially the same molar
concentrations, in Fig. 3. The strong patterns obtained in all cases
establish unequivocally that aggregates are present. When the scattering-length density of water and of oil are approximately matched
(see Fig. 3, $C_{16}D_{34}/D_2O$), scattering is still observed; further,
there is a broad peak at intermediate Q, an observation consistent
with scattering from a surfactant/cosurfactant shell between water
droplets and a continuous oil phase, if spherical aggregates are
present. A transient bicontinuous phase, in which the water droplets fluctuate and may coalesce for short times, would look similar
and provide a path for electrical conduction at the water volume
fraction here, with resistance dependent on the frequency of the
fluctuations [2].

The two extreme models posed are therefore locally similar. Consequently, we can approximate the local geometry around aqueous

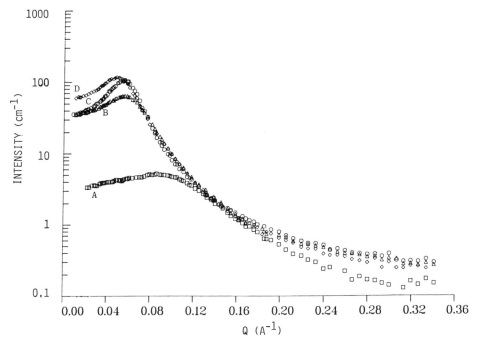

FIG. 2 SANS curves for KOl w/o microemulsions having D_2O cores.
Compositions are given in Table 1.

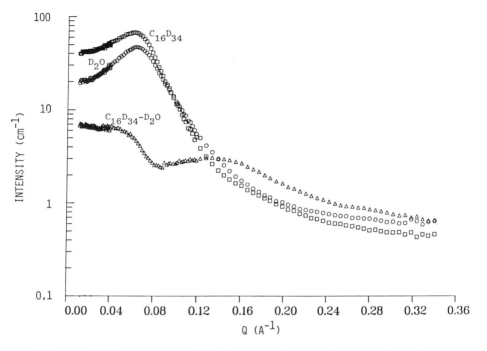

FIG. 3 SANS curves for KO1/1-pentanol w/o microemulsions formulated with various H/D contrasts. Compositions are given in Table 1.

regions by using simple water droplet-in-oil models, for which we can calculate scattering patterns at various levels of approximation. The results should not, however, be taken to indicate that the transient alternative necessarily does not exist. Dense fluids such as these will in any case have significant local neighbor order.

A. Analysis of the Scattering Patterns

The scattered intensity for these concentrated dispersions of particles may be expressed by Eq. (1),

$$I(Q) = N_p <|F(Q)|^2> + N_p [S(Q) - 1] |<F(Q)>|^2 \qquad (1)$$

assuming that there is little correlation between the particles' size and orientation and the interparticle distance. This so-called decoupling approximation [29–31] is valid for Q values equal to and greater than that near the first maximum in S(Q). However, at least for the case of polydisperse hard spheres at small Q, it does deviate significantly form the exact calculations of van Beurten and Vrig [32].

The quantity $\langle F(Q) \rangle$ is the Fourier transform of the average radial distribution of scattering length within one particle; $\langle |F(Q)|^2 \rangle$ is the Fourier transform of the average distribution of distances $P(r)$ within one particle; $S(Q)$ is the Fourier transform of the radial distribution function $g(r)$ for the centers of mass of the particles; and N_p is the number particle density.

Our task is thus to determine the form factor, $F(Q)$, and structure factor, $S(Q)$, which reproduce the observed scattering patterns. We have discussed at length elsewhere [33] the various models tested, so we shall present only a summary here.

B. The Structure Factor

A hard-sphere interaction potential was assumed initially; the analytical solution for $S(Q)$, due to Hayter and Penfold [34], which has the advantage of allowing an additional (either repulsive or attractive) term to be included in the interaction potential, was used.

Two parameters are important in the computation of structure factors for uncharged systems such as those of interest here: a radius, commonly referred to as "hard sphere," here designated as R_{sf} for structure factor, describing the closest approach of the particles to one another; and the volume fraction, ϕ, of the disperse phase, here the fraction of volume occupied by the particles. Notice that the position of the first peak in $S(Q)$ has the following dependence on R_{sf} and ϕ: At constant ϕ, the maximum shifts to lower Q as R_{sf} increases; at constant R_{sf}, it shifts to higher Q as ϕ increases. Increasing ϕ, of course, also decreases $S(Q)$.

Initially, in fitting results in which heavy water was the only deuterated component and the scattering resulted primarily from the core, we computed R_{sf} from the value obtained for the radius of the core, R_{core} (see details on the form factor), plus the volumes of the hydrocarbon parts of the surfactant and alcohol assigned to the shell, a total radius of R_{tot} of the particle was implied. The volume fraction ϕ was obtained from stoichiometry and the fraction of alcohols assigned to the disperse phase; R_{tot} and the number of KOl per micelle, AGG, were computed from R_{core}, and R_{sf} was set equal to R_{tot}. Independent variation of R_{sf}, rather than assignment of the value of R_{tot}, also gave poor fits. Similarly, the inclusion of an attractive Yukawa tail in the interaction potential did not improve the fits.

If instead the determination of R_{sf} and ϕ are uncoupled from the form factor and from the stoichiometry of the dispersed phase (i.e., R_{sf} and ϕ are treated as adjustable parameters), good fits are obtained. We shall present them in the next section.

C. The Form Factor

Figure 4 presents scattering length density (ρ) profiles for the three contrasts used in this work. When the only deuterated component is D_2O, ρ_{shell} is approximately equal to $\rho_{nC_{16}}$, so that the form factor is dominated by contrast between the aqueous cores (to which the $-COO-$, K^+, and $-OH$ moieties are assigned as well; half of the alcohol is assumed to be present in the dispersed phase, *vide infra*) and the surrounding hydrocarbon milieu.

These cores may be disposed in principle as mono- or polydispersed spheres or as prolate or oblate ellipsoids in solution. In fact, the best fits were obtained for monodispersed oblate ellipsoidal cores.

For nonspherical particles it is necessary to average the form factor over the orientations of the particles with respect to Q. Thus we write

$$< |F(Q)|^2 > = \int_0^1 |F(Q, a, ECC, t, \mu)|^2 \, d\mu \qquad (2)$$

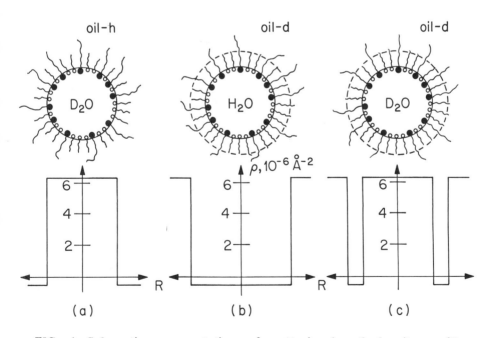

FIG. 4 Schematic representations of scattering length density profiles for (a) $D_2O/nC_{16}H_{34}$; (b) $H_2O/nC_{16}D_{34}$; (c) $D_2O/nC_{16}D_{34}$.

$$| <F(Q)> |^2 = | \int_0^1 F(Q, a, ECC, t, \mu) |^2 \tag{3}$$

The function F is given by

$$F(Q, a, ECC, a', ECC', t, \mu) = V_C \cdot (\rho_c - \rho_{sh}) \tag{4}$$

$$\cdot \phi(u') + V_T \cdot (\rho_{sh}$$

$$- \rho_{solv}) \cdot \phi(u)$$

where the ellipsoid's core (of volume V_C) has axes a', a', and ECC' · a' (ECC' > 1 for prolate, ECC' < 1 for oblate), and the overall ellipsoid (of volume V_T) has axes a, a, and ECC · a. The thickness of the shell region is given by t; ECC is the eccentricity (axial ratio) of the ellipsoid. The quantity u(a, ECC) is Q · $[(ECC \cdot a \cdot \mu)^2 + a^2(1 - \mu^2)]^{1/2}$; $\phi(x)$ is $3(\sin x - x \cos x)/x^3$. When $\rho_c \neq \rho_{sh} \sim \rho_{solv}$, the second term of Eq. (4) is zero, and the form factor is that for the cores only. R_{core} is expressed as $[(a')^3 \cdot ECC']^{1/3}$. The integrals in Eqs. (2) and (3) were evaluated numerically.

D. The Fitting Protocol

Zero-angle intensity is fixed by the particle concentration (N_p), contrasts, and dimensions of the various regions of the particles, and the structure function at zero angle. The particle dimensions selected by the least-squares fit, constrained by stoichiometry, determine the value of N_p. Absolute intensities are therefore derived from the fit. However, these values are experimentally less precise than relative intensities as a function of angle. The Center estimates that the calibration standards and procedures are good to −10% in absolute intensity. In addition, residual errors in the correction for differences in coherent scattering between sample and background [35], and errors in transmission measurements, add to uncertainties in normalization of intensities to absolute values. To avoid distortions of other parameters to compensate for errors in absolute intensity, we include an adjustable parameter A ("scale") multiplied by the terms in Eq. (1), and an adjustable parameter B to account for residual incoherent scattering. If there were no errors from these sources, the value of A should be 1, but if within −15% of unity, we do not infer problems with the model under test. This point is discussed further in Ref. 31.

Figure 5 compares three fits (all assuming monodisperse oblate ellipsoids) for sample E. Figure 5II is for ϕ computed from stoichiometry, with R_{sf} as an adjustable parameter; in 5I, R_{sf} is set equal

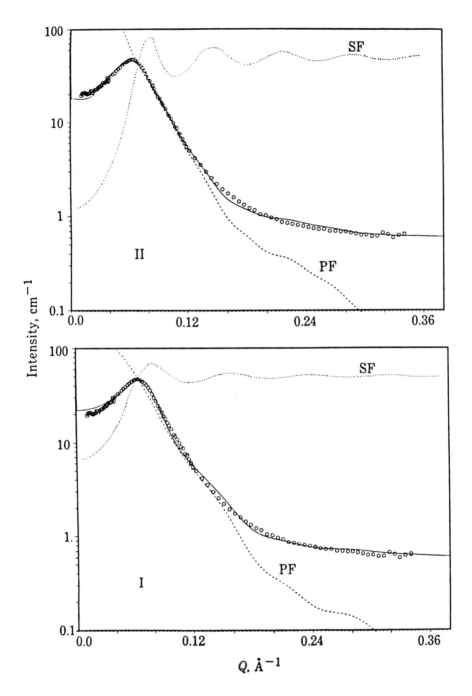

FIG. 5 Results of nonlinear least-squares fitting for composition E: effect of various assumptions about R_{sf} and ϕ. PF denotes the quantity $N_p \cdot \langle |F(Q)|^2 \rangle$, SF denotes the structure factor.

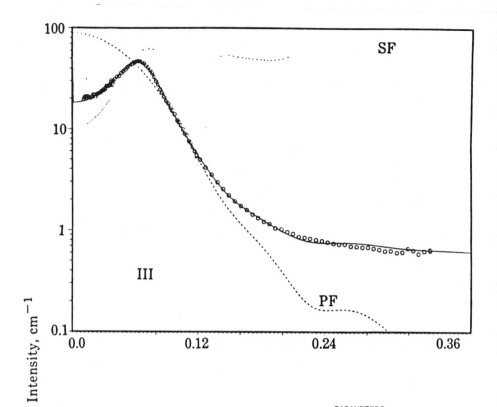

	PARAMETERS					
	I $R_{sf} = R_{tot}$		II No penetration		III R_{sf} by LSTSQ	
		±		±		±
R_{core}	33.4	21.4	47.9	4.1	28.2	0.6
R_{sf}	(39.5)		43.7	8.9	42.0	0.8
Ax Rat	0.40	0.96	0.30	0.14	0.51	0.03
Apparent ϕ	0.26	0.52	[1]		0.21	0.009
Scale	0.97	1.35	1.39	0.42	0.87	0.02
AGG	(108)		(319)		(65)	
B	0.57	3.25	0.56	0.58	0.59	0.06
CHI	11.3		6.5		3.0	

( ) Value computed from other parameters.

$Q,\ \text{Å}^{-1}$

FIG. 5 (Continued)

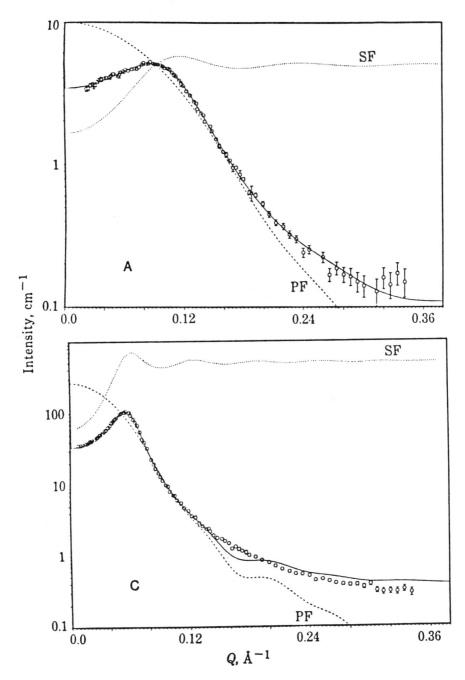

FIG. 6 Results of nonlinear least-squares fitting for compositions A-D (D_2O cores). See Table II for a compilation of the parameters.

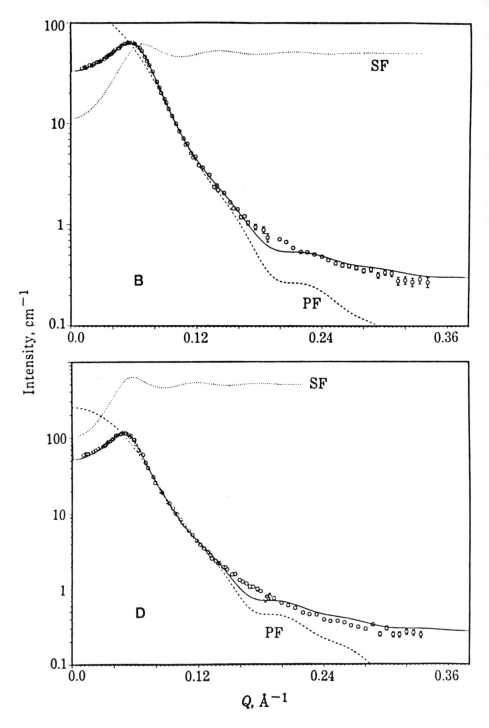

FIG. 6 (Continued)

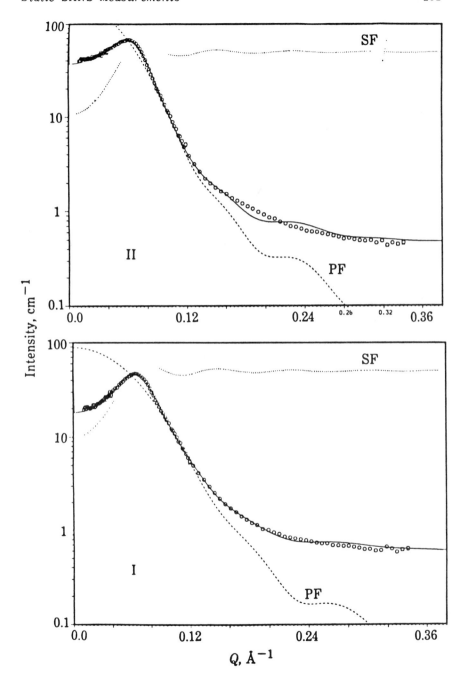

FIG. 7 Results of nonlinear least-squares fitting for composition E, F and G (core, particle and shell contrasts).

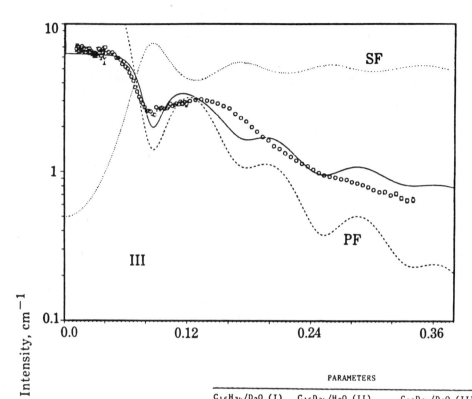

	PARAMETERS					
	$C_{16}H_{34}/D_2O$ (I)		$C_{16}D_{34}/H_2O$ (II)		$C_{16}D_{34}/D_2O$ (III)	
		±		±		±
R_{core}	28.2	0.6	27.8	5.9	29.6	62.6
R_{tot}	(33.4)		(33)		(35.2)	
R_{sf}	42.0	0.8	41.1	8.0	36.4	182.8
Ax Rat	0.51	0.03	0.54	0.22	0.58	2.39
Apparent ϕ	0.21	0.009	0.20	0.09	0.30	1.57
Scale	0.87	0.02	0.81	0.19	1.89	16.18
AGG	(65)		(64)		(77)	
B	0.58	0.06	0.44	0.53	0.58	8.86
CHI	3.0		6.9		13.7	

( ) Value computed from other parameters.

$$Q, \text{Å}^{-1}$$

FIG. 7 (Continued)

to R_{tot} and ϕ is allowed to vary; in 5III, both are allowed to vary. The fitting residuals summarized by CHI values confirm the visual impression that III is the best fit, and also indicate the I is better than II. Also persuasive are the values of the parameters and their uncertainties. With no reduction in ϕ allowed, the core radius is larger than R_{sf}, a physically unreasonable dimension.

The value of R_{sf} is larger in comparison to the total particle dimension computed from disperse-phase constituents only; the difference, $R_{sf} - R_{core}$, is also substantially larger than the extended chain length of 1-C_5OH, which is 7.8 Å. We shall return to this latter point in Sec. IV.

This visible shell model gives essentially the same fit as the invisible one did for sample E; see Fig. 7I. The scattering-length densities for the core + shell fit were 3.55×10^{-7} Å^{-2} for the shell -3.11×10^{-7} Å^{-2} for the continuous phase, in comparison with 6.01×10^{-6} Å^{-2} for the core. The fit to the case of deuterated hexadecane (Fig. 7II), in which dominant contrast is between the total aggregate and the continuous medium, is good, though not as good as for D_2O. The magnitudes of the parameters in the two fits are very close. The agreement increase confidence in the basic features of the model.

The fit (Fig. 7III) to the case in which there is strong contrast between the shell and both the core and medium reproduces only general aspects of the experiment. Part of the problem here may arise from instrumental smearing, to which this case would be particularly vulnerable, although it seems unlikely that loss of resolution could be this large. Details of structure to which the other two are indifferent may be important here.

IV. DISCUSSION

The scattering patterns establish, without recourse to models, that structures characterized by dimensions of tens of angstroms are present in these solutions. Hence, water cores are present in the 1-C_5OH cases as well as in the 1-C_6OH case; communication between the cores must be more facile in the pentanol case. The present observations are consistent with a distribution of particles of either a narrow range of sphere sizes, or of a slightly anisotropic shape. Oblate ellipsoids gave clearly the best fits, but we are not certain they represent the actual configuration. Since the volume fraction of the disperse phase is so high in these microemulsions, perhaps local, transient ordering of hydrocarbon tails on adjacent aggregates causes distortion from a spherical shape.

Monodisperse oblate ellipsoids give reasonable fits to the other compositions reported, all having only water deuterated (Fig. 6,

parameters in Table 2). Thus there are aggregates present in all
the microemulsions studied; the variations in core radius are not
surprising. Core radius and aggregation number increase as water/
amphiphile increase, for constant amphiphile/hydrocarbon ratios (A,
B, and D). Not quite as expected for an increase in disperse phase/
hydrocarbon ratios, at constant water/amphiphile, is the decrease in
aggregate size of B versus E; possibly the distribution of alcohol
into the aggregates is greater in this case. Although the molar
concentrations are not quite the same, it appears that substitution
of hexanol for pentanol increases aggregate size (B versus C).
Axial ratios are similar for the different concentrations, clustering
about 0.5.

A. Other Contrasts; The Effect on the Form Factor

Varying the visibility to neutrons of different regions of the aggregates can be expected to illuminate further aspects of their structure. It is clear that for some contrasts (see Fig. 3) an invisible
shell model is not appropriate, and the fits to be presented of the
systems E-G having different components deuterated are carried out
with a core + shell model. For computation of the particle scattering
function, R_{tot} (and the shell thickness and scattering length density
consistent with R_{tot}) are computed by assigning to the shell the hydrocarbon parts of the surfactant and aggregated cosurfactant; the
value of AGG is obtained as before from the least-squares-derived
core radius. With deuterated continuous phase, there is a logical
discrepancy in the independent variation of R_{sf}, in that solvent penetration should affect the scattering length density of the shell. The
effect will be compensated to some extent in that lowering of intensity, because of lower shell-solvent contrast, will be counteracted by
increased scattering as a result of greater shell thickness.

It is curious that we apparently can distinguish between geometries causing such slight departures from monodisperse spheres,
in view of the fact that the distribution of orientations of asymmetric
particles with respect to direction of the neutron beam can to a first
approximation be represented by an assembly of spheres of differing
size [31]. Perhaps the main reservation stems from the use of
structure functions based on spheres of equivalent size. It may be
that deficiencies of this procedure, particularly in the test of polydisperse models, explain the apparent sensitivity to different particle geometries [32].

The assignment of half of the stoichiometric alcohol to the disperse phase perhaps merits comment. The selection was made partly
on the basis of getting some sort of agreement to fit and experiment
with solution G, which was particularly sensitive to this parameter.
The number used had little effect on the D_2O-only microemulsions.
For example, with solution E, monodisperse oblate ellipsoid/invisible

TABLE 2 Parameters from Least-Squares Fits to Scattering Patterns from W/O Microemulsions of Different Compositions (Monodisperse Oblate Ellipsoid, Invisible Shell Model)[a]

Mole/Mole KOleate	A		B		C		D		E	
1-Pentanol	6		6		—		6		6	
1-Hexanol	—		—		5.2		—		—	
Hexadecane	5.5		5.5		5.5		5.5		4.8	
D_2O	18		45		45		57		45	
Parameters		±		±		±		±		±
R_{core}, Å	18.2	0.2	32.7	1.2	37.9	6.6	37.1	8.5	28.2	0.6
R_{tot}, particle function	(25.0)		(38.9)		(45.1)		(42.8)		(33.4)	
R_{sf}, structure function	26.1	0.2	44.2	1.3	49.8	4.8	52.5	12.0	42.0	0.8
Axial ratio (ECC)	0.46	0.02	0.47	0.04	0.50	0.16	0.47	0.22	0.51	0.03
Apparent ϕ	0.142	0.004	0.19	0.01	0.27	0.06	0.21	0.09	0.211	0.009
Scale	0.89	0.02	0.98	0.03	1.11	0.19	0.96	0.20	0.87	0.02
AGG	(42)		(104)		(163)		(120)		(65)	
B	0.09	0.02	0.27	0.11	0.36	0.63	0.24	0.78	0.59	0.06
CHI	1.2		3.0		5.3		6.3		3.0	

[a] ( ) Computed from other parameters.

shell model, putting all the alcohols in the shell, instead of three per KOl, affected only the scale and penetration factors appreciably. Scale was changed from 0.87 to 0.80, because of slightly modified scattering densities in the various regions, and the aggregation number changed from 65 to 63 because of more alcoholic $-OH$ groups in the core.

Cebula et al. [11], from SANS measurements performed on dilute W/O microemulsions of KOl/1-C_6OH/water/toluene, inferred a shell thickness of 8 Å: the effect on the form factor of changing from D_2O/h-toluene to H_2O/d-toluene dominated their scattering curves. Ottewill and co-workers [10] found that using SANS for concentrated W/O microemulsions in the system sodium dodecylbenzene-sulfonate/1-C_6OH/water/xylene, R_{sf} could be set equal to R_{core} + 8 Å. Our results, obtained using R_{sf} as an adjustable parameter (see Table 2), give values for $R_{sf} - R_{core}$ ranging from 7.9 to 15.4 Å. The extended length of KOl's 17-carbon tail (recall that the double bond is cis) is 18 Å; thus one can say that a portion of the particles' shells are not rigid, but instead interpenetrate when the particles approach each other. However, in n-hexadecane this interpenetration is less than it is when the oil is a shorter-chain n-alkane.

We believe that the apparent reduction in volume fraction below that calculated from stoichiometry may also be a manifestation of the particles' soft outer shells. However, the physical origin of this is less clear than for R_{sf} being less than R_{core} plus the extended length of the oleate tail.

V. ACKNOWLEDGMENT

Work at the University of Tennessee was sponsored by the National Science Foundation (CHE-8308362).

Small-angle neutron measurements were carried out at the National Center for Small-Angle Scattering Research (NCSASR). NCSASR is funded by National Science Foundation Grant No. DMR-77-244-58 through Interagency Agreement No. 40-637-77 with the Department of Energy (DOE) and is operated by the U.S. Department of Energy under contract DE-AC05-840R21400 with Martin Marietta Energy Systems, Inc.

VI. REFERENCES

1. Bellocq, A. M., Biais, J., Bothorel, P., Clin, B., Fourche, G., Lalanne, P., Lemaire, B., Lemanceau, B., and Roux, D., *Adv. Colloid and Interface Sci.* 20, 167 (1984).

2. Cazabat, A. M., Langevin, D., Meunier, J., and Pouchelon, A., *Adv. Colloid and Interface Sci.* *16*, 175 (1982).
3. Dvolaitzky, M., Guyot, M., Lagues, M., LePesant, J. P., Ober, R., Sauterey, C., and Taupin, C., *J. Chem. Phys.* *69*, 3279 (1978).
4. Lagues, M., Ober, R., and Taupin, C., *J. Physique Lett.* *39*, L-487 (1978).
5. Dvolaitzky, M., Lagues, M., LePesant, J. P., Ober, R., Sauterey, C., and Taupin, C., *J. Phys. Chem.* *84*, 1532 (1980).
6. Auvray, L., Cotton, J.-P., Ober, R., and Taupin, C., *J. Physique* *45*, 913 (1984).
7. Auvray, L., Cotton, J.-P., Ober, R., and Taupin, C., *J. Phys. Chem.* *88*, 4586 (1984).
8. De Geyer, A., and Tabony, J., *Chem. Phys. Lett.* *113*, 85 (1985).
9. Cabos, C., and Delord, P., *J. Appl. Cryst.* *12*, 502 (1979).
10. Cebula, D. J., Ottewill, R. H., Ralston, J., and Pusey, P. N., *J. Chem. Soc., Faraday Trans. 1*, *77*, 2585 (1981).
11. Cebula, D. J., Harding, L., Ottewill, R. H., and Pusey, P. N., *Colloid Polymer Sci.* *258*, 973 (1980).
12. Kotlarchyk, M., Chen, S. H., Huang, J. S., and Kim, M. W., *Phys. Rev. A29*, 2054 (1984).
13. Chen. S.-H., Lin, T. L., and Kotlarchyk, M., paper to be published in the Proceedings, 5th International Symposium on Surfactants in Solutions, Bordeaux, July 1984.
14. Robinson, B. H., Toprakcioglu, C., and Dore, J. C., *J. Chem. Soc., Faraday Trans. 1*, *80*, 13 (1984).
15. Toprakcioglu, C., Dore, J. C., Robinson, B. H., and Howe, A., *J. Chem. Soc., Faraday Trans. 1*, *80*, 413 (1984).
16. Shah, D. O., and Hamlin, R. M., *Science* *177*, 483 (1971).
17. Falco, J. W., Walker, R. D., and Shah, D. O., *AICHE J.* *20*, 510 (1974).
18. Shah, D. O., Bansal, V. K., and Hsieh, W. C., in *Improved Oil Recovery by Surfactant and Polymer Flooding*, Academic Press, New York, 1977, p. 293.
19. Sjöblom, E., and Friberg, S., *J. Colloid and Interface Sci.* *67*, 16 (1978).
20. Bellocq, A. M., and Fourche, G., *J. Colloid and Interface Sci.* *78*, 275 (1980).
21. Boussaha, A., Djermouni, B., Fucugauchi, L. A., and Ache, H. J., *J. Am. Chem. Soc.* *102*, 4654 (1980).
22. Atik, S. S., and Thomas, J. K., *J. Phys. Chem.* *85*, 3921 (1981).
23. Caljé, A. A., Agterof, W. G. M., Vrij, A., in *Micellization, Solubilization and Microemulsions*, Mittal, K. L. (ed.), Plenum Press, New York, 1977, vol. 2, pp. 779-90.

24. Lemaire, B., Bothorel, P., and Roux, D., *J. Phys. Chem.* 87, 1023 (1983).
25. Brunett, S., Roux, D., Bellocq, A. M., Fourche, G., and Bothorel, P., *J. Phys. Chem.* 87, 1028 (1983).
26. Huang, J. S., Safran, S. A., Kim, M. W., Grest, G. S., Kotlarchyk, M., and Quirke, N., *Phys. Rev. Lett.* 53, 592 (1984).
27. Huang, J. S., *J. Chem. Phys.* 82, 480 (1985).
28. Triolo, R., Magid, L. J., Johnson, J. S., Jr., and Child, H. R., *J. Phys. Chem.* 86, 2389 (1982).
29. Hayter, J. B., and Penfold, J., *Colloid Polymer Sci.* 261, 1022 (1983).
30. Kotlarchyk, M., and Chen, S.-H., *J. Chem. Phys.* 79, 2461 (1983).
31. Hayter, J. B. in *Physics of Amphiphiles: Micelles, Vesicles and Microemulsions*, Degiorgio, V., and Corti, M. (eds.), North Holland, Amsterdam, 1985, p. 59.
32. Van Beurten, P., and Vrij, A., *J. Chem. Phys.* 74, 2744 (1981).
33. Caponetti, E., Magid, L. J., Hayter, J. B., and Johnson, J. S., Jr., *Langmuir, in press.*
34. Hayter, J. B., and Penfold, J., *J. Mol. Phys.* 42, 109 (1981); *J. Chem. Soc., Faraday Trans. 1*, 77, 1851 (1981).
35. Akcasu, A. Z., Summerfield, G. C., Jahshan, S. N., Han, C. C., Kim, C. Y., and Yu, H., *J. Polymer Sci., Polymer Phys.* 18, 863 (1980).

18

Structural Evolution in Concentrated Zones of Ternary and Quaternary Microemulsions

F. C. LARCHE, P. DELORD, and J. L. DUSSOSSOY*
University of Montpellier 2, Montpellier, France

I. INTRODUCTION

The structure of microemulsions in regions of the phase diagram where neither water nor oil are major constituents has been investigated by x-ray and neutron scattering. The systems studied were the ternary one comprised of sodium p-octylbenzene sulfonate/pentanol/water, and the quaternary one obtained by the addition of decane. In the treatment, the peak we observed in the spectra resulted from a preferred distance in the solutions. In a central region of the quaternary system, the variation of this distance with water volume fraction agrees with the model of Jouffroy et al. [6]. A structure made of a dispersion of droplets is also considered in this chapter. Together with previously published data, these results seem to be more in agreement with a bicontinuous structure for the regions with the highest surfactant/decane ratios. Results on the isotropic alcohol-rich phase of the ternary system are also discussed.

In this chapter we use the term microemulsion in its broadest sense to mean all isotropic solutions having a supramolecular structure that contains at least water, a surfactant, and a hydrophobic compound. The structure of these microemulsions, in regions of the phase diagram either very rich in water or very rich in oil, are reasonably well understood [1]. As the zone corresponding to equal amounts of oil and water is approached, the structure becomes complicated and the analysis of experimental results, such as x-ray

---

Present affiliation: Centre d'Etudes Nucléaires de Cadarache, Saint Paul-Les-Durance, France.

scattering, becomes difficult. Recently, several theoretical developments have emerged whose predictions can be checked against experimental data. Parodi [2] postulated a locally lamellar structure. Consequently, X-ray (or neutron) diffraction spectra are expected to show a broad peak whose maximum will vary linearly with the volume fraction of water. Data on the sodium p-octylbenzene sulfonate/pentanol/water system, where the alcohol in part plays the role of oil, follow such predictions [3]. The X-ray spectra predicted by the Talmon-Prager model [4] have been computed by Kaler and Prager [5]. This highly random structure produces a continuously decreasing small-angle scattering. Jouffroy et al. [6] modified this model by introducing a characteristic distance ξ, imposed by the properties of the interfacial film, In the limit of small film volume fractions, ξ is proportional to the product $\phi_o\phi_w$ of the oil and water volume fractions. We have extended the X-ray data on the previously mentioned ternary system to higher water and alcohol content, and present neutron scattering data on the quaternary system so as to test the relevance of these models.

II. EXPERIMENTAL

The octylbenzene sulfonate (OBS) was synthetized according to Gray et al. [7] from 1-phenyloctane (Fluka, 99% pure). It was purified by several recrystallizations in water and ethanol. The water was doubly distilled, the 1-pentanol (Merck, p.a.) and decane (Fluka, purum) were used as received.

The low-angle X-ray equipment has been described previously [8]. It was equipped with a position-sensitive detector. The samples were contained in 0.3-mm-diameter sealed glass capillaries. The temperature was controlled at $25 \mp 0.1°C$.

To obtain the neutron data, water was replaced by heavy water. The samples were contained in a cell with quartz windows, separated by a 1-mm precision polished washer. The low-angle neutron scattering instrument (D11 at the Laüe-Langevin Institute in Grenoble) was operated with 0.485-nm wavelength neutrons. The 2-D spacial detector was placed at 2.5 m, and the results radially averaged. Standard corrections were made, and the intensities were relative to that of water.

All the quaternary samples were made with 0.5 wt% NaCl brine, so as to make possible a direct comparison with previously published results [9]. The two-dimensional composition cut investigated was at a fixed pentanol/OBS ratio of 2.1 by weight (or 7 by mole).

III. RESULTS

The neutron data were taken from samples located on three straight
lines going through the water corner of the phase diagram (Fig. 1).
Along any of these lines, only the water content is changed. A
typical spectrum taken in the center of the homogeneous zone is
shown in Fig. 2. Over the scattering-vector range $(0.3 < Q <
2 \, nm^{-1})$, the spectra show a well-developed maximum and no other

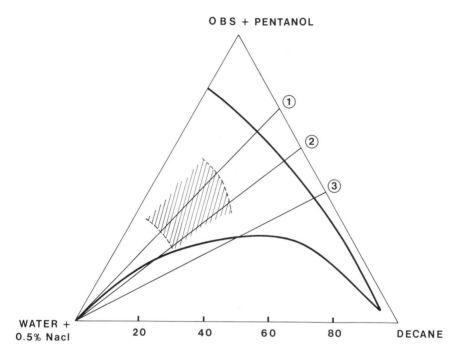

O B S + PENTANOL

① ② ③

**WATER +
0.5% Nacl**   20   40   60   80   **DECANE**

FIG. 1 The isotropic one-phase region of the system OBS-pentanol-
decane-brine (0.5% by weight) at a molar ratio pentanol/OBS = 7.
The neutron scattering data were taken from samples along the three
lines passing through the water corner. The replacement of water
by heavy water for contrast enhancement has a negligible effect
on the diagram. The decane/(OBS + pentanol) weight ratios for
lines 1, 2, and 3 are 0.25, 0.39, and 0.55 respectively. The dashed lines indicate approximatively the extent of the bicontinuous zone
(see text).

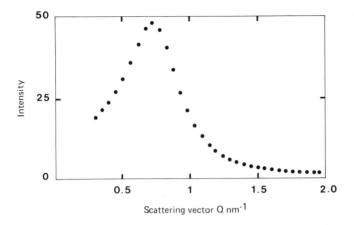

FIG. 2 The neutron scattered intensity of a sample of line 1, containing 44.1 vol% water, as a function of the scattering vector Q. The well-defined maximum is a typical feature of the spectra in the central region of the isotropic phase.

fine structure. As the interpretation is not unique, to use such data requires further information on the structure of the solutions. These are provided by the conductivity and self-diffusion measurements previously published on the same system [9]. They indicate that, in the central zone, the structure of the oil-rich and water-rich regions of these microemulsions should be open. Recent X-ray data on a similar system [10] have shown the existence of a well-defined surfactant film. This is corroborated in our system by the surfactant diffusion coefficient, which is constant over practically all the isotropic region of the phase diagram. These data are in favor of a bicontinuous structure first proposed by Scriven [11].

Talmon and Prager [4] made the first quantitative description of a possible bicontinuous geometry. The random nature of the Voronoï polyhedra used to generate the structure clearly appears in the scattering calculated by Kaler and Prager [5]. No maximum is expected from this model, and thus it cannot be applied to our data. The modification introduced by Jouffroy et al. [6] postulated the existence of a characteristic length in the system. This persistence length ξ is a direct consequence of the mechanical properties of the film alone. A recent theoretical development by Widom [12] would make this length the result of the minimum of the free energy of the whole system. In both models, the structure is generated by a random mixing of a cubic array stacking of cubic cells of size ξ.

In the limit of low film volume fractions, the relevant geometric parameters of the system are related by

$$\xi = \frac{j\phi_o\phi_w}{\Sigma\rho}$$ (1)

where ϕ_o and ϕ_w are the volume fraction of oil and water, respectively, ρ is the surfactant number density, Σ is the film area per surfactant molecule, and j is a geometric constant, equal to 6 for the cubic array mentioned earlier.

In our system, the surfactant film contains some alcohol. The rest of the alcohol can be considered to be mostly in the oil (pentanol solubility in water is small, and the presence of sodium chloride tends to make it even smaller). Without precise knowledge of the partition coefficient, ϕ_o and Σ are not known. However, when the composition is changed by changing only the water content, the relative proportion of all the other constituents remains the same. In a systematic study of the evolution of ξ in lamellar structures of the same systems [3,13], it was found that the partition coefficient is controlled entirely by the amount of alcohol in the interface. The same behavior is expected in the microemulsion phase, and Σ would also remain constant upon addition of water. Therefore, even if we don't exactly know the product $j\phi_o/\Sigma\rho$, it remains constant when only the water content is changed. According to Eq. (1), if we plot ξ as a function of ϕ_w along such a composition path, we should find a straight line passing through the origin. We assumed that the maximum in the neutron spectrum provides a good measure of this characteristic length. The plots obtained appear in Fig. 3. In series 1 and 2, they show a sigmoidal shape. A straight line passing through the origin fits very well the points located between 20 and 45 vol% water. Its slope increases from series 1 to series 2, as predicted by Eq. (1), since the amount of surfactant and alcohol are decreased, decreasing both Σ and ρ. A dispersion of droplets has previously been proposed as a possible structure for a similar microemulsions [15,16]. In such concentrated solutions a crystalline-like stacking of droplets is usually assumed. We shall take a face-centered cubic lattice as the most probable local structure in a system of attracting spherical droplets [17,18].

Our measurements have been made along lines where the ratio (pentanol + OBS)/decane is constant. In this system it fixes the surface per polar head Σ. The total film surface per unit volume is given by

$$S = 4\pi N R_w^2$$

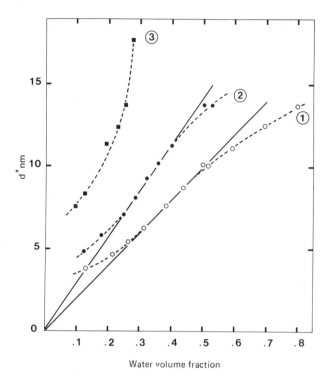

FIG. 3 The characteristic length d* of the microemulsions computed from the maximum of the neutron scattering intensity data, as a function of the water volume fraction, for series 1 to 3 (cf. Fig. 1). Dashed lines through the experimental points are only visual aids.

where N is the number of particles per unit volume and R_W is the radius of the water core. R_W and ϕ_W are related by

$$\phi_W = \frac{4}{3} \pi N R_W^{\,3}$$

The water volume fraction is also related to the unit cell parameter $\bar{a}$ by

$$\phi_W = \frac{16}{3\bar{a}^{-3}} R_W^{\,2}$$

After elimination of N and R_W among these three relations,

$$\bar{a} = \frac{(144)^{1/3}}{S} \phi_W^{2/3}$$

$\bar{a}$ is also related to the maximum Q_m of the scattering peak by

$$\bar{a} = \frac{2\pi\sqrt{3}}{Q_m}$$

Since S is a constant, a plot of $2\pi/Q_m$ versus $\phi_W^{2/3}$ should give a straight line passing through the origin. The results appear in Fig. 4. For series 1 and 2, a straight line can be drawn through

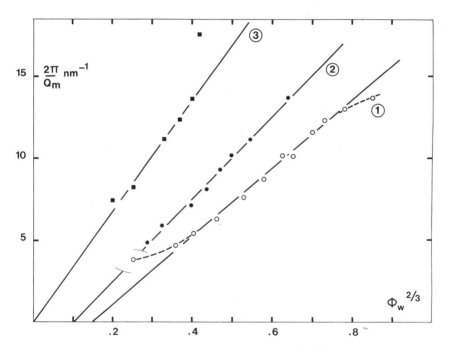

FIG. 4 The droplets model. The parameter $\bar{a}/\sqrt{3}$ of a face-centered cubic unit cell versus $\phi_W^{2/3}$.

the inflection point, but it does not pass through the origin; on the other hand, the results agree with series 3.

Comparing the two models, it seems that a bicontinuous structure gives a better fit to the experimental data for the region corresponding to series 1 and 2.

It is interesting to note that the center of this region corresponds to 30 vol% water. In our experiments, however, the film volume fraction is not negligible. A rough estimate, assuming that 50% of the pentanol is in the film, gives the volume fraction of the film as 30% for the series 1 samples in the middle of the linear zone. This correction returns the linear zone to where it is expected, i.e., nearly equal volume fractions of oil and water [1].

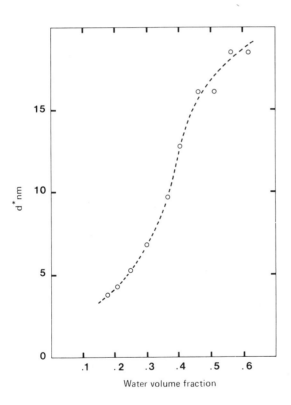

FIG. 5 The characteristic length d* obtained from the maximum of the X-ray spectrum as a function of the water volume fraction, in the ternary system OBS-pentanol-water, along a line at constant ratio pentanol/OBS = 4 by weight.

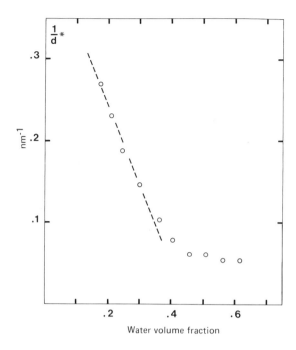

FIG. 6 The inverse of the characteristic length d* as a function of the water volume fraction in the ternary system pentanol/OBS of 4 by weight.

In the ternary system, the L_2 isotropic zone extends to relatively high water concentrations. Small-angle X-ray scattering spectra were again obtained along a straight line passing through the water corner. The pentanol/OBS ratio (4 by weight) was chosen so as to span the largest possible water content. Once again the spectra show a well-developed maximum, which is again assumed to provide a good measure of the characteristic length ξ. The data are plotted in Fig. 5. Although the general shape is still sigmoidal, the tangent through the point of inflection does not pass through the origin. Self-diffusion data [14] may be in favor of open structure, but because of the time scale involved, it may be a dynamic feature. In this region, the data are not entirely consistent with the Jouffroy et al. model. We have also tested Parodi's model, by plotting 1/d* as a function of ϕ_W. Figure 6 shows that at the lowest water volume fraction such a locally lamellar structure would be in agreement with the data; but at $\phi_W > 0.3$, there is a marked discrepancy with the expected behavior.

In contrast, at much lower alcohol/OBS ratio, the linear relation between $1/d*$ and ϕ_W is in agreement with the data [3]. The difference in behavior between these two regions is not explained by the existing models at this time.

IV. DISCUSSION AND CONCLUSION

Several models have been tested to explain the presence of a peak in X-ray and neutron scattering spectra of concentrated microemulsions and to relate quantitatively its evolution with the composition of the samples.

To relate the peak position to a characteristic length requires a model of structure from which scattering intensities could be computed and fitted.

The Kaler-Prager model does not provide such a framework, since there is no favored length in it; indeed, it does not show a maximum in the scattering intensity.

The Jouffroy-Levinson-de Gennes model has been tested with the assumption that the position of the maximum of the peak gives a good measure of the characteristic length.

In this case, the spacings $d*$ corresponding to the peaks are somewhat shorter and those found in Winsor microemulsions by Auvray et al. [10]. Apart from the difference in chemical composition, their samples contained a lower amount of surfactant, and this alone, according to Eq. (1), could explain the difference. The center of the linear region for $d*$ and for the conductivity as a function of ϕ_W is not centered around $\phi_W = 0.5$ but more toward the oil-rich side. We calculated the volume fraction of water with respect to the total volume of the sample and not with respect to the oil and water volume. This latter procedure requires a knowledge of the alcohol partition coefficient between the film and the oil. Assuming equal amounts in each location would change the sample $\phi_W = 0.315$ of the series 1 into $\phi_W = 0.458$. However, in Eq. (1) the volume of film is neglected. In view of this approximation, and because of the lack of knowledge of the pentanol partition coefficient, we have chosen to use a definition that does not require an assumption on unknown parameters. In a first approximation, the functional dependence of ξ and ϕ_W would be the same. Obviously, there is a need for a geometric model closer to the volumic constrains of the molecules involved.

Keeping in mind that we are judging here data that are orders of magnitude, we can say that this model gives more convincing results than that of a droplet model in the region concentrated both in water and oil.

We have also tested a model of droplets with a water core in a face-centered cubic structure; in this model, the peak is due to first-order reflection on (111) planes; the results seem to agree with experimental results for high-oil-volume fractions in the quaternary system.

Another model, proposed by Parodi, presupposes a lamellar local order and gives satisfactory results in the ternary system without salt and oil, over 15 wt% of OBS. This model does not agree with experimental results in the other part of the L_2 isotropic phase of the ternary system, nor in the large isotropic phases of the quaternary system.

V. AKNOWLEDGMENTS

We would like to thank an anonymous referee for very valuable and positive criticisms. This work was partially supported by PIRSEM (C.N.R.S.) AIP 2004.

VI. REFERENCES

1. De Gennes, P. G., and Taupin, C., *J. Phys. Chem. 86*, 2294 (1982).

2. Parodi, O., in *Colloïdes et Interfaces*, Summer School, Aussois, Les Editions de Physique, Paris, 1983, p. 355.

3. Marignan, J., Delichere, A., and Larche, F. C., *J. Physique . Lett., 44*, 609 (1983).

4. Talmon, Y., and Prager, S., *J. Chem. Phys. 69*, 2984 (1978).

5. Kaler, E. W., and Prager, S., *J. Colloid and Interface Sci. 86*, 359 (1982).

6. Jouffroy, J., Levinson, P., and de Gennes, P. G., *J. Physique 43*, 1241 (1982).

7. Gray, F. W., Gerecht, J. F., and Krems, I. J., *J. Org. Chem. 20*, 511 (1955).

8. Assih, T., Larche, F. C., and Delord, P., *J. Colloid and Interface Sci. 89*, 35 (1982).

9. Larche, F., Rouviere, J., Delord, P., Brun, P., and Dussossoy, J. L., *J. Physique Lett. 41*, 437 (1980).

10. Auvray, L., Cotton, J. P., Ober, R., and Taupin, C., *J. Physique 45*, 913 (1984).

11. Scriven, L. E., in *Micellization, Solubilization and Microemulsions*, Mittal, K. L., (ed.), Plenum Press, New York, 1977, vol. 2.

12. Widom, B., *J. Chem. Phys. 81*, 1030 (1984).

13. Larche, F. C., Quebbaj, S. El., and Marignan, J., Proceedings of the Symposium of Magnetic Resonance and Scattering in Surfactant Systems, Magid, L. (ed.), *to be published*.

14. Larche, F. C., Dussossoy, J. L., Rouviere, J., and Marignan, J., *J. Colloid and Interface Sci. 94*, 564 (1983).

15. Lagües, M., Ober, R., and Taupin, C., *Physique Lett. 39*, L487 (1978).

16. Cebula, D. J., Ottewill, R. H., Ralston, J., and Pusey, P. N., *J. Chem. Soc. Faraday Trans. 1 77*, 2585 (1981).

17. Kotlarchyk, M., Chen, S. H., Huang, J. S. and Kim, M. W., *Phys. Rev. Lett. 53*(9), 941 (1984).

18. Kotlarchyk, M., Chen, S. H., and Kim, M. W., *Phys. Rev. A 29*, 2054 (1984).

19

Glass-Forming Microemulsions and Their Structure

D. R. MACFARLANE, I. R. McKINNON, E. A.
HILDEBRAND *Monash University, Clayton,
Victoria, Australia*

C. A. ANGELL *Purdue University, West Lafayette,
Indiana*

I. INTRODUCTION

The recent development of glass-forming microemulsions has allowed
the direct probing of their structure using electron microscopic
techniques without the danger of the structure being modified by
crystallization. The possibility of trapping a variety of simple
molecular liquids in arrested, glassy states has also been investigated via their incorporation as the nonpolar phase in these
microemulsions.

II. DISCUSSION

Of the four components in the classical microemulsion, invariably one
or more crystallizes easily if the temperature is lowered sufficiently
below room temperature. Unless the microemulsion as a whole has
become immobile, this event will probably cause severe disruption
of the microemulsion structure. On the other hand, if the components are separately or collectively glass-forming, then the crystallization of all possible crystalline phases can be bypassed, perhpas
by rapid cooling, to form a totally arrested amorphous solid. This
vitreous microemulsion is much more likely to retain the structure
of the room-temperature microemulsion. The advent of vitrifiable
microemulsions thus opens the way to the direct examination of this
structure by classical quench-fracture electron microscopic techniques.
Several years ago it was reported [1, 2] that glass-forming microemulsions could be prepared by choosing a cosurfactant component
of the oil-in-water microemulsion that would also inhibit the nucleation

and growth of ice in the microemulsion when the temperature was
lowered below 273 K. In the cases studied to date, this cosurfactant component has always been propane-1,2-diol (propylene glycol,
PG) although recent work [3] has suggested that other simple poly-
ols act similarly. Propylene glycol is known [1] to render aqueous
solutions glass-forming at concentrations in excess of 12 mol% PG.
As long as the dispersed oil phase is glass-forming and provided that
there is no lower temperature phase instability with respect to a two-
phase mixture, the microemulsion as a whole can hence be vitrified.
No such phase separation has been observed in the PG-based systems
studied to date; for example, a sample cooled quite slowly was obtained in vitrified form without any increase in turbidity [2], and a
H_2O/PG/Tween 80/CCl_4 (Tween 80 = polyoxyethylene sorbitan

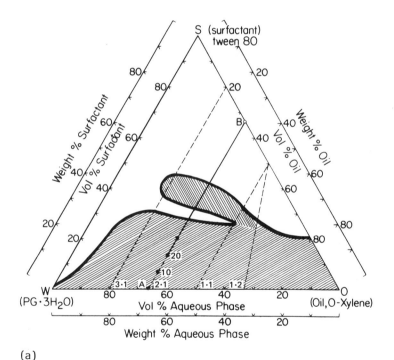

(a)

FIG. 1   (a) Pseudo-three component phase diagram for the system
propylene glycol·$3H_2O$/Tween 80/o-xylene.   (b) Thermal analysis
(DSC) evidence for vitrification of o-xylene and other normally un-
vitrifiable liquids in microemulsion state; note closeness of melting
points to bulk crystal values, consistent with very small surface free
energies for droplet phase.

monooleate) microemulsion has been found to be stable to at least −90°C for a number of hours [4]. However, at temperatures near the glass transition temperature, T_g, (at which point equilibration ceases on a given time scale) of the higher-T_g component, kinetic slowing down of equilibration processes could inhibit any thermodynamically expected phase change.

The pseudo-ternary phase diagram for the system H_2O/PG/Tween 80/o-xylene is shown in Fig. 1. All compositions in the quaternary are glass-forming as long as the H_2O/PG mole ratio is less than 7. The phase diagrams for o-xylene and a number of other simple molecular liquids (most of which are not glass-forming and are of interest for this reason—see below) are similar. Recent work in the Monash laboratories has shown that H_2O/PG/Tween 80/o-xylene has a continuous region of clear, stable, one-phase compositions from the finger region indicated in Fig. 1 to the H_2O-free ternary.

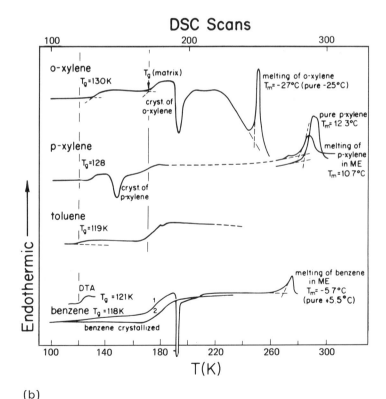

(b)

FIG. 1 (Continued)

This suggests that a three-component microemulsion has been formed (which is incidentally strongly glass-forming), in which the polar phase (PG) is also acting at the surface as cosurfactant, if such a description continues to have any meaning in the ternary.

A recent article [4] reported the vitrification in glassy micro-emulsions of three molecular liquids (CS_2, benzene, and CCl_4) that had not previously been vitrified and that, on the basis of bulk glass-forming trends, would not be expected to vitrify. The glass transition temperatures T_g (an iso-structural relaxation time point) for these liquids were as expected on the basis of extrapolations from binary-solution T_g data, indicating that the microemulsion droplets exhibit largely unaltered relaxation properties with respect to their bulk liquids.

The observation of glass formation in these simple liquids entrapped in the microemulsion (although nonetheless retaining unit activity), simply because of the small volume of the liquid droplet, indicates another fascinating application of these glass-forming micro-emulsions in the elucidation of the dynamics of these simple liquids.

Nonetheless, the major interest in these microemulsions remains their ability to provide a static structure that is respresentative of the ambient-temperature microemulsion. Especially attractive is the possibility of trapping the microemulsion in the transition region of compositions between what are thought to be clearly water-in-oil and clearly oil-in-water droplet structures. The bicontinuous and lamel-lar structures postulated for this region of composition should be quite distinguishable under microscopic examination. Three low-temperature electron microscopic techniques have been applied over the last few years. In one series of experiments [4], the electron micrographs were obtained by the quenching of a thin (~ 0.1 µm) film of the microemulsion on an electron microscope sample screen. These samples, examined by transmission electron microscopy (TEM), revealed a clear droplet structure in all the compositions studied and, at the higher dispersed phase contents, a hexagonally ordered packing of the droplets.

In the most recent work [5], attention was focused on a quench-fracture technique in which the sample of microemulsion is supported between two metal disks, the assembly quenched and the disks separated, fracturing the sample. The fracture faces thus produced should be truly representative of the bulk microemulsion. The fracture face is then coated with a platinum-carbon film while being held at liquid nitrogen temperatures. Finally, the Pt-C replica is removed and examined by TEM.

In Fig. 2a, an example of the micrographs obtained for micro-emulsions in the H_2O/PG/Tween 80/o-xylene systems is shown. An obvious droplet structure is evident, the droplets having diameters of the order of 50 nm. For comparison a micrograph, at similar

magnification, of a simple H_2O/PG solution is shown in Fig. 2b.
Comparison of Figs. 2a and 2b reveals a similarity between the continuous region of the microemulsion micrograph and the micrograph
of the propylene glycol/water solution. This essentially establishes
a control for the microemulsion micrographs by indicating that the
freeze-facture procedure does not produce artifactual surface features, nor does it produce devitrification in the bulk continuous phase
toward crystalline ice. A repeat of this experiment utilizing a different sample container revealed similar results except that the droplet diameters were, on average, smaller.

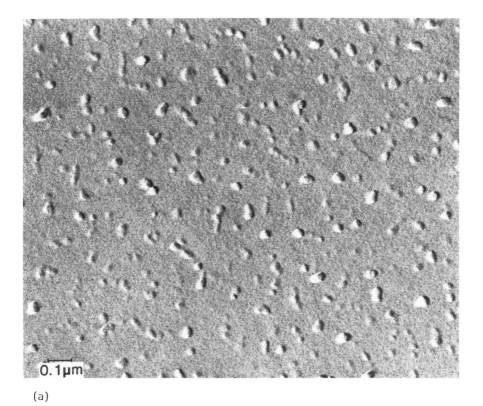

0.1μm

(a)

FIG. 2 (a) Transmission electron micrographs of a 40 vol% $PG \cdot 3H_2O$/
27 vol% o-xylene/33% Tween 80 microemulsions obtained by the double-
replica freeze-fracture technique at 50,000X magnification. All micrographs were recorded on a JEOL 100S TEM. (b) Transmission electron micrograph of a $PG \cdot 3H_2O$ sample at the same 50,000X magnification.

(b)

FIG. 2   (Continued)

     The techniques used in this work are essentially guaranteed to
produce a representation of the microemulsion in the fluid state. It
is, however, not clear to which region of temperature this representation applies. Since the droplet size is a result of a rather critical
free energy balance, it would be expected that the droplet size distribution would change as a function of temperature. Hence the
droplets observed in the vitreous microemulsion are representative
of the structure of the microemulsion at the lowest temperature (de-
noted $T_{g,m}$) at which the microemulsion was able to maintain complete equilibration during cooling. $T_{g,m}$ is likely to be related to
the T_g's of the continuous phase and the droplet surface (whichever
is higher) and would intuitively be expected to be higher than both.
Since T_g is always a kinetic event, it is markedly dependent on
cooling rate, and hence the difference between the results obtained

using the two freeze-fracture techniques (of intrinsically different cooling rates) may well represent a strong temperature dependence of the microemulsion structures.

It is clear from all the results, and particularly from Fig. 2, that the microemulsion "droplets" represented there vary in both size and shape. Chains of 2–12 predominate over single droplets which, of course, may be only the exposed portion of a more complex structure.

The polydispersity seems greater than the $\sim 20\%$, which is predicted as a maximum value compatible with stability in a three-component microemulsion at ambient temperature [6], a value that should decrease with decreasing T. Again it must be recognized that apparent size variations may be exaggerated by different extents of exposure at the fracture surface of more monodisperse droplet distributions. The stringing together of droplets is interesting and could represent the manner in which the system moves toward the cylindrical phase predicted by Safran et al. [7] for certain regions in the parameter space of the three-component microemulsion theories. The *shape* variations seem real and cannot be explained by variable degrees of exposure of uniform droplets. Variations in shape of droplets are predicted [6,8] and can become larger near a phase boundary. The predicted shape variations may be observed in Fig. 2, though the extent to which shape variations may be imposed on a droplet phase by stress as the matrix phase (continuously) vitrifies around them, as in the present case, have not been evaluated. Since changes in composition of the matrix phase should make it possible to vary the value $T_{g,m}/T_{g,d}$ through unity, such questions should receive experimental answers in the near future.

In summary, the possibilities for direct investigations of microemulsion structure opened up by the development of glass-forming microemulsion compositions are considerable and deserve to be exploited. In particular, parallel studies by neutron scattering and electron microscopic methods are to be sought, and may be available in the near future [9].

III. ACKNOWLEDGMENTS

This work has been supported by the U.S. National Science Foundation (DMR 8304887).

IV. REFERENCES

1. MacFarlane, D. R., and Angell, C. A., *J. Phys. Chem.* 86, 1927 (1982).

2. Angell, C. A., Kadiyala, R. K., and MacFarlane, D. R., *J. Phys. Chem. 88*, 4593 (1984).
3. MacFarlane, D. R., McKinnon, I. R., and Hildebrand, E. A., *to be submitted to J. Phys. Chem.*
4. Dubochet, J., Adrian, M., Teixeira, J., Alba, C. M., Kadiyala, R. K., MacFarlane, D. R., Angell, C. A., *J. Phys. Chem. 88*, 6727 (1984).
5. Hildebrand, E. A., McKinnon, I. R., and MacFarlane, D. R., *J. Phys. Chem. 90*, 2784 (1986).
6. Safran, S. A., in *Surfactants in Solution*, ed. Mittal, K. L., and Lindman, B. (eds.), Plenum Press, New York, 1984, vol. 3, p. 1781.
7. Safran, S. A., Turkevich, L. A., and Pincus, P., *J. Phys. Lett., 45*, L-69 (1984).
8. Safran, S. A., *J. Chem. Phys. 78*, 2073 (1983).
9. Alba, C. M., and Teixeira, J., *private communication.*

20

Oxygen as an Electrochemical Probe in Studies of Physico-Chemical Structures of Microemulsions

ALAIN BERTHOD   *Université Claude Bernard, Lyon, Villeurbanne, France*

I. INTRODUCTION

Polarographic reduction of dissolved oxygen in an air-saturated microemulsion system composed of water, pentanol, dodecane, and sodium dodecylsulfate was compared with a similar study in methylene chloride microemulsions [1]. In both systems, oxygen was more soluble in microemulsion than in each separated component. The excess of solubility can reach 200% (3.5 mmoles/liter) in a methylene chloride system and 60% (2.4 mmoles/liter) in a dodecane system. The study of the diffusion coefficient of hydrogen peroxide (electrogenerated at the mercury electrode by oxygen reduction) gave an insight into the physicochemical structures of the systems. The first polarographic maximum, observed in oil-rich microemulsions of both systems, allowed us to delimit the L2 structure area.

    In a recent work [1], we have shown that well-known electrochemical techniques, such as DC and AC polarography, are able to bring out some information about microemulsions that agree with results obtained using more sophisticated techniques, such as high-resolution NMR [2], quasielastic light scattering [3], radioactive tracers [4], electron microscopy [5], conductivity or viscosity measurements [6], small-angle neutron scattering [7], and microwave dielectric measurements [8] as nonexhaustive examples.

    At room temperature, oxygen dissolves naturally in any liquid medium. In the case of microemulsions, the oxygen dissolution does not modify either the compound proportion or the ionic strength. This chapter presents the polarographic study of oxygen in several compositions for a water/dodecane/n-pentanol/sodium dodecylsulfate (SDS) (mass ratio 2/1) microemulsion system and compares it with the results obtained in a water/methylene chloride/n-pentanol/sodium paraoctylbenzenesulfonate (SOBS) (mass ratio 2/1) system [1]. This work attempts to obtain information about the physicochemical structure of these two microemulsion systems.

II. MATERIAL AND METHODS

The direct current (DC) polarograms were obtained with a PRG5 (Solea Tacussel, Villeurbanne, France) or a PAR 174 (Princeton Applied Research, Princeton, N.J.), two polarographic analyzers. The AC polarograms were obtained with a PRG3 AC polarograph (Solea Tacussel); the amplitude of the applied alternating voltage was 10 mV, the frequency was 60 Hz, and the demodulation phase angle was 90° (i.e., capacitive current). The working electrode was a dropping mercury electrode (DME), the reference electrode was a saturated calomel electrode (SCE) with a KCl agar bridge to limit water or salt diffusion in the microemulsion studied. All experimental potentials were referred to this electrode. The auxiliary electrode was a platinum wire. This three-electrode system compensated electronically for the iR drop between the working and reference electrodes (provided that the microemulsion conductivity was greater than about 1 mS/cm).

As described in a recent publication [1], we ensured that potential shifts in oxygen reduction did not originate from a variation in liquid junction potentials, which were always below 10 mV in the whole range of microemulsions studied. The coulometric procedure has already been described [1]. Gasometric measurements were performed with a Van Slyke apparatus [9]. A known volume of the microemulsion under examination was put in the sample chamber and fully isolated from the atmosphere by mercury. After a complete vacuum extraction from the liquid phase (Winsor IV) of the pure gases, mainly N_2, CO_2 and O_2, a measurement of gas pressure at constant volume and temperature was performed then, a O_2 adsorber (anthraquinone-2-sulfonic acid, sodium salt) was added and the new gas pressure, at the same volume and temperature, allowed us to calculate the O_2 concentration in the liquid sample [10]. The addition of an ionic O_2 adsorber often broke the microemulsion, but due to the fact that the whole experiment took place without any gas exchange with the atmosphere, this addition did not modify the initial O_2 concentration. This procedure did not work with methylene chloride microemulsions because of the high vapor pressure of methylene chloride at 25°C (Table 1).

SOBS (MW 292) was synthesized by sulfonation of n-octylbenzene (Fluka). SDS (MW 288.4), methylene chloride, and n-dodecane were analytical-grade reagents (Merck), and n-pentanol was pro-analysis reagent (Rhone Poulenc Prolabo). Water was deionized and distilled.

TABLE 1 Physicochemical Constants and Oxygen Solubility at P_{O_2} = 160 mm Hg and 25°C for the Microemulsion Components[a]

Compound	MW	d	Vapor Pressure mm Hg (25°C)	bp	O$_2$ solubility mmole/liter	mole fraction × 10^5
Water	18	1.00	24	100	0.28	0.50
Methylene chloride	84.9	1.33	415	40.7	1.9	12.1
Dodecane	216.3	0.749	1	170.3	2.1	60.6
Pentanol	88.2	0.814	5	137.3	1.5	16.3
SDS micelles	288.3	1.17	–	–	0.85	21.0

[a]MW, d, vapor pressure, and bp from Ref. 23. O$_2$ solubility from Refs. 1, 16, and 17.

All concentrations were expressed as moles per liter of microemulsion unless otherwise indicated. All measurements were performed in a 20-ml cell thermostated at 25 ± 0.5°C by water circulation.

III. RESULTS

A. Oxygen Concentration

The polarographic reduction of molecular oxygen, at pH about 7, is known to produce two waves, corresponding to the reactions

$$O_2 + 2H^+ + 2e^- \; \rightleftharpoons \; H_2O_2 \tag{1}$$

$$H_2O_2 + 2H^+ + 2e^- \; \rightleftharpoons \; 2H_2O \tag{2}$$

The tests for the diffusion-controlled process, already described [1] (plots of i, the limiting current, versus $h^{0.5}$, the mercury height, versus C_{O_2}, the oxygen concentration, and versus the temperature), indicated a mainly diffusion-controlled current.

Water, pentanol, and oil (dodecane or methylene chloride) were air-saturated and then mixed with the surfactant (SDS or SOBS) to obtain a given composition of microemulsion spontaneously. The DC and AC polarograms were recorded and the coulometric cell was filled with the microemulsion. The oxygen concentration C was calculated from $C = Q/nFV$, where Q is the quantity of electricity (coulombs) for a complete oxygen reduction, n is the number of exchanged electrons, F is the faraday (96,500 C), and V is the cell volume (0.772 µl). When the conductivity was too slight (<1 mS/cm), the procedure with hydrogen peroxide, described elsewhere [26], was used with methylene chloride microemulsions and gasometric measurements were performed with dodecane microemulsions.

Oxygen concentration being dependent on atmospheric pressure (weather and altitude), the measured concentration was corrected by the ratio 760/actual pressure. Figures 1 and 2 show the isoconcentration curves for oxygen in the Winsor IV area of the two systems studied.

B. Diffusion Coefficients

The variations of the diffusion coefficients of oxygen and hydrogen peroxide were evaluated from the value of the limiting current at the DME. For this comparative study in microemulsions of various compositions, the Ilkovic equation related adequately the diffusion current i to the diffusion coefficient D:

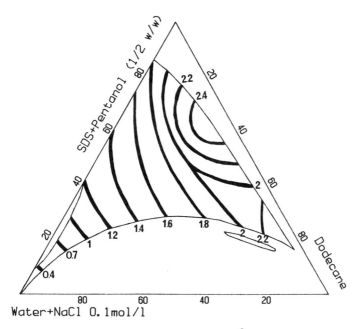

FIG. 1 Isoconcentration curves $(10^{-3}$ mole/liter) for oxygen at 25°C and P_{O_2} = 160 mm Hg in dodecane microemulsion. Accuracy 10%.

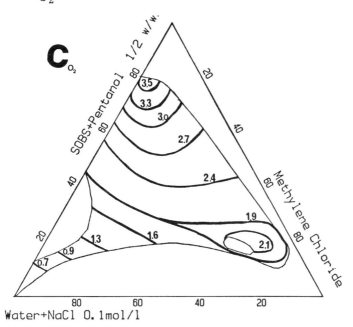

FIG. 2 Isoconcentration curves $(10^{-3}$ mole/liter) for oxygen at 25°C and P_{O_2} = 160 mm Hg in methylene chloride microemulsion. Accuracy 10%.

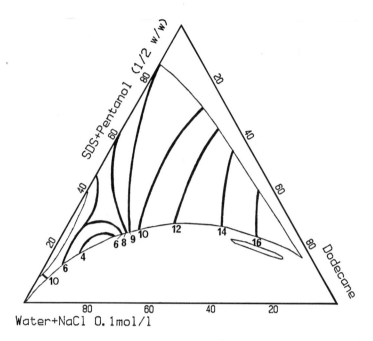

FIG. 3 Diffusion coefficient (10^{-6} cm^2/s) of molecular oxygen in dodecane microemulsion. Accuracy 20%.

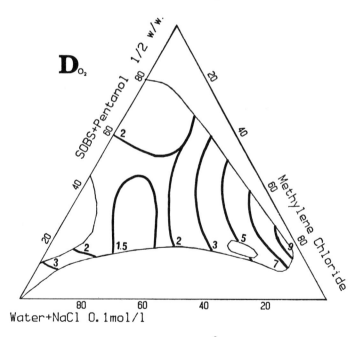

FIG. 4 Diffusion coefficient (10^{-6} cm^2/s) of molecular oxygen in methylene chloride microemulsion. Accuracy 20%.

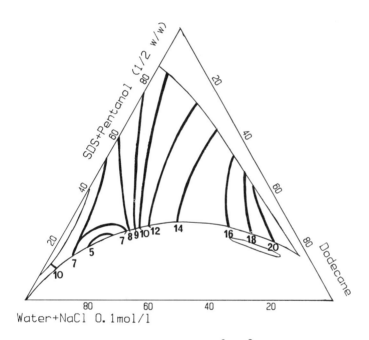

FIG. 5 Diffusion coefficient (10^{-6} cm^2/s) of hydrogen peroxide in dodecane microemulsion. This coefficient is theoretical; see text and Fig. 9. Accuracy 20%.

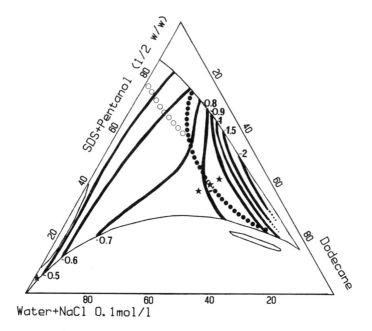

FIG. 6 Isopotential curves (V versus SCE) of the first reduction
step of oxygen in dodecane microemulsion. Accuracy 5%. (●) Area
inside which a polarographic maximum is present. (○) Γ_0 conductivity lines (Ref. 21). (⋆) Composition corresponding to the polarograms of Fig. 11.

$$i = nFaC \left(\frac{7D}{3\pi t}\right)^{1/2}$$ (3)

where a and t denote the mercury drop area and time, respectively,
and C is the oxygen concentration. Results are presented in Figs.
3 and 4 for oxygen diffusion coefficients D_{O_2} in both systems. The
diffusion coefficient of hydrogen peroxide, $D_{H_2O_2}$, was always from
1 to 25% greater than D_{O_2} (Fig. 5), but the general appearance was
similar (Figs. 3 and 5).

C. Half-wave Potentials

The half-wave potentials $E_{1/2}$ were measured directly on polarograms
for the two reduction steps. Figures 6 and 7 show the isopotential
curves corresponding to the first step (Eq. 1). In the area limited

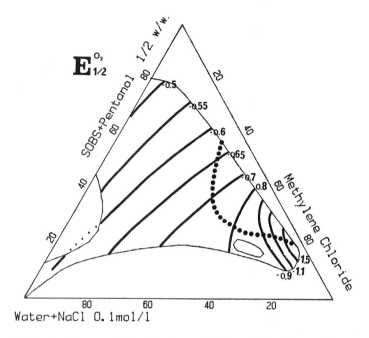

FIG. 7 Same as Fig. 6, in methylene chloride system.

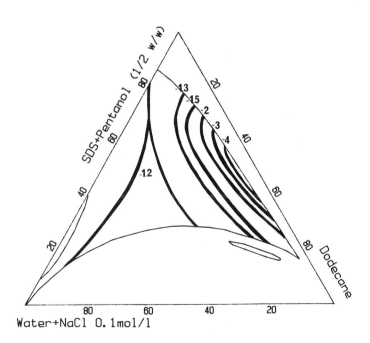

FIG. 8 Isopotential curves (V versus SCE) of the second step of oxygen reduction (H_2O_2) in dodecane microemulsion. Accuracy 5%.

by dots, the first wave presented a strong polarographic maximum (Fig. 11b); this was very surprising given the proportion of surfactant present. Figure 8 shows the isopotential curves corresponding to the second (H_2O_2) step of oxygen reduction (Eq. 2).

IV. DISCUSSION

A. Oxygen Solubility

Given the oxygen solubility in each component of both microemulsion systems (Table 1), oxygen seems to present an excess of solubility (synergy) in the microemulsions. In active blend-rich methylene chloride microemulsions, the excess of oxygen solubility can reach 200% (3.5 mmoles/liter in microemulsion and about 1.2 mmoles/liter as the sum of oxygen concentration in each component). This synergy is less important in dodecane microemulsions; the excess of oxygen solubility reaches at the most 60% (2.4 mmoles/liter in microemulsions and about 1.5 mmoles/liter as the sum in each component). In both systems, the excess of O_2 solubility disappears in oil or in water-rich microemulsions. Oxygen seems to have a preferential solubility in the interphase that incorporates the greatest part of surfactant.

Our results are consistent with those of Almgren et al. [11], who have studied in water-rich dodecane microemulsion; they have noted a local oxygen solubility in the microemulsion that is much higher than that in water. Wong et al. [12] have determined a local oxygen concentration, in water bubbles of an Aerosol OT-heptane system, two times higher than in pure water. These results were confirmed by Matheson and Rodgers [13] but questioned by Gaudin et al. [14]. In a micellar solution of SDS, Abuin and Lissi [15] have shown that the local O_2 concentration increased when the amount of alcohol added was greater, which supports our results about the preferential solubility of oxygen in the interphase.

As stated by King et al. [16,17], the observed synergy may be explained by the microemulsion structure [1]. Delpuech et al. [18, 19] have shown that the existence of microcavities between the perfluorinated molecules was the cause of the great solubility of oxygen in perfluorocarbons. The possible existence of such cavities, in the interfacial film of surfactant and alcohol, may explain the oxygen solubility in the interphase.

B. Diffusion Coefficients

Even if it is not entirely correct [27], by using the bulk viscosity η of the microemulsion (Ref. 20 for methylene chloride and Refs. 21 and 22 for dodecane microemulsions) and the diffusion coefficient D of oxygen (Fig. 3 or 4), the Stokes-Einstein law,

$$D = \frac{KT}{6\pi\eta R} \tag{4}$$

allows us to obtain a rough estimate of the radius R of the species diffusing in the bulk microemulsion. The calculated radius R spread out from 1.3×10^{-8} cm to 1.8×10^{-8} cm. The radius of molecular oxygen being 1.46×10^{-8} cm [23], and taking into account that this calculation cumulates errors in O_2 concentration and in microemulsion viscosity, the diffusing species is certainly molecular oxygen, whatever the microemulsion composition. This means that O_2 can move with great ease in the whole microemulsion, undergoing unrestricted diffusion.

As we have shown [1], the diffusion of hydrogen peroxide is different: This molecule is much more soluble in the aqueous phase than in the oil phase. Then, assuming that the whole H_2O_2 is located in the aqueous phase of the microemulsion, one can evaluate a local H_2O_2 concentration $C^W_{H_2O_2}$:

$$C^W_{H_2O_2} = \frac{C_{H_2O_2}}{dW} \tag{5}$$

where d denotes the density of the microemulsion versus water, W is the weight fraction of water, and $C_{H_2O_2}$ is the hydrogen peroxide concentration in the diffusion layer of the mercury electrode, equal to the oxygen concentration in the microemulsion (Eq. 1). Equations (3) and (5) give an apparent diffusion coefficient $D^W_{H_2O_2}$ in the aqueous phase as

$$D^W_{H_2O_2} = D_{H_2O_2} \times (dw)^2 \tag{6}$$

Figure 9 represents the map of $D^W_{H_2O_2}$, which is very different from the map of $D_{H_2O_2}$ (Fig. 5).

The plot of $D^W_{H_2O_2}$ versus $(dw)^{2.5}$, according to Mackay and Agarwal [24], was not linear for the dodecane system (Fig. 10). Since the mobility of a compound solubilized in water can be related to the mobility of water molecules [1,17], the values of $D^W_{H_2O_2}$ are related to the physicochemical structure of the dodecane microemulsion. At a very low water content, an oil continuous phase consists of microdroplets of water (L2 structure), which explains the very slight values of $D^W_{H_2O_2}$ ($< 10^{-17}$ cm^2/s). At high water content,

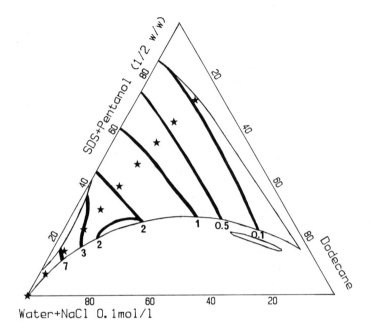

FIG. 9 Diffusion coefficient (10^{-6} cm^2/s) of H$_2$O$_2$ in aqueous phase as calculated by Eq. (5). ($\star$) Compositions corresponding to the points of Fig. 10.

the straight line obtained (Fig. 10) indicates a behavior of a concentrated suspension of polydisperse hard spheres [24]. As stated by the theory in case of L1 structure [24], the slope of the straight line (1.9×10^{-5} cm^2/s) corresponds to the diffusion coefficient of H$_2$O$_2$ in pure water [11] (1.6×10^{-5} cm^2/s). The intermediate values of $D_{H_2O_2}^W$ might correspond to a bicontinuous structure [25] (Fig. 10).

The use of the Ilkovic equation (Eq. 3) with a local concentration of $C_{H_2O_2}^W$ that is different from the bulk concentration may be questionable. The diffusion coefficients of water-soluble compounds have been studied by Mackay [8,24,27] in microemulsions. The measured diffusion coefficient D, for a species entirely in the aqueous phase in an O/W system, is given by

$$D = D^\circ(1 - \phi')^{3/2} \underset{\sim}{\sim} D^\circ(1 - a\phi_{comp})^{5/2} \tag{7}$$

where D° is the diffusion coefficient in pure water, ϕ' is the actual phase volume, ϕ_{comp} is the compositional phase volume ($= 1 - dW$), and a is a system-dependent semiempirical constant (about 1.2). Since D° is known (1.6×10^{-5} cm^2/s), ϕ' can be calculated for each measured value of D (Fig. 5), and then it should be possible to relate ϕ' to the structure. In the dotted area of Fig. 9, the calculated ϕ' values ranged from 0 (water apex) to 0.42; they were smaller than 0.66 and the ϕ_{comp} values. These two conditions correspond to a random-sphere L1 structure. For the other microemulsion compositions, the calculated ϕ' values were greater than the ϕ_{comp} values, so these compositions do not correspond to O/W system.

C. Half-wave Potentials

The half-wave potentials $E_{1/2}$ depend on the rate constant k_s of the electrochemical reduction at the electrode surface:

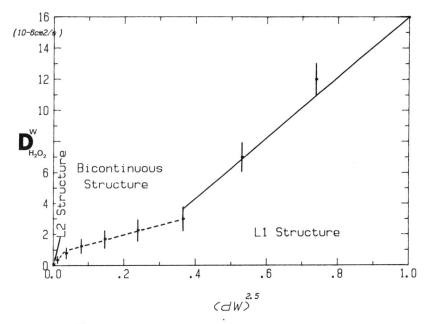

FIG. 10 $D^W_{H_2O_2}$ versus $dW^{2.5}$ for a water-dilution line of initial composition 20% dodecane and 80% w/w active blend.   Full line: L1 structure; dashed line: bicontinuous structure.

$$E_{1/2} = E_0 + \left(\frac{RT}{\alpha nF}\right) \log \left[1.35k_s \left(\frac{t}{D}\right)^{1/2}\right] \tag{8}$$

where E_0 is the standard potential of the redox couple (+0.031V versus SCE and + 1.107V versus SCE in water for the first and second step, respectively [26]; α denotes the charge transfer co-efficient (about 0.1) [26], t is the mercury drop time (2s), and D is the diffusion coefficient of the reduced species. Figure 11A shows the polarogram of O_2 in a water-rich (94%) microemulsion. The first, incomplete wave corresponds to the oxygen reduction on a naked mercury drop. The second wave corresponds to the oxygen reduction on a surfactant-covered mercury drop. We have fully reported this phenomenon elsewhere [26].

In the area bounded by dots in Figs. 6 and 7, a polarographic maximum was obtained on the first reduction wave of O_2 as illustrated by Fig. 11B. Given the characteristics of polarographic

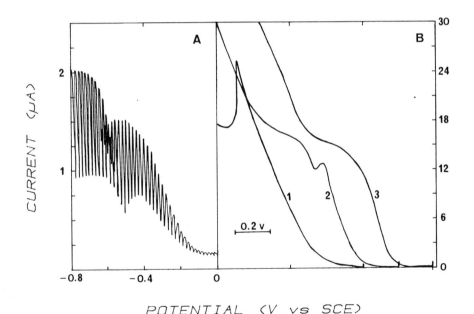

FIG. 11 A. DC polarogram of oxygen in a water-rich microemulsion. B. DC polarograms of oxygen showing the appearance of a polarographic maximum. Water content: 1, 18%; 2, 19.5%; 3, 22%. Each polarogram's start is at −0.4 V versus SCE. Mercury drop time: 2s; mercury flow: 1.04 mg/s; cathodic scan rate: 10 mV/s.

maxima [1], dots on Figs. 6 and 7 seem to delimit the L2 areas of the two systems. In dodecane microemulsions, Zradba and Clausse [6,21] have defined, by conductivity measurements, a Γ_0 line. They have indicated [21] that the Γ_0 line bounds an area in which water microdroplets (i.e., L2 structure) are present. Their Γ_0 line corresponds exactly to our dotted line of Fig. 6 in microemulsions containing more than 30% w/w dodecane. In active blend-rich microemulsions, the absence of polarographic maximum is not a point in favor of L2 structure. Bicontinuous structure [25] or even, given the water and dodecane solubility in pentanol, quaternary solution of hydrated surfactant molecules might correspond to the pentanol-rich microemulsion.

If the sudden appearance (Fig. 11B) of a polarographic maximum allows us to delimit precisely the L2 area of a microemulsion system, the polarographic study of molecular oxygen is a means to get a physicochemical structures of ionic microemulsions.

V. ACKNOWLEDGMENTS

This work was supported by the Centre National de la Recherche Scientifique, U.A. 0435, Pr. Porthault.

VI. REFERENCES

1. Berthod, A., and Georges, J., *J. Colloid and Interface Sci.* *106*, 194 (1985).
2. Lindman, B., Stilbs, P., and Moseley, M. E., *J. Colloid and Interface Sci.* *83*, 569 (1981).
3. Cazabat, A. M. and Langevin, D., *J. Chem. Phys.* *74*, 3148 (1981).
4. Lindman, B., Kamenka, N., Kathopoulis, T. M., Brun, B., and Nilson, P. G., *J. Phys. Chem.* *84*, 2485 (1980).
5. Bellocq, A. M., Biais, J., Clin, B., Lalanne, P., and Lemanceau, B., *J. Colloid and Interface Sci.* *70*, 524 (1979).
6. Boned, C., Peyrelasse, J., Heil, J., Zradba, A., and Clausse, M., *J. Colloid and Interface Sci.* *88*, 602 (1982).
7. Bendedouch, D., Chen, S. H., and Koehler, W. C., *J. Phys. Chem.* *87*, 153 (1983).
8. Epstein, B. R., Foster, K. R. and Mackay, R. A., *J. Colloid and Interface Sci.* *95*, 218 (1983).
9. Van slyke, D. D., and Neill, J. M., *J. Biol. Chem.* *61*, 523 (1924).
10. Zander, R., and Euler, D., in *Measurement of Oxygen*, Degn, H., Balslev, I., and Brook, R., (eds.), Elsevier, Amsterdam, 1976, p. 271.

11. Almgren, M., Grieser, F., and Thomas, J. K., *J. Am. Chem. Soc. 102*, 3188 (1980).
12. Wong, M., Thomas, J. K., and Grätzel, M., *J. Am. Chem. Soc. 98*, 2391 (1976).
13. Matheson, I. B. C., and Rodgers, M. A. J., *J. Phys. Chem. 86*, 885 (1982).
14. Gaudin, E., Lion, Y., and Van De Vorst, A., *J. Phys. Chem. 88*, 281 (1984).
15. Abuin, E. G., and Lissi, E. A., *J. Colloid and Interface Sci. 95*, 198 (1983).
16. Bolden, P. L., Hoskins, J. C., and King, A. D., Jr., *J. Colloid and Interface Sic. 91*, 454 (1983).
17. Matheson, I. B. C., and King, A. D., Jr., *J. Colloid and Interface Sci. 66*, 464 (1978).
18. Delpuech, J. J., Hamza, M. A., Serratrice, G., and Stebe, M. J., *Ann. Chim. France 9*, 763 (1984).
19. Hamza, M. A., Serratrice, G., Stebe, M. J., and Delpuech, J. J., *J. Am. Chem. Soc. 103*, 3733 (1981).
20. Berthod, A., and Georges, J., *J. Chim. Phys. 80*, 245 (1983).
21. Zradba, A., Thesis, Université de Pau et des pays de l'Adour, 1983.
22. Chen, J. W., and Georges, J., *personal communication*.
23. Weast, R., in *Handbook of Chemistry and Physics*, 64th ed., CRC Press, Boca Raton, Fla., 1984.
24. Mackay, R. A., and Agarwal, R., *J. Colloid and Interface Sci. 65*, 225 (1978).
25. Bennett, K. E., Hatfield, J. C., Davis, H. T., and Scriven, L. E., in *Microemulsions*, Robb, I. D. (ed.), Plenum Press, New York, 1982, p. 65.
26. Berthod, A., and Georges, J., *Anal. Chim. Acta 147*, 41 (1983).
27. Mackay, R. A., in *Microemulsions*, Robb, I. D. (ed.), Plenum Press, New York, 1982, p. 207.

21

Is the Low Interfacial Tension Observed in Two-Phase Microemulsion Systems Related to the Composition of the Adsorbed Interfacial Layer?

J. S. ZHOU, M. KAMIONER, and M. DUPEYRAT *Université Pierre et Marie Curie, Paris, France*

I. INTRODUCTION

This study shows that the low interfacial tensions observed between the microemulsion and the excess phase in two-phase systems are related to the composition of the adsorbed interfacial layer. A branched dodecyl benzene sulfonate and butanol were used as surfactant and cosurfactant and studied at the interface dodecane/water + NaCl (40 g/liter). From Gibbs' relationship, the interfacial densities of the solutes in the adsorbed layer at the critical micelle concentration (cmc) were determined for butanol varying from 0% to 6% by weight. A maximum packing and a minimum interfacial tension (10^{-2} mN/m) were observed when the ratio sulfonate/alcohol in the adsorbed layer tends toward unity. Some preliminary evaluation of free enthalpy, enthalpy, and entropy of micellization under the same conditions are reported and discussed in this chapter.

Works about the microemulsions, especially anionic systems, have been published regularly for many years. These works deal with phase diagrams and structural studies but also with interfacial tensions between the microemulsion and the excess phase in two- and three-phase systems. The low or very low interfacial tensions observed in these cases are believed to be relevant to two mechanisms. in three-phase systems, the very low tension is ascribed to a critical behavior of the microemulsion with respect to one of the excess phases [1,2]. Several authors found that dilute water/surfactant/oil two-phase systems could present low interfacial tension provided that the cmc is reached in the aqueous phase [3,4]. Moreover, Pouchelon et al. [5] point out that a low interfacial tension is observed no matter what the concentration of the micelles. They proposed that this low interfacial tension is due to the layer adsorbed at the interface. Thus, we wished to determine if the composition of the condensed, interfacial layer, just before the cmc is reached, is more or less related to the ability of such systems to give low interfacial tension and microemulsions.

The composition of the adsorbed layer for any composition of bulk phases was determined from Gibbs' relationship using excess variables defined by Hansen [6] by means of Motomura's derivation [7]. If the solvents A and B are immiscible, the following expression is obtained at constant pressure and temperature:

$$\left(\frac{\partial \gamma}{\partial x_i^A}\right)_{T,P,x_j^A} = -\sum_{j=1}^c \Gamma_i^H \left(\frac{\partial \mu_j^A}{\partial x_i^A}\right) \qquad i = 1, \ldots, c$$

where γ is the interfacial tension, Γ_i^H is the interfacial density according to Hansen, μ_i is the chemical potential, and x_i^A is the mole fraction in phase A of ith$_2$ component.

II. MATERIAL AND METHODS

The surfactant was the 4p-phenylsulfonate dodecane sodium salt, a branched dodecyl benzene sulfonate, referred to in the following as sulfonate. The cosurfactant was n-butanol (Merck 99.7%). Their interfacial properties were investigated at the interface dodecane/water with NaCl (40 g/liter), a NaCl concentration slightly larger than the optimal salinity according to Reed and Healy [8]. Dodecane (99%) supplied by SDS was purified by removing polar impurities with alumina so that the interfacial tension dodecane/water was 51 mN/m. The temperature was 25 ± 1°C. The composition of the samples was such that the percentages (in weight) of the aqueous phase and dodecane were always the same. The percentage of butanol and the initial concentration of sulfonate varied. The measurements were performed at equilibrium, after stirring for 24 h. The interfacial tensions were measured either by a detachment method or by the spinning drop method. Under these conditions, the hypothesis of inmiscibility of the solvents holds because their mutual solubility is very low and the variations of activity coefficient are negligible in the solute concentration range that was studied. Moreover, we can assume that there is no change in activity coefficients of alcohol with surfactant concentration and in activity coefficients of sulfonate with alcohol concentration because alcohol is a nonelectrolyte and the surfactant concentrations are very low (10^{-6} to 10^{-4} mol/liter). Besides, the adsorption of NaCl is considered negligible, and $\Gamma_{NaCl} = 0$. Thus, the equations are simplified to:

$$\Gamma_{sulf} = - \frac{1}{RT} \left(\frac{\partial \gamma}{\partial \ln a_{sulf}^A} \right)_{T, P, x_{alc}^A, NaCl}$$

$$\Gamma_{alc} = - \frac{1}{RT} \left(\frac{\partial \gamma}{\partial \ln a_{alc}^A} \right)_{T, P, x_{sulf}^A, NaCl}$$

Therefore the Γ_i's are evaluated from the slopes of the $\ln a_i^A$ curves.

III. EXPERIMENTAL RESULTS

The variations of the interfacial tension γ as a function of the initial aqueous concentration of sulfonate are plotted in Fig. 1 for

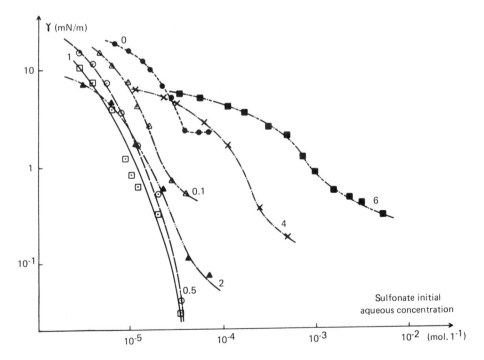

FIG. 1 Variations in interfacial tension as a function of the log of sulfonate initial concentration for various concentrations in butanol: ● 0%; △ 0.1%; ⊙ 0.5%; ▫ 1%; ▲ 2%; X 4%; ■ 6%.

various concentrations in butanol. The minimum interfacial tension and the cmc decrease from 0% to 1% butanol, reaching 2×10^{-2} mN/m when the sulfonate concentration is about 3×10^{-5} mol/liter. Then, the interfacial tension increases and reaches 0.8 mN/m at a sulfonate concentration of 4×10^{-3} mol/liter when the butanol concentration is 6%. Thus we can observe low interfacial tensions at very low sulfonate concentration.

IV. DISCUSSION

The aqueous phase was chosen as the reference A phase, and the aqueous concentration of the surfactant and cosurfactant were measured by ultraviolet spectroscopy and gas chromatography, respectively. The distribution coefficients of the two solutes are reported in Table 1. The plots of γ/log sulfonate aqueous concentration and γ/log butanol aqueous activity are represented in Figs. 2a and 2b, respectively. The curves γ/log sulfonate concentration, in the concentration range we investigated, are straight lines, meaning that the adsorbed layers are condensed as far as possible. Thus we calculated the interfacial Γ_i's of the two solutes in such layers for various concentrations of butanol. The results are plotted in Fig. 3. It is interesting to note that up to 1% of butanol, Γ_{sulf} is constant while $\Gamma_{butanol}$ is increasing. That means that a layer of sulfonate condensed as far as possible is really not very condensed, since alcohol molecules can penetrate it. Indeed, the area per molecule of sulfonate alone is 0.39 nm^2, in agreement with the area per molecule at the cmc, 0.415 nm^2, found by Pethica [9] with sodium dodecyl sulfate at the interface decane/water + 6% NaCl. This area is large compared to the cross section of the alkyl chain, which is about 0.20 nm^2. This can be explained by the repulsions between the

TABLE 1 Distribution Coefficient Between Dodecane and Aqueous Phase of Sulfonate and Alcohol for Various Concentrations in Alcohol

% alcohol	0.1	0.5	1	2	4	6
K^{dod}_{eau} sulfonate	<0.1	0.4	1	5	~60	400-600
K^{dod}_{eau} butanol	0.15	0.15	0.17	0.22		0.92

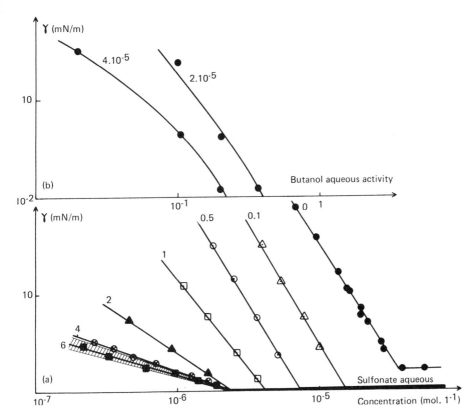

FIG. 2 (a) Interfacial tension as a function of the log of sulfonate aqueous concentration. (b) Interfacial tension as a function of the log of butanol aqueous activity.

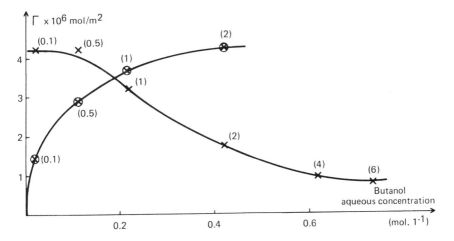

FIG. 3 Variation in interfacial excess as a function of butanol aqueous concentration: X sulfonate; ⊗ alcohol.

sulfonate ions. Thus, the penetration of alcohol comes to a screen
effect that favors an increase in density and a lowering of interfacial
tension. Then the layer becomes really condensed, and more butanol
can penetrate the interface only if the sulfonate molecules leave it.
But the interfacial activity of butanol is lower than that of sulfonate,
and the interfacial tension decreases. In Table 2 we summarize the
interfacial densities of surfactant and cosurfactant; the composition
of the condensed adsorbed layers expressed as the mole fraction of
alcohol; and the mean area as a function of the butanol concentration. In order to get more information about the packing of molecules in the interface, we plotted (Fig. 4) the mean area as a function of the mole fraction of butanol in the adsorbed layer. We can
observe a strong deviation from ideality, which is a maximum for
0.5, that is, a ratio alcohol/sulfonate = 1. We also reported, as a
function of the alcohol mole fraction, the logarithm of the lowest
tension we measured. This tension also tends to a minimum for 0.5,
that is, for a maximal packing. It is possible that this strong modification in the packing of the molecules in the adsorbed layer is related to some interaction between them. Micellization enthalpy (ΔH)
measurements could give us that information. Calorimetric measurements are difficult, due to the very low concentration (10^{-5} mol/
liter). We performed only some preliminary evaluation from classical
thermodynamics:

$$\frac{\partial}{\partial 1/T} \left(\frac{\Delta G}{T} \right) = \Delta H$$

where ΔG is the micellization free enthalpy. Thus, the slope of the
curve $\Delta G/T$ as a function of $1/T$ gives ΔH. The problem is to evaluate ΔG. The basic equation, from the mass action law model, is

$$\frac{\Delta G}{RT} = \ln \text{ cmc} + \frac{m}{n} \ln (X^-)$$

where (X^-) is the counterion activity, m is the number of fixed
counter ions, and n is the aggregation number. The second term
is negligible with respect to the first, and we used, as a first approximation for preliminary experiments, the equation

$$\frac{\Delta G}{T} = R \log \text{ cmc}$$

The experimental results are plotted in Fig. 5 for two extreme
cases, 0% and 3% butanol, and three temperatures, 25, 35, and 45°C.
They each show that the cmc varies with temperature in inverse
order, depending on whether the butanol concentration is 0% or 3%.
The values of ΔG, ΔH, and ΔS (the micellization entropy) deduced

TABLE 2 Γ_i = interfacial excess $\times 10^6$ mol/m^2; x_{alc} = mole fraction of alcohol in the adsorbed layer; A = area per molecule (Å^2)

% buta--nol	0	0.1	0.5	1	2
Γ but.	0	1.5	3	3.7	4.2
Γ sulf.	4.2	4.2	4.2	3.2	1.75
x alc.	0	0.26	0.42	0.54	0.7
A	0.39	0.29	0.23	0.24	0.28

from these experiments, assuming no change in the distribution coefficients with increasing temperature, are reported in Table 3. Note that these values result from preliminary and rough measurements and have to be considered cautiously. However, we can make some observations. They are high with respect to the values reported in

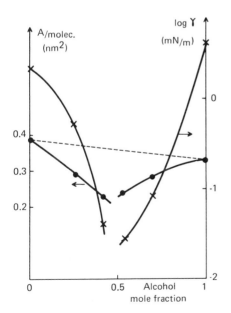

FIG. 4 • mean area per molecule as a function of alcohol more fraction in the adsorbed layer; X $\log_{10}$ of the minimum interfacial tension.

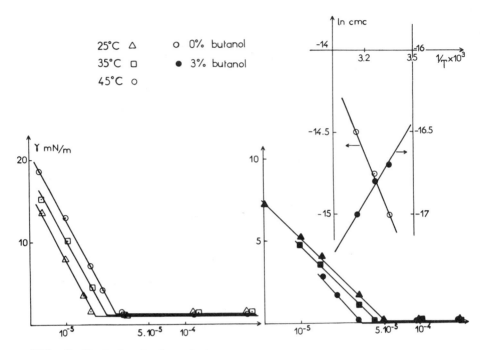

FIG. 5 Variation in interfacial tension as a function of the sulfonate concentration at Δ 25°C, □ 35°C, ○ 45°C. Open symbols are for 0% butanol, closed symbols are for 3% butanol.

other studies (−5 to +5 kJ/mol) for micellization enthalpies in water [10]. We would like to compare them to the ones reported by Rosano et al. [11] for other two-phase systems, obtained by a very different approach. These authors determined the free enthalpy "accompanying the adsorption [of cosurfactant] at the interface during the

TABLE 3 Micellization Free Enthalpy, Enthalpy, Entropy at 0% and 3% Butanol

Butanol	ΔG kJ mol^{-1}	ΔH kJ mol^{-1}	ΔS kJ mol^{-1}K^{-1}
0%	− 37	− 18,5	+ 0,062
3%	− 41	+ 12	+ 0,177

transformation of [a] primary coarse emulsion into a transparent system" (that is, a microemulsion). The reported values of enthalpy and entropy vary from -25 to $+18.7$ kJ/mol and from -18 to $+82$ J/mol K, respectively. It can indeed be noticed that the cmc in water is different from the cmc in water saturated with dodecane. But this slight difference cannot account for the difference in the enthalpies. Thus, some interaction with dodecane should be taken into account, probably due to the penetration of dodecane into sulfonate and alcohol chains. The entropic term would be responsible only for micellization with 3% butanol, meaning the absence of attraction between surfactant chains due to the numerous short alcohol chains, while such an attraction between long chains could be added to an entropic term, less important than previously because of repulsion between polar heads, with 0% butanol.

V. CONCLUSION

Low interfacial tensions observed in some two-phase systems seem to be due to the composition of the interface. There is a clear correlation between a compact packing in the interface and a low interfacial tension. The maximum effect is observed for a ratio alcohol/surfactant about unity, which corresponds just to a sulfonate distribution coefficient unity. This means that under these conditions, the chemical potential of sulfonate is the same in each phase, as previously mentioned by Shah [3] for other systems, which would favor a critical behavior. To relate the composition of the interface to a type of microemulsion is very difficult when the sulfonate concentration is so close to the cmc. We can only observe (Fig. 1) that while the cmc is well marked for low alcohol concentrations corresponding to direct micelles and to oil/water (o/w) microemulsion, a slow variation of γ with the sulfonate concentration is observed at higher alcohol concentrations, suggesting the presence of inverse micelles and the formation of W/O microemulsion.

VI. ACKNOWLEDGMENTS

The authors thank the Institut Francais du Pétrole for financial support.

VII. REFERENCES

1. Bellocq, A. M., Bourbon, D., Fourche, G., and Lemanceau, B., *Adv. Colloid Interface Sci. 20*, 261 (1984).

2. Cazabat, A. M., Langevin, D., Meunier, J., and Pouchelon, A., *Adv. Colloid Interface Sci. 16*, 175 (1982).
3. Chan, K. S., and Shah, D. O., *J. Dispersion Sci. Technol. 1*, 55 (1980).
4. Morgan, J. C., Schechter, R. S., and Wade, W. H., in *Improved Oil Recovery by Surfactant and Polymer Flooding*, Shah, D. O., and Schechter, R. S. (eds.), Academic Press, New York, 1977, p. 101
5. Pouchelon, A., Chatenay, D., Meunier, J., and Langevin, D., *J. Colloid and Interface Sci. 82*, 418 (1981).
6. Hansen, R. S., *J. Phys. Chem. 66*, 410 (1962).
7. Motomura, K., *Adv. Colloid & Interface Sci. 12*, 1 (1980).
8. Reed, R. L., and Healy, R. N., in *Improved Oil Recovery by Surfactant and Polymer Flooding*, Shah, D. O., and Schechter, R. S. (eds.), Academic Press, New York, 1977, p. 383.
9. Pethica, B. A., *Trans. Faraday Soc. 50*, 413 (1954).
10. Lindman, B., and Wennerström, H., in *Topics in Current Chemistry No. 87*, Springer Verlag, 1980, p. 16.
11. Rosano, H. L., Cavallo, J. L., and Lyons, G. B., *this volume*.

22

Mechanism of Formation of Water-in-Xylene Microemulsions and Their Characterization

R. C. BAKER* and TH. F. TADROS   *Jealotts Hill Research Station, Bracknell, Berkshire, United Kingdom*

I. INTRODUCTION

Interfacial tension measurements using both the Wilhelmy plate and spinning drop techniques showed that the major factors responsible for microemulsion formation are fluctuation in interfacial tension, a negative transient interfacial tension, and formation of a "fluid"-like boundary with some liquid crystalline phase formation. Under these conditions, spontaneous expansion of the interface occurs with the formation of small droplets. The droplet size of the resulting microemulsions can be determined from time-averaged light-scattering measurements using a hard-sphere model to calculate the structure factor (for concentrated dispersions). The results showed that, at any given surfactant concentration, the droplet radius increases with increase of volume fraction of water, ϕ_{H_2O}. The radius estimated

from the interfacial area, assuming a molar ratio of hexanol to sodium dodecyl benzene sulfonate of 2:1 at the interface, was in good agreement with the value obtained from light scattering.

The term microemulsion has been adopted to describe isotropic and thermodynamically stable systems consisting of water, oil, and amphiphile (which may be a single surfactant or a combination of surfactants) [1]. There has been much debate about the use of the term microemulsion to describe such systems, since structural studies [2] indicated that in many systems of bicontinuous solutions, i.e., both oil and water, continuous and easily deformable and

---

*Present address: Coopers Animal Health Ltd., Berkhampsted, Herts, United Kingdom.

flexible interfaces exist. In other words, in many systems large
aggregates are not present. However, in many other microemulsion
systems, distinct separation into hydrophilic and hydrophobic domains exist, with definite "cores" surrounded by surfactant molecules [2]. It must then be acknowledged that the term microemulsion indeed covers a wide range of systems, which may vary significantly in their structure.

Having now recognized the problem of definition, the next points
that must be considered are the following. First, what is the driving force for the thermodynamic stability of microemulsions? The
answer to this question has been adequately put forward by Ruckenstein and collaborators [3] and Overbeek and his co-workers [4,
5]. These authors considered the free energy of formation of microemulsions, ΔG_m, which was considered to consist of three main contributions: ΔG_1, an interfacial free energy term; ΔG_2, an energy
of interaction between the droplets term; and ΔG_3, an entropy term
accounting for the dispersion of the droplets into the continuous
phase. The main conclusion from these theoretical treatments is
that a low interfacial tension, $\gamma_{o/w}$, appears to be essential for the
spontaneous formation of a microemulsion. This results in a low interfacial free energy term, which is compensated for by the negative
entropy term for dispersion of the droplets into the continuous phase,
thus resulting in a net negative free energy for formation of the
microemulsion.

The second and most important question to answer about microemulsions is the mechanism of formation. For that purpose, interfacial tension measurements are required under conditions where
microemulsions are formed. This is the purpose of the present study,
in which we have combined the Wilhelmy plate technique with the
spinning drop method to gain some insight into the mechanism of
microemulsification. Finally, the resulting water-in-xylene microemulsions were characterized using time-averaged light scattering and
the average droplet size compared with that calculated from the
interfacial area (which can be obtained from a knowledge of the
amount of surfactant at the interface assuming reasonable values for
the area per molecule).

II. EXPERIMENTAL

A. Materials

Water was twice distilled in an all-glass apparatus. Xylene was an
Analar-grade material (ex. Koch Light) and used as received. The
surfactant was a commercial product, Nansa 1106, supplied by Al-
bright and Wilson (Whitehaven, U.K.). This material, which is
usually referred to as sodium dodecyl benzene sulfonate (NaDBS)

has been purified by the method described before [6]. The linear hydrocarbon chain consists of a distribution from C_9 to C_{14} and its composition was given before [6]. n-Hexanol was a B.D.H. material and was used as received.

B. Interfacial Tension Measurements

Two techniques were used for interfacial tension measurements. The first method was the Wilhelmy plate technique, in which a roughened platinum plate was used in conjunction with a microbalance (CI Robal Electronics, Salisbury, U.K.) as described before [7]. For measurement of low interfacial tension (<2 mN m^{-1}), a spinning drop apparatus constructed by Bailey Engineering (Windsor, U.K.) was used. The apparatus was essentially the same as described by Cayias et al. [8]. The basic procedure was described before [7]. Basically, a droplet contained in a capillary tube and equilibrated with the solution under study is rotated at various speeds, resulting in its deformation from spherical shape. When the ratio of the length of the droplet to its width is greater than 4, achieved by adjusting the speed of the motor, the interfacial tension can be calculated from the simple equation given by Vonnegut [9],

$$\gamma = \frac{(\rho_1 - \rho_2)\omega^2 r^3}{4} \tag{1}$$

where ρ_1 and ρ_2 are the densities of the aqueous and oil phase, respectively, r is the drop radius in centimeters, and ω is the angular velocity in radians per second.

C. Light-Scattering Measurements

A Sofica Photo-Gonio Diffusometer (Model 42,000) instrument was used for light-scattering measurements. The technique was described in detail before [10]. The intensity of scattered light at 90° to the direction of the incident beam is measured, allowing one to calculate the Rayleigh ratio, R_{90}. The latter is related to the molecular mass of the scattering units and their concentration C by the equation

$$R_{90} = K_0 \, MC \, P(90) \, S(90) \tag{2}$$

where K_0 is an optical constant, $P(90)$ is the particle form factor, and $S(90)$ is the structural factor, both at an angle of 90°. Since $M = (4/3)\pi R^3 \rho N_A$, where R is the radius of the scattering unit, then

$$R_{90} = K_1 \, \phi_{H_2O} \, R^3 \, P(90) \, S(90) \qquad (3)$$

Where $R \ll \lambda_0$ (the wave length of incident light), $P(90)$ is approximately equal to unity and, therefore, Eq. (3) reduces to

$$R_{90} = K_1 \, \phi_{H_2O} \, R^3 \, S(90) \qquad (4)$$

Thus, R can be calculated from R_{90} using Eq. (4), provided that $S(90)$ can be estimated as a function of ϕ_{H_2O}. $S(90)$ was calculated using the hard-sphere model suggested by Percus and Yevic [11] and applied by Ashkroft and Lekner [12]. Details of the calculations were given before [10].

III. RESULTS AND DISCUSSION

A. Interfacial Tension Measurements

1. Dynamic Measurements

Figure 1 shows the $\gamma_{o/w} - t$ curves for 0.05 and 0.5 wt% n-hexanol in xylene against 5×10^{-4} mol dm^{-3} NaDBS. In both cases $\gamma_{o/w}$ increases slowly with time and finally reaches a constant value. This points toward some slow diffusion of the NaDBS into the oil

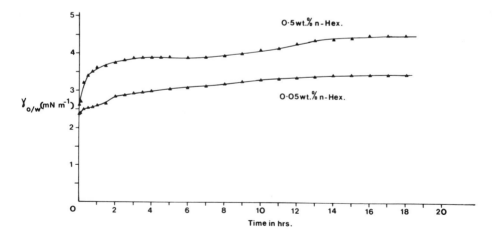

FIG. 1 $\gamma_{o/w} - t$ curves for 0.05 and 0.5 wt% hexanol in xylene against 5×10^{-4} mol dm^{-3} NaDBS.

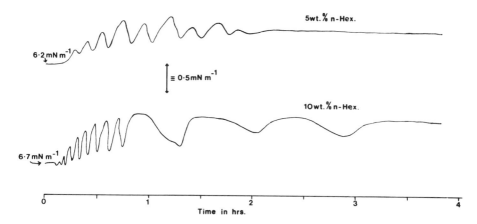

FIG. 2 $\gamma_{o/w}$ – t curves for 5 and 10 wt% hexanol in xylene against 5×10^{-4} mol dm^{-3} NaDBS.

(note that NaDBS is soluble in xylene). When the NaDBS concentration was increased to 10^{-3} mol dm^{-3} (i.e., above the critical middle concentration (cmc) of the surfactant) and the n-hexanol concentration was maintained below 3%, such time variation disappeared. This may be taken as an indication that the presence of NaDBS micelles enhances diffusion from one phase to the other. However, when the n-hexanol concentration in the oil phase was increased above 3%, considerable fluctuations in $\gamma_{o/w}$ were observed. This is illustrated in Fig. 2, which shows that $\gamma_{o/w}$ passes through a series of maxima and minima, at least during the first 2 to 3 h. When the NaDBS concentration was increased to 10^{-3} mol dm^{-3}, such oscillations were damped considerably. No measurements could be made using the Wilhelmy plate technique above 10^{-3} mol dm^{-3}, as the values of γ were too low. These fluctuations suggest instability of the interface, resulting from the disordering of the mixed film, and/or exchange of NaDBS and n-hexanol between oil and water phases. Such instability, which occurs at high n-hexanol concentrations, may be responsible for the interface expansion during microemulsion formation.

2. Equilibrium Measurements

Equilibrium measurements of $\gamma_{o/w}$ were made at low NaDBS concentration ($\leqslant 10^{-3}$ mol dm^{-3}) using both Wilhelmy plate and spinning drop techniques (Fig. 3) and at high NaDBS concentrations using the spinning drop technique only (Fig. 4). Figure 3 shows that the results obtained using the spinning drop technique are significantly lower than these using the Wilhelmy plate technique, but the

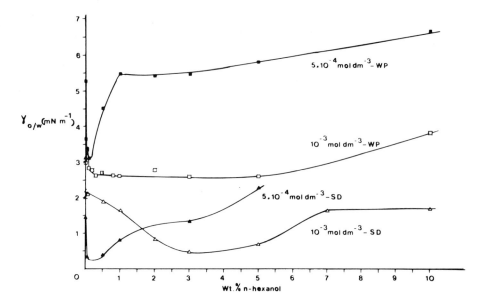

FIG. 3 $\gamma_{o/w}$ – $C_{hexanol}$ using the Wilhelmy plate (WP) and spinning drop (SD) techniques at 5×10^{-4} and 10^{-3} mol dm^{-3} NaDBS.

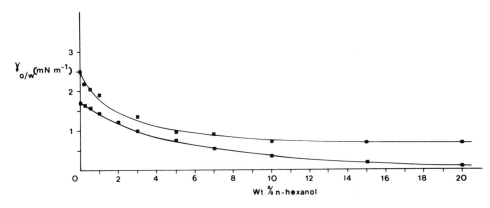

FIG. 4 $\gamma_{o/w}$ – $C_{hexanol}$ using the spinning drop technique at 10^{-2} (top curve) and 10^{-1} mol dm^{-3} (bottom curve) NaDBS.

general trend is the same. This variation between the two methods is due to the different procedures adopted for sample preparation in the two types of measurements [7]. At 5×10^{-4} mol dm^{-3}, the results show an initial drop in $\gamma_{o/w}$ reaching a minimum of ~ 3 mN m^{-1}, using the Wilhelmy plate, and ~ 0.3 mN m^{-1} using the spinning drop technique, after which $\gamma_{o/w}$ increases sharply with increase of n-hexanol concentration, but above 1% n-hexanol, $\gamma_{o/w}$ increases much more slowly. The minimum in $\gamma_{o/w}$ may be due to the coopera-tive adsorption of n-hexanol at the O/W interface and, moreover partitioning of some of the n-hexanol into the aqueous phase would cause a reduction in the cmc of NaDBS [13]. The cooperative ad-sorption of NaDBS and n-hexanol results in further lowering of $\gamma_{o/w}$, particularly when the two surfactants are different in nature [4]. However, with further increase in the n-hexanol concentra-tion, partitioning of NaDBS into the oil phase may take place, thus reducing its activity in the aqueous phase. Thus, the contribution of surface excess from NaDBS is reduced and as a result $\gamma_{o/w}$ in-creases. This explains the increase in $\gamma_{o/w}$ with increasing n-hexanol concentration after reaching a minimum. As the NaDBS concentration is increased to 10^{-3} mol dm^{-3}, $\gamma_{o/w}$ shows a more gradual decrease with increase in n-hexanol, producing only a shal-low minimum. This is due to the fact that one is above the cmc of the surfactant, and hence any partitioning of NaDBS into the oil phase has a relatively smaller effect on $\gamma_{o/w}$ than the case below the cmc. At NaDBS concentrations, well above the cmc (10^{-2} and 10^{-1} mol dm^{-3}, Fig. 4), $\gamma_{o/w}$ decreases slowly with increase in hexanol concentration, reaching a constant value of ~ 0.7 mNm^{-1} at n-hexanol $\geqslant 10\%$ for 10^{-2} mol dm^{-3} NaDBS and ~ 0.1 mN m^{-1} at n-hexanol $\geqslant 15\%$ for 10^{-1} mol dm^{-3} NaDBS. The latter concentration was the highest that could be reached while maintaining a stable droplet in the spinning drop apparatus. Above 10^{-1} mol dm^{-3}, the droplets became unstable. This is illustrated in Fig. 5, which shows photographs of the drops at various NaDBS concentration while maintaining the n-hexanol concentration at 10%. It is clear that be-yond 10^{-1} mol dm^{-3} NaDBS, the droplet became unstable forming a viscous gel (Figs. 5f and 5g). An attempt was made to inject the droplet while the tube was spinning, but the droplet broke into a large number of small droplets (Fig. 5h).

B. Injection Experiments

These experiments were carried out using the Wilhelmy plate techni-que, following a similar procedure as described by Rosano and co-sorkers [14,15]. In the first experiment, 90 cm^3 of xylene were placed on 90 cm^3 of water and 10 cm^3 of n-hexanol were injected as one aliquot in ~ 1 min into the xylene phase. The $\gamma_{o/w}$ was moni-tored as a function of time using the Wilhelmy plate-microbalance

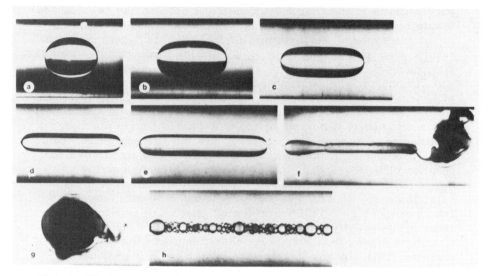

FIG. 5 Photographs of the drops of n-hexanol in xylene formed at various NaDBS concentrations (mol dm^{-3}): (1) 10^{-4}; (b) 5×10^{-4}; (c) 10^{-3}; (d) 10^{-2}; (e) 10^{-1}; (f) 3×10^{-1}; (g) 5×10^{-1}; (h) 5×10^{-1} (drop injected while tube spinning).

arrangement. The results showed a great deal of fluctuations similar to those described in Fig. 2 but with much higher amplitude, which continued for ~ 0.5 min, after which they disappeared and eventually a constant value was reached. The interfacial tension was reduced from 36.4 to 12.9 mN m^{-1}. Similar fluctuations were produced when the n-hexanol was injected into the water phase, although in this case the xylene became turbid, indicating that some water molecules had partitioned into the oil phase. In this case the interfacial tension was reduced from 36.5 to 7.2 mN m^{-1}.

In the second experiment, 10 cm^3 of n-hexanol were injected into 90 cm^3 of xylene which was in equilibrium with 90 cm^3 of 3×10^{-1} mol dm^{-3} NaDBS. Immediately after the injection of n-hexanol, $\gamma_{o/w}$ fell from its initial value of ~ 0.2 mN m^{-1} to a small negative value of ~ -0.5 mN m^{-1}. No fluctuations in $\gamma_{o/w}$ were observed in this case (the time duration of fluctuations was probably too short). It should be mentioned that this small negative interfacial tension, which is difficult to measure, is only approximate and was a transient measurement. When the system was left standing, $\gamma_{o/w}$ became positive although it was too small to measure. Moreover, a blue tinge was observed at the aqueous phase of the interface, whereas a liquid crystalline phase formed at the oil side of the interface. Such fluctuations, transient negative interfacial tension, and liquid crystalline phase formation all give a picture of micro-emulsion formation, in which partitioning of oil, water, and surfactant occurs between the two phases, with a relatively diffuse and disordered interface being the main cause of low interfacial tension and expansion of the interface. Miller et al. [16] have observed liquid crystalline phases when an aqueous petroleum sulfonate solution was brought into contact with an oil. It is concluded from these injection experiments that microemulsion formation is spontaneous and, when the conditions are right, does not require any mechanical work.

C. Characterization of the Microemulsion Using Light Scattering

Figure 6 shows plots of R_{90} versus the volume fraction of water, ϕ_{H_2O} for the four NaDBS concentrations of the samples studied.

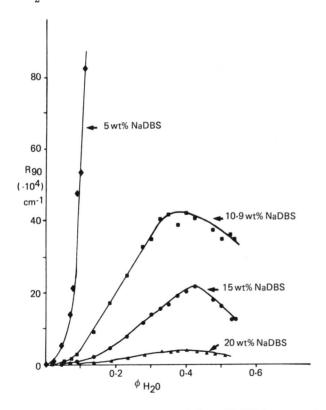

FIG. 6 R_{90} versus ϕ_{H_2O} at four NaDBS concentrations.

Except for the 5% NaDBS curve, all other systems show a maximum in the $R_{90}-\phi_{H_2O}$ curves. Similar results were obtained by Agterof et al. [17] for the system water benzene/potassium oleate/n-hexanol. As mentioned before, for calculation of the droplet radius R from the Rayleigh ratio one needs to calculate the structure factor S(90) [see Eq. (4)]. The latter was calculated using the hard-sphere model as given by Ashkroft and Lekner [12].

$$S(Q) = \frac{1}{1 - NC(2QR_{HS})} \tag{5}$$

where N is the number of particles per cubic centimeter, Q is the scattering vector, and C is a function of the hard-sphere radius R_{HS}, and the hard-sphere volume fraction ϕ_{HS}. The hard-sphere radius R_{HS} is related to the core radius R by the equation

$$R_{HS} = R + t \tag{6}$$

where t is the thickness of the surfactant layer, which was taken to be in the region of 3 to 8 Å. Details of the calculations were given before [10], and the results of R versus ϕ_{H_2O} at the four NaDBS concentrations studied are shown in Fig. 7. These results show an increase in R with increase in ϕ_{H_2O}, as expected, since the surfactant concentration was kept constant in each series. It also shows that R decreases with increase of NaDBS concentration, as expected. The droplet radius at any given surfactant concentration can be calculated if one is able to estimate the interfacial area A (m^2/kg^{-1}), since

$$A = \frac{3}{R\rho} \tag{7}$$

where ρ is the density $(kg\ m^{-3})$ and R is given in meters. The interfacial area A can be estimated if one knows the amount of surfactant at the interface and the area per molecule. If all surfactant molecules were adsorbed at the interface,

$$A = \frac{n_s N_A A_s + n_a N_A A_a}{\phi_{H_2O}} \tag{8}$$

where n_s and n_a are the number of moles of surfactant and alcohol per kilogram of microemulsion and A_s and A_a are the areas per

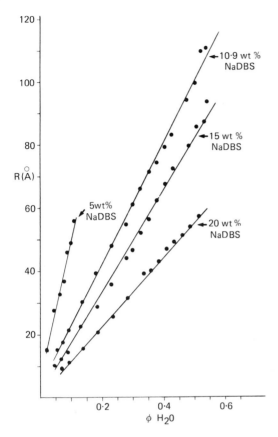

FIG. 7 Variation of droplet radius (R) with ϕ_{H_2O} at four NaDBS concentrations.

molecules occupied by each respectively. Using reasonable values of the area per molecule of 30 and 20 $Å^2$ for NaDBS and n-hexanol, respectively, R was calculated using Eqs. (7) and (8). The results showed that the calculated R from the interfacial area was significantly lower than that determined from light scattering. However, when it was assumed that all NaDBS was adsorbed, whereas n-hexanol is only partly adsorbed to give a molar ratio of hexanol to NaDBS of 2:1 as compared to a molar ratio of 3.24:1 in the microemulsion, the calculated R values were in reasonable agreement with those estimated from light-scattering measurements.

IV. REFERENCES

1. Danielsson, I., and Lindman, B., *Colloids and Surfaces 3*, 391 (1981).
2. Lindman, B., Stilbs, B., and Mosely, M. E., J. *Colloid and Interface Sci. 83*, 569 (1981).
3. Ruckenstein, F., and Chi, J. C., *J. Chem. Soc. Faraday Trans. 71*, 1690 (1975); Ruckenstein, E., J. *Colloid and Interface Sci. 66*, 369 (1978); *Chem. Phys. Lett. 57*, 517 (1978); Ruckenstein, E., and Krishnan, R., *J. Colloid and Interface Sci. 71*, 321 (1979); *75*, 476 (1980); *76*, 188, 201 (1980).
4. Overbeek, J. Th. G., *Faraday Dis. Chem. Soc. 65*, 7 (1978).
5. Overbeek, J. Th. G., de Bruyn, P. L., and Verhoeckx, F., in *Surfactants*, Tadros, Th. F. (ed.), Academic Press, London, 1984, p. 111.
6. Baker, R. C., Florence, A. T., Tadros, Th. F., and Wood, R. M., *J. Colloid and Interface Sci. 100*, 311 (1984).
7. Baker, R. C., and Tadros, Th. F., in *Surfactant Solution*, Mittal, K. L. (ed.), Plenum Press, New York, 1986, *in press*.
8. Cayias, J. L., Schechter, R. S., and Wade, W. H., *ACS Symp. Ser. 8*, 234 (1975).
9. Vonnegut, R., *Rev. Sci. Instrum. 13*, 6 (1942).
10. Baker, R. C., Florence, A. T., Ottewill, R. H., and Tadros, Th. F., *J. Colloid and Interface Sci. 100*, 332 (1984).
11. Percus, K. J., and Yevic, G. J., *Phys. Rev. 110*, 1 (1958).
12. Ashkroft, N. W., and Lekner, J., *J. Phys. Rev. 145*, 83 (1966).
13. Zana, R., Tio, S., Strazielle, C., and Lianos, P., *J. Colloid and Interface Sci. 80*, 208 (1981).
14. Rosano, H. L., *J. Soc. Cosmet. Chem. 25*, 609 (1974).
15. Gerbacia, W., and Rosano, H. L., *J. Colloid and Interface Sci. 44*, 142 (1974).
16. Miller, C. A., Hawen, R., Benton, W. J., and Tommlinson, F., *J. Colloid and Interface Sci. 61*, 554 (1977).
17. Agterof, W. G. M., van Zomeren, J. A. J., and Vrij, A., *Chem. Phys. Lett. 43*, 363 (1976).

23

Formamide as a Water Substitute IX: Waterless Microemulsions 6; A New Type of Water-Insoluble Surfactants and Nonaqueous Microemulsions

I. RICO and A. LATTES   *Université Paul Sabatier, Toulouse, France*

I. INTRODUCTION

Molecular aggregation phenomena in structured nonaqueous solvents have received little attention compared to the number of investigations carried out in water.

We demonstrate here for the first time surfactant properties of water-insoluble molecules (long-chain phosphonium salts) in a structured nonaqueous solvent (formamide).

We also demonstrate that microemulsions can be prepared in formamide instead of water with various surfactants, particularly with phosphonium salts for which it is impossible to prepare aqueous microemulsions. Winsor systems for these waterless microemulsions will be discussed.

These results show the general nature of micellization and microemulsion obtainment.

Molecular aggregation phenomena is structured nonaqueous solvents have received little attention compared to the number of investigations carried out in water. But water's usual properties as a solvent may not be unique. If hydrophobic interactions are required for certain systems (micellar solutions or microemulsions), solvophobic bonding tendencies analogous to water can be found in other solvents. By studying the effect of water and other solvents on the structure of biopolymers, Sinanoglu and Abdulnur [1] have determined a descending order of solvophobic force for pure solvents:

water > (glycerol, formamide) ≫ glycol

$$> \frac{\text{(n-butanol, methanol}}{\text{n-propanol, ethanol)}} > \text{t-butanol}$$

TABLE 1 Physicochemical Characteristics of Water and Formamide
(25°C)

Formula	Boiling point (°C)	Freezing point (°C)	$\varepsilon°$	μ (D)	Surface tension (mN m^{-1})
H_2O	100	0	80	1.80	70
$H-C\diagup^{O}_{\diagdown NH_2}$	210	2.5	110	3.37	58.5

The same order was obtained by Wacker and Lodemann [2] in a
study of the influence of the interfacial energy of organic solvents
on the photochemical dimerization of thymidylyl (3'-5')-thymidine.
Therefore, the comparison of physical properties of solvents shows
that formamide is quite similar to water. For example, some physicochemical characteristics of these both solvents are summarized in
Table 1. It therefore occurred to us that it should be possible to
study molecular aggregation phenomena (micellar solutions and microemulsions) in formamide rather than water.

II. MICELLAR SOLUTIONS IN FORMAMIDE

As shown in Table 2, in contrast to water, formamide solvates large
ions preferentially. We have therefore carried out a study of molecular aggregation of large ions that are insoluble in water but
that can aggregate in formamide. We chose the long-chain substituted phosphonium salts (2), which were prepared by heating tri-
phenylphosphine and various alkyl or fluoroalkyl iodides (1) to
95°C in the absence of solvents:

$$(C_6H_5)_3P + ICH_2CH_2R \xrightarrow{95°C} (C_6H_5)_3\overset{+}{P}CH_2CH_2R,I^-$$

$$(1) \qquad\qquad (2) \qquad\qquad \text{yields} = 60-90\%$$

a: R = C_4F_9; b ÷ R = C_6F_{13}; c: R = C_8F_{17}
d: R = C_4H_9; e ÷ R = C_6H_{13}; f ÷ R = C_8H_{17}

the phosphonium salts (2) are insoluble in water even at high
temperatures. They are soluble, however, in formamide, and

micellization was observed by conductimetry. The plots of confidency versus concentration for compounds (2) show a sharp change in slope at a given concentration. We considered this to be the point of micelle formation (Fig. 1) [3].

It is noteworthy that the form of discontinuity shown in Fig. 1 is not the usual form obtained for micellar aggregation in water. This phenomenon is probably due to the fact that micellar aggregation in formamide is different from that in water. In formamide, with its high dielectric constant ($\varepsilon = 110$), the micelles are probably more dissociated than in water ($\varepsilon = 80$), and so the conductivity increases beyond the critical micelle concentration (cmc). We are now undertaking a study of this phenomenon.

However, measurement of surface tension also shows a sharp change at the same concentration (Fig. 2) [4]. In this case, the form of discontinuity is usual for micellar aggregation and indicates the tensioactive properties of the molecules. The yields and cmc's for the different derivatives (2) are shown in Table 3.

The results indicate the following: As in the case of classical surfactants in water, the cmc decreases with increasing chain length, for both alkyl and fluoroalkyl compounds. In formamide, however, the fluorinated compounds have higher cmc's than their alkyl homologs. The reverse is found in water, where the long-chain

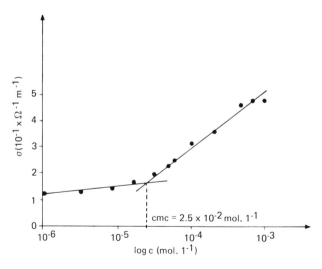

FIG. 1 Specific conductance (σ) versus concentrations (c) at 64°C for formamide solutions of (2c).

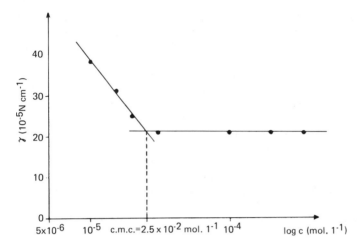

FIG. 2 Surface tension (γ) versus concentration (c) at 64°C for formamide solutions of (2c).

TABLE 2 Solubility of Various Salts in Formamide (T = 25°C)

Type	Formula	Solubility (g/liter)	Solubility (mol/liter)
Small cation, small anion	LiF	16	0.62
	NaCl	8	0.39
	NaF	15	0.36
Small cation, large anion	LiI	380	2.83
	NaI	1000	6.67
Large cation, small anion	KF	60	1.03
	KCl	30	0.40
Large cation, large cation	KI	400	2.41

TABLE 3   Characteristics of Compounds (2)

Compound	Yield (%)	cmc (mol dm^{-3})
(2a)	85	0.034
(2b)	88	0.030
(2c)	60	0.025
(2d)	90	0.030
(2e)	90	0.022
(2f)	90	0.019

fluorinated compounds have markedly lower cmc's than their alkyl homologs (up to 1.5 times lower for perfluorinated derivatives) [5]. The differences are due to the increased hydrophobic nature of the fluorinated compounds.

It would appear from our results that the fluorinated derivatives are considerably less "solvophobic" in formamide than in water. We have demonstrated for the first time surfactant properties of water-insoluble molecules in a structured nonaqueous solvent (formamide) and hence have shown the general nature of the micellization phenomenon.

Before this work, Gopal and Singh [6a,6b] investigated a number of surfactants in formamide, including CTAB. Contrary to the results of the present study, the former studies show no anomalous behavior of the electrical conductivity.

Recently, Almgren et al. [6c] demonstrated that SDS does not form micelles in formamide at 20°C.

In both cases, authors worked at temperatures below the Krafft points of surfactants in formamide. These points are higher than in water (55°C for SDS), and their results must be reexamined.

III.   WATERLESS MICROEMULSIONS IN WHICH FORMAMIDE SUBSTITUTES FOR WATER

Two definitions of microemulsions can be proposed, depending on either thermodynamic stability or the kinetics of formation:

1. "A system of water, oil, and amphiphiles which is a single optically isotropic and thermodynamically stable liquid solution" [7].

2. Friberg's definition, which considers that the thermodynamic
stability appears to be an exception. Therefore, a change re-
quiring spontaneous formation has been proposed [8].

Macroscopically, microemulsions are transparent dispersions of
two immiscible fluids (water and oil), which are formed spontaneously by admixing certain amphiphilic molecules. A typical system
contains four constituents: water, oil (hydrocarbon, fluorocarbon),
surfactant (ionic, zwitterionic, nonionic detergent), and cosurfactant (short-chain alcohols, amines, and also oximes). Such a system has a complex phase diagram, but single-phase domains can be
found for microemulsions.

Some ternary mixtures of n-hexane, water, and 2-propanol exhibit the properties of a microemulsion; "detergentless microemulsions" can be formed [9]. Three components that do not require
a cosurfactant have been known for a long time for non-ionic surfactants [10,11]; recently reports have also been published for
ionic systems.

Nevertheless, in all cases, water seemed to be required. Therefore, it occurred to us that it should be possible to prepare microemulsions with formamide instead of water [12-14].

Very recently, and independently of us, Friberg and colleagues
described microemulsions in which glycerol [15] and ethylene glycol
[16] replaced water. However, phase separations occurred after
some days.

Reported here are some four-component systems that utilize formamide rather than water. All the systems were stable for many
months. In drawing pseudo-ternary phase diagrams, we used transparency criteria to determine the microemulsion monophasic area.
Transparency was determined by a visual method that is generally
used for such systems [17].

A. Monophasic Areas of Formamide Microemulsions
   (at 25°C)

1. Ionic Microemulsions

a. Perhydrogenated Microemulsions. The oil used was cyclohexane
or isoctane. The cosurfactant was 1-butanol, and two surfactants
were studied: cetyltrimethyl ammonium bromide (CTAB) and cetyl
tributyl phosphonium bromide (CT_bPB).

It is noteworthy that no microemulsion was obtained with aqueous systems and CT_bPB. As for micellar solutions, in contrast to
water, formamide solvates large ions preferentially. So CT_bPB,
which is slightly soluble in water, is largely soluble and can aggregate in formamide. Therefore, it can lead to microemulsions systems
with formamide in place of water.

Figures 3 and 4 show the pseudo-three-component phase diagrams of (1-butanol/surfactant = K, formamide, cyclohexane) mixtures. For CTAB, we found similar monophasic areas to those obtained with water.

Figure 5 shows different types of pseudo-three-component phase diagrams for (formamide/CTAB = 2, 1-butanol, cyclohexane, or iso-octane) mixtures. In these cases, microemulsions of the phase boundary may be obtained by the addition of detergent to biphasic mixtures of formamide, 1-butanol, and cyclohexane. Thus, these are not three-component solvent mixtures, but are really microemulsions.

b. Perfluorinated Microemulsions. Production of aqueous four-component perfluorinated microemulsions has been reported recently [18]. However, the authors obtained only two separate narrow

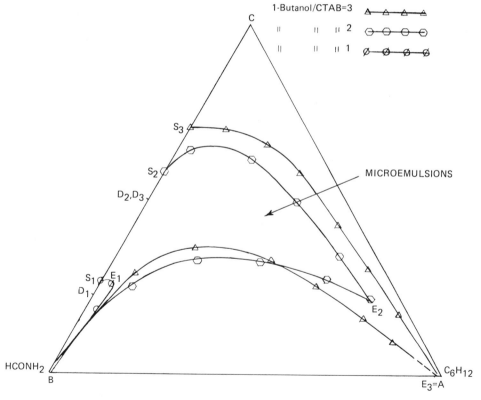

FIG. 3 Pseudo-ternary phase diagram of (1-butanol/CTAB = 3, 2, 1, formamide, cyclohexane).

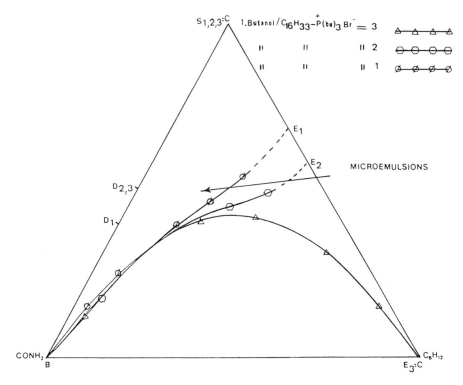

FIG. 4 Pseudo-ternary phase diagrams of (1-butanol/CT$_b$PB = 3, 2, 1, formamide, cyclohexane).

monophasic areas O/W and W/O microemulsions. The systems studied consisted of quaternary mixtures of formamide, 2,2,3,4-tetrahydro-perfluoroundecanoate [(3), surfactant], 1,1,2,2-tetrahydroperfluoro-hexanol [(4), cosurfactant] and perfluorinated oils: perfluorode-calin (5a), 1,1,2-trihydroperfluoro 1-decene (5b), and 1,1,1,2,2-pentahydroperfluorodecane (5c).

Figure 6 shows the pseudo-ternary phase diagrams for a sur-factant/cosurfactant weight ratio of 1/2. This ratio was chosen so as to obtain the largest monophasic areas.

Surfactant effect: Microemulsions were not obtained with per-hydrogenated surfactants. This may be due to segregation phenome-non between the hydrogenated tails of the surfactant and the per-fluorinated chains of the oil components. This has been established in the case of other perhydrogenated and perfluorinated compounds [19]. Microemulsions were not obtained with the lithium salts of the perfluorinated surfactants due to their poor solubility in formamide.

Cosurfactant effect (with $C_8F_{17}C_2H_4CO_2K$ as surfactant): Only narrow monophasic areas were obtained with perhydrogenated cosurfactants, such as 1-butanol, even when using large quantities of surfactant ($C_8F_{17}C_2H_4CO_2$ K/1-butanol = 2). This is most probably due to segregation between the tails of the surfactant molecules, the oil chains, and the alcohol. 1-Butanol is insoluble in perfluorinated oils, in contrast to perhydrogenated oils, in which it is very soluble. It does not appear to be a good cosurfactant for perfluorinated oils.

With 1,1,2,2-tetrahydroperfluorohexanol as cosurfactant, large monophasic areas were obtained (Fig. 6). This perfluoroalcohol is completely soluble in perfluorinated oils and is also slightly soluble in formamide. The areas obtained were similar to those obtained with aqueous perhydrogenated systems. In these latter systems, 1-butanol as cosurfactant is completely soluble in oil and slightly soluble in water.

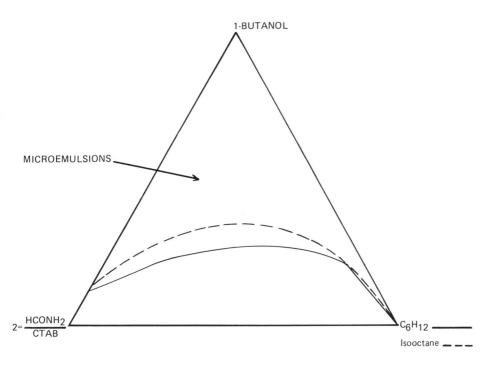

FIG. 5 Pseudo-ternary phase diagrams of (formamide/CTAB = 2,1-butanol, cyclohexane or isooctane).

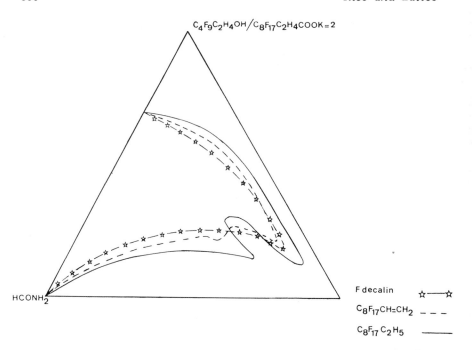

FIG. 6 Pseudo-ternary phase diagrams of the systems Formamide-
($C_4F_9C_2H_4OH/C_8F_{17}C_2H_4CO_2K$ = 2)-F-decalin; Formamide-($C_4F_9C_2H_4OH/$
$C_8F_{17}C_2H_4CO_2K$ = 2)-C_8F_{17}-CH=CH$_2$; Formamide-($C_4F_9C_2H_4OH/$
$C_8F_{17}C_2H_4CO_2K$ = 2)-$C_8F_{17}C_2H_5$.

For aqueous perfluorinated microemulsions, the monophasic areas
are narrow [18] because the fluoroalcohol is insoluble in water.

Oil effect: The Microemulsion monophasic area was slightly larger for $C_8F_{17}C_2H_5$ than for $C_8F_{17}CH=CH_2$ and F-decalin. In the
formamide-rich region especially, the phase boundary was lower for
$C_8F_{17}C_2H_5$ than for $C_8F_{17}CH=CH_2$ and F-decalin. This is probably
due to the fact that (5c) is more hydrogenated (and so less "solvophobic" for formamide) than (5b) and (5a), which are completely
perfluorinated.

In order to study this phenomenon in more detail, we are preparing mixed oils R_F-R_H (with R_F and R_H of various lengths) to
test the formamide microemulsions.

Because of the importance of aqueous perfluorinated O/W microemulsions as blood substitutes [20], we intend to test such mixed
oils in aqueous microemulsions. Due to dilution of the microemulsion
in blood, the phase boundary must be low in the water-rich region.

2. Nonionic Microemulsions

The nonionic surfactants Pluronic L-64 (MW = 2900) and Pluronic F-88 (MW = 10,800) were used. These agents are trisequenced co-polymers of ethylene polyoxide and propylene oxide, and they have been widely used for the production of aqueous microemulsions [21]. The interest of these microemulsions is that the surfactants can be easily eliminated by ultrafiltration, which makes them particularly useful as reaction media.

These zones were determined on the following systems:

A. Formamide or water/Pluronic L-64 = 2, toluene (oil)-isopropanol (cosurfactant)
B. Formamide or water/Pluronic F-88 = 2, toluene (oil)-isopropanol (cosurfactant)

The water systems have been studied by Riess et al. [21].

We used a solvent/surfactant ratio of 2, as this produces larger microemulsion zones than those with isopropanol/surfactant = 2.

Figures 7 and 8 show the pseudo-ternary phase diagrams for both systems. The curves at the top of the diagram indicate the limits of miscibility of water (for formamide), toluene, and isopropanol. Miscibility is observed above the line and emulsion below. The curves at the bottom indicate the limits of miscibility in the presence of surfactants.

Comparison between the aqueous and formamide systems show that:

1. The microemulsion regions are larger in formamide than in water.
2. Less isopropanol is needed for production of microemulsions in formamide than in water.

B. Phase Behavior of Formamide Microemulsions

We have examined in more detail the polyphasic zones in formamide microemulsions (demonstration of Winsor I, II, and III systems). The different Winsor systems were determined as for aqueous micro-emulsions [22-24] by the progressive addition of an electrolyte to the system.

1. Choice of electrolyte

Sodium chloride is usually used as electrolyte in aqueous microemul-sions. However, this salt is not very soluble in formamide, and so we investigated various other halogenides (see Table 2).

Iodides (large anion) were found to be quite soluble in forma-mide, and we therefore used sodium iodide for these experiments.

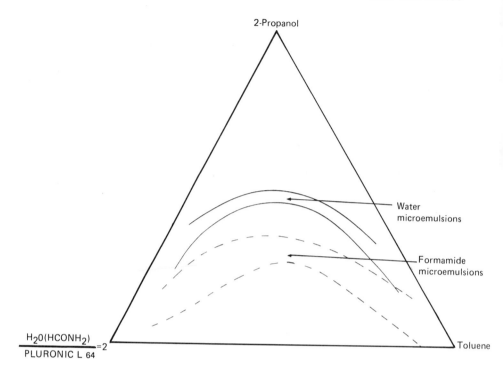

FIG. 7 Pseudo-ternary phase diagram for aqueous and formamide microemulsions (with Pluronic L-64 as surfactant).

2. Phase Behavior in Ionic Microemulsions

Two types of system, using either hydrogenated or fluorinated constituents, were used.

a. Hydrogenated Systems. We used isooctane as oil, ctyltrimethyl ammonium bromide (CTAB) as surfactant, and 1-butanol as cosurfactant. As for aqueous system [5], we kept the mass ratio of isooctane/formamide at unity.

Figure 9 shows a diagram of the various Winsor systems for these microemulsions. It represents the percentage of surface-active components as a function of salt concentration in formamide. It has a similar appearance to that observed for aqueous microemulsions [23]. A large Winsor III zone, characterized by its extent in the electrolyte-rich medium, can be seen. These systems are stable only above 80°C. If the electrolyte concentration is above 100 g/liter, the constituents tend to separate out below this temperature.

b. Fluorinated Systems. For oil we used 1,1,2-trihydroperfluoro 1-decene, potassium 2,2,3,3-tetrahydroperfluoroundecanoate as surfactant, and 1,1,2,2-tetrahydroperfluorohexanol as cosurfactant. The monophasic zones for this system have been reported before. Keeping the oil/formamide ratio at unity, as for the hydrogenated system, we observed Winsor systems of types I, II, III, and IV. In these microemulsions, systems containing less electrolyte than for the hydrogenated microemulsions are stable at 30 to 40°C. The phase diagram is shown in Fig. 10.

It should be noted that due to the increased density of fluorinated oils, the Winsor phases are reversed. The oil phase is at the bottom, and the formamide phase is at the top.

The optimal salinity (S^* = intersection of Winsor I, II, III, and IV zones) is found at 25% surface-active agents and 35 g/liter NaI. These values are considerably lower than those found for the hydrogenated microemulsions (S^* = 43% surface-active agents, and

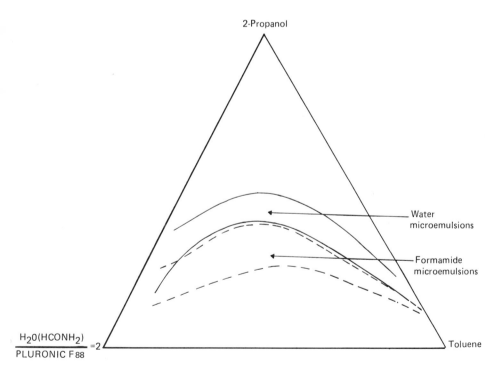

FIG. 8 Pseudo-ternary phase diagram for aqueous and formamide microemulsions (with Pluronic F-88 as surfactant).

100 g/liter NaI). The fluorinated systems are thus more sensitive
to salt effects than the hydrogenated systems.

Due to increased stability of these fluorinated systems, we were
able to measure the interfacial surface tensions between the different phases of the Winsor I, II, and III systems. We used the whirl-
ing drop method [25] at a temperature of 30°C. The results are
shown in Fig. 11 (interface surface tension as a function of salinity
at different percentages of surface-active agents).

As for aqueous microemulsions, we observed very low interfacial
surface tensions [26], especially in the Winsor III zone, where they
were lower than 10^{-4} nM/m.

Overall, these results show that the phase behavior of these
nonionic formamide microemulsions is in agreement with the presence
of an organized microemulsion, rather than a simple cosolubilization
phenomenon.

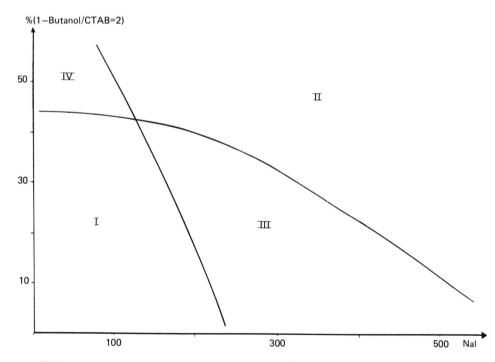

FIG. 9 Phase behavior for the system (formamide, isooctane,
1-butanol, and CTAB) and NaI.

%($C_4F_9C_2H_4OH/C_8F_{17}C_2H_4CO_2K = 2$)

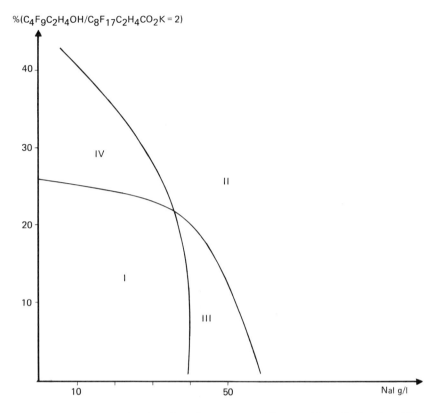

FIG. 10 Phase behavior for the system (formamide, $C_8F_{17}CH=CH_2$, $C_4F_9C_2H_4OH$, and $C_8F_{17}C_2H_4CO_2K$) and NaI.

3. Phase Behavior of Nonionic Microemulsions

Figure 12 shows the phase diagram for the nonionic microemulsion [toluene/formamide = 1, formamide/Pluronic L-64 = 2, (isopropanol) fn (salt)]. The appearance of the diagram is similar to that for the ionic systems except that a Winsor III zone is not observed between the Winsor I and II zones.

The optimal salinity S* (intersection between the Winsor I, II, and IV zones in this case) corresponded to 44% alcohol and 200 g/liter NaI. This latter value is much higher than that found for the ionic surfactants. This indicates a weak salt effect for nonionic systems. A similar phenomenon has been found for aqueous microemulsions. As for the ionic microemulsions, we carried out these experiments at a temperature of about 80°C.

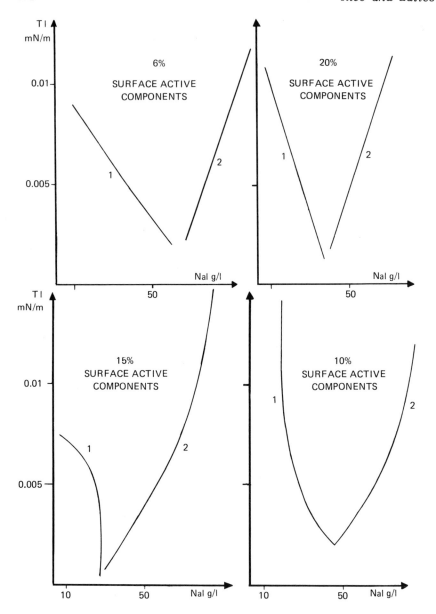

FIG. 11 Interfacial surface tensions between the different phases
of the Winsor I, II, and III systems for fluorinated microemulsions
of Fig. 10.

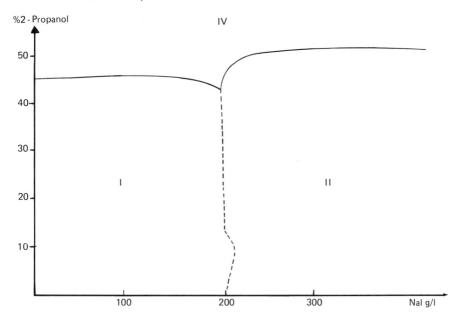

FIG. 12 Phase behavior for the system (formamide, toluene, Pluronic L-64, and isopropanol).

IV. CONCLUSION

In conclusion, these results show that these waterless microemulsions have similar structures to their aqueous counterparts.

Phase behavior of nonaqueous microemulsions, with a large Winsor III zone in the case of ionic surfactants, is good evidence of the analogy of structure with aqueous microemulsions.

Change of conductivity, analogous to the change seen by Crausse [27] with water, is also evidence against a cosolubilized system.

We are pursuing investigations of various chemical reactions in these formamide microemulsions in view of their potential applications. Compounds reacting with water can be studied, and the use of a solvent such as formamide with its high dielectric constant ($\varepsilon =$ 110) could provide a valuable medium for the investigation of many chemical and photochemical reactions.

Other studies by NMR, X-ray and light-scattering experiments are also in progress.

V. ACKNOWLEDGMENT

We are grateful to Atochem for the Providing of perfluorinated compounds. We thank Greco Microemulsions (PIRSEM AIP No. 2201,
CNRS, France) for financial support.

VI. REFERENCES

1. Sinanoglu, O., and Abdulnur, S., *Fed. Proc. Am. Soc. Exp.
    Biol. 24*, 5 (1965).
2. Wacker, A., and Lodemann, E., *Angew. Chem. Int. Ed. Engl.
    4*, 150 (1965).
3. Conductivity measurements were carried out at 64°C using a
    Tacussel CD 6N-G conductimeter (the Krafft points of the de-
    rivatives were around 60°C, the breaks in slope were clear-
    cut and gave the values of cmc's with a 5% uncertainty). It
    should be pointed out that the exact nature (true spheres or
    otherwise) of the aggregates obtained is not known. Morpho-
    logical studies would be required to provide this information.
4. Surface tension measurements were carried out at 64°C using a
    Prolabo tensiometer (Tensimat n 3) with the ring detachment
    method.
5. Kuneda, H., and Shinoda, K., *J. Phys. Chem. 80*, 2468 (1976).
6. (a) Gapal, R., and Singh, J. R., *Kolloid Z. 239*, 699 (1970);
    (b) Gopal, R., and Singh, J. R., *J. Phys. Chem. 71*, 554
    (1973); (c) Almgren, M., Swarup, S., and Lofroth, J. E.,
    J. Phys. Chem. 89, 4621 (1985).
7. Danielson, I., and Lindman, B., *Colloids Surfaces 3*, 381 (1981).
8. Friberg, S. E., *Colloids Surfaces 4*, 201 (1982).
9. Smith, G. D., Donelan, C. E., and Barden, R. E., *J. Colloids
    and Interface Sci. 60*, 488 (1977).
10. Shindo, K., *J. Colloids and Interface Sci. 24*, 4 (1967).
11. Friberg, S. E., and Lapczynska, I., *Progs. Colloid Polymer
    Sci. 56*, 16 (1975).
12. Rico, I., and Lattes, A., *Nouveau J. Chimie 8*(7), 429 (1984).
13. Rico, I., and Lattes, A., *J. Colloids and Interface Sci. 102*,
    285 (1984).
14. Rico, I., and Lattes, A., *Proceedings of 5th International Sym-
    posium on Colloids*, Mittal, K. L., (ed.), *in press*.
15. Friberg, S. E., and Wohn, C. S., *Colloid and Polymer Sci.
    263*, 156 (1985).
16. Friberg, S. E., and Podzimek, M., *Colloid and Polymer Sci.
    262*, 252 (1984).
17. Cayias, J. L., Schechter, R. S., and Wade, W. H., *J. Colloids
    and Interface Sci. 59*, 31 (1977).

18. Oliveros, E., Maurette, M. T., and Braun, A. M., *Helv. Chim. Acta 66*(4), 1183 (1983).
19. (a) Mukerjee, P., *J. Am. Oil. Chem. Soc. 59*(12), 573 (1982); (b) Shinoda, K., and Namura, T., *J. Phys. Chem. 84*, 365 (1980).
20. Riess, J. G., and Le Blanc, M., *Pure Appl. Chem. 54*, 2383 (1982).
21. Riess, G., and Nervo, J., *Information Chimie 18* (1977).
22. Winsor, P. A., *Trans. Faraday Soc. 44*, 376 (1948).
23. Bourrel, M., and Chambu, C., *Soc. Pet. Eng. J.* 17 (1982).
24. Bourrel, M., Salager, J. L., Schechter, R. S., Wade, W. W., Colloques Nationaux: CNRS No. 938, Physicochimie des Composés Amphiphiles, June 1978.
25. Bellocq, A. M., Biais, J., Bothorel, P., Clin, B., Fourche, G., Lalanne, P., Lemaire, B., Limanceau, B., and Roux, D., *Adv. Colloid Interface Sci. 20*, 167 (1984).
26. Healy, R. N., and Reed, R. L., *Soc. Pet. Eng. J.* 129 (1977).
27. (a) Clausse, M., *Nature 293*, 636 (1981); (b) Clausse, M., and Heil, *Nuovo Cimento Lett. 36*, 369 (1983).

24

Formamide as a Water Substitute X: Waterless Microemulsions 7; A Study of Cyclocondensation Reactions in Waterless Microemulsions

A. LATTES and I. RICO   *Université Paul Sabatier, Toulouse, France*

I. INTRODUCTION

Aqueous microemulsions have attracted considerable interest as media for chemical reactions. We have demonstrated that it is possible to prepare microemulsions using formamide instead of water.

The waterless systems are very interesting as reaction media: Compounds insoluble or reacting with water can be studied, and the use of a solvent such as formamide with its high dielectric constant ($\varepsilon_{HCONH_2} = 110$; $\varepsilon_{water} = 80$) provides a valuable medium for many reactions. In particular, cyclocondensation reactions in various formamide microemulsions have been studied and the results interpreted in terms of medium structure and polarity.

There have been a large number of studies of aqueous microemulsions over the last few years [1]. Microemulsions have been used as reaction media [2], although their more widespread use has often been limited by inadequate substrate solubility in water.

Recently, we have reported that it is possible to prepare micro by replacing water with the highly structured solvent formamide [3]. In this study we report the first use of such nonaqueous microemulsions as reaction media. We carried out a Diels-Alder addition of methyl acrylate to cyclopentadiene in various formamide microemulsions. This type of reaction, which is highly sensitive to the polarity of the medium [4], can be used as a chemical probe of microemulsion.

II. RESULTS

A. Preparation of the Reaction Media: Monophasic Areas of Formamide Microemulsion

We replaced water with formamide, and we used a visual transparency criterion to determine the phase diagram [5]. The spontaneity of the change between milky emulsion and transparent solution conformed to Friberg's definition [6]. None of the microemulsions prepared separated even after some months.

An ionic system was used with isooctane as oil, 1-butanol as cosurfactant, and cetyltrimethylammonium bromide (CTAB) as surfactant.

Figure 1 shows the pseudo-ternary phase diagram for (formamide/CTAB = 2, 1-butanol, isooctane) mixtures, where ratio and compositions are given in weight percent.

It should be pointed out that the microemulsions at the phase boundary could be obtained by addition of the detergent to biphasic mixtures of formamide, 1-butanol, and isooctane. They are not, therefore, three-component solvent mixtures, but are true microemulsion.

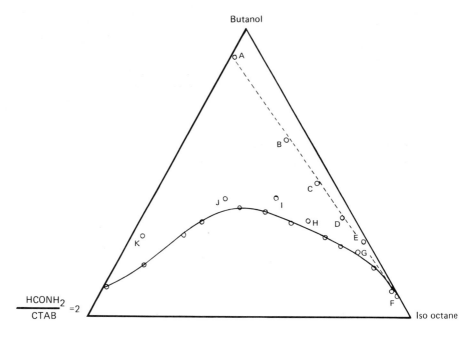

FIG. 1 Pseudo-ternary phase diagram of the system formamide/ CTAB = 2, 1-butanol, isooctane.

B. Probe Reaction

We chose the Diels-Alder addition reaction of methyl acrylate to cyclopentadiene:

It is known that the endo/exo selectivity increases with increasing polarity of the solvent [4]. We carried out a preliminary study of the influence of the various constituents of the microemulsion on the reaction.

C. Effect of Microemulsion Components on the Probe Reaction

Table 1 summarizes the results obtained in different solvents. We have included results obtained in water and N-methyl acetamide for comparison.

The use of water as a reaction medium for Diels-Alder additions has been reported previously [7-11]. For example, Breslow et al. [10] and Grieco et al. [11] have shown that high endo/exo selectivities can be obtained under these conditions. Laszlo et al. [12] have recently shown that high selectivities can also be achieved for Diels-Alder reactions carried out in the presence of clay.

Our results suggest the following:

1. In agreement with Berson's explanation [4], the solvents with the highest dielectric constant (most polar) favor formation of the endo isomer. This can explain the high stereoselectivity in both water and formamide.

2. Nevertheless, the above parameter along cannot fully explain the stereoselectivities observed. For example, in N-methyl acetamide (ε = 182.0), the endo/exo selectivity was less than in water or formamide. As suggested by Breslow [10c], "solvophobic"-type interactions would seem to play an important role. Therefore, in formamide, which is highly structured, the stereoselectivity was similar to that observed in water; whereas in N-methyl acetamide, which is less structured than water, the selectivity fell to that found in alcoholic solvents such as 1-butanol.

Overall, our results confirm Breslow et al.'s explanation for the stereoselectivity of Diels-Alder reactions. For high endo/exo selectivity, "solvophobic" interactions (and to a lesser extent solvent polarity) are fundamental.

Formamide, which strongly favors solvophobic interactions [3,13], would therefore seem to be an excellent medium for Diels-Alder

TABLE 1   Endo/Exo Product Ratios in Organic Media and in Water

Solvent	Dielectric constant ε (20-25°C)	Concentration of diene and dienophile (M)	ENDO/EXO
Isooctane	2.0	0.15	2.3
1-Butanol	17.1	0.15	5.0
Water	78.5	0.15	7.4
Water	78.5	0.30	5.3
Formamide	109.0	0.07	6.9
Formamide	109.0	0.15	6.7
Formamide	109.0	0.30	6.7
N-Methyl acetamide	183.0	0.15	4.7

reactions, especially at high concentrations. By using formamide, we have obtained considerably improved stereoselectivities, at high concentrations, over those obtained by Breslow [10b] (for a concentration of 0.30 mol/liter, endo/exo = 5.3 for water, and 6.7 for formamide). It is interesting that, contrary to the results in water, where the reaction takes place in a heterogeneous medium (because of the low solubility of products), concentration effects were not observed in formamide. The same stereoselectivity was obtained whatever the substrate concentration. This is probably due to the fact that in formamide the reaction takes place in a homogeneous medium.

In addition, the use of formamide as a homogeneous reaction medium significantly improves yield. For water, the yields of product (after extraction) do not exceed 40%, whereas in formamide, the yields are quantitative whatever the initial concentrations of the reactants. The yields of isolated product (after extraction) are then close to 85%.

We are carrying out further investigations on the use of formamide in other preparative Diels-Alder reactions.

D. Diels-Alder Reactions in Formamide Microemulsions

As opposed to results obtained for the same type of reaction (cyclocondensation between cyclopentadiene and methylmethacrylate) in

TABLE 2 Microemulsion Compositions (by weight) of the System (1-butanol, formamide/CTAB = 2, isooctane), and Selectivity of the Diels-Alder Probe Reaction ($C_{substrate}$ = 0.15 M)

Microemulsion	Percent isooctane	Formamide		Percent 1-butanol	Percent CTAB	Endo/exo
		Percent	ϕ*			
A	1	6	0.07	90	3	4.50
B	32	5	0.04	60.5	2.5	4.50
C	48	4	0.03	46	2	4.27
D	62	3	0.02	34.5	1.5	3.87
E	72	1	0.01	26.5	0.5	3.16
F	92	0.2	0.003	7.7	0.1	2.85
G	72	4	0.03	22	2	3.76
H	54	10	0.08	31	5	3.96
I	38	14	0.15	41	7	4.09
J	20	26	0.28	41	13	4.12
K	3	46	0.51	28	23	4.55

*ϕ = volumetric fraction calculated from the relative densities of formamide and the microemulsions.

aqueous microemulsions (water, toluene, 2-propanol as cosurfactant both in the absence and in presence of surfactant: SDS and CTAB) studied only in the continuous oil phase [14], we have been able to exploit the high solubility of substrates in formamide to investigate the complete microemulsion phase diagram (continuous oil phase and continuous formamide phase). Representative points of the micro-emulsions studied are shown in Fig. 1. The microemulsion compositions and reaction selectivities are shown in Table 2.

III. DISCUSSION

The results demonstrated the presence of two zones in the microemul-sions studied: (1) a zone of relatively low selectivity in the oil-rich part of the diagram, and (2) a much wider zone of higher selectivity. Good agreement between the changes in conductivity and endo/exo selectivity are found along the lines AF and FK (demixing line) on a plot of these changes as a function of the formamide volumetric frac-tion ϕ (Figs. 2 and 3).

The abrupt changes in selectivity and conductivity around C on line AF, and around G on the demixing line, can be attributed to

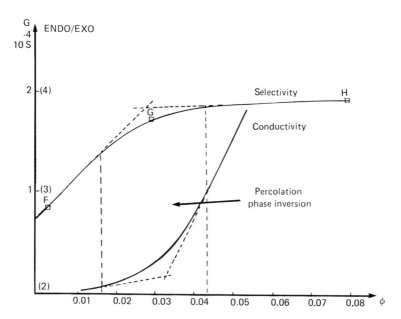

FIG. 2 Plot of conductivity and endo/exo ratios as a function of the formamide volumetric fraction ϕ, demixing line.

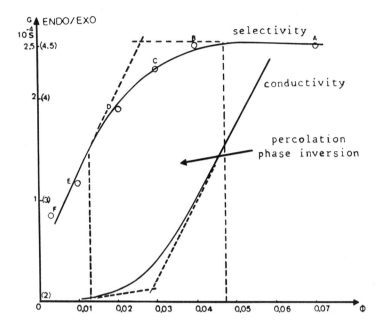

FIG. 3 Plot of conductivity and endo/exo ratios as a function of the formamide volumetric fraction ϕ, along the line AF.

phase inversion from reverse micelles (formamide in oil) to direct micelles (oil in formamide) or, taking into account the low percentage of formamide at the compositions C and G, to transition from inverse micelles (no free formamide, all bound) to formamide/oil microdroplets (inverse micelles with an oil "core").

A. Conductivity Change

This type of change in conductivity has been demonstrated, for aqueous microemulsions, to be due to percolation, in which there is a concatenation of reverse micelles of water in oil, followed almost immediately by a phase inversion from reverse (low conductivity zone) to direct micelles (higher conductivity zone, on the demixing line [15]. Clausse et al. [16] have demonstrated analogous results with nonsaturated microemulsions in all parts of the phase diagram.

B. Selectivity Change

1. Reverse Micelles

In the continuous oil phase of microemulsion G, it appears that the reaction takes place in the bicontinuous phase. The endo/exo ratio (= 2.85, Table 2) is close to that obtained for pure isooctane (= 2.3, Table 1).

2. Direct Micelles

On the other hand, in the formamide-rich zone, the endo/exo ratio (= 4.5) obtained in microemulsion K is virtually the same as that found in the alcohol-rich microemulsion A (Table 2). This value is also close to that found in pure butanol (= 5.0, Table 1). It seems that the reaction takes place at the micelle interface in the continuous formamide phase.

IV. CONCLUSION

This Diels-Alder reaction is a valuable probe reaction for a preliminary study of formamide microemulsions. It is able to discriminate the continuous oil phase from structured phases: formamide/oil microdroplets or inverse micelles.

We are carrying out kinetic studies of this reaction in the constituent solvents of the microemulsions. This should enable us to localize the reaction site at the microscopic level, in the heterogenous medium. We intend, in the first instance, to study the reaction in pure formamide, which appears to be an excellent medium for Diels-Alder reactions.

V. ACKNOWLEDGMENT

We thank Greco Microemulsions (PIRSEM AIP No. 2201, CNRS, France) for financial support.

VI. REFERENCES

1. Bellocq, A. M., Biais, J., Bothorel, P., Clin, B., Fourche, G., Lalanne, P., Lemaire, B., Lemanceau, B., and Roux, D., *Adv. Colloid & Interface Sci.* 20, 167 (1984) and cited refs.
2. (a) Mackay, R. A., *Adv. Colloid & Interface Sci.* 15, 131 (1981); (b) Whitten, D. G., Russell, J. C., and Schmehl, R. H., *Tetrahedron* 38, 2455 (1982); (c) Rico, I., Maurette, M. T., Oliveros, E., Riviere, M., and Lattes, A., *Tetrahedron Lett.* 48, 4795 (1978); (d) Fargues, R., Maurette, M. T., Oliveros, E., Riviere, M., and Lattes, A., Nouveau *J. Chim.* 3, 487 (1979); (e) Rico, I., Maurette, M. T., Oliveros, E., Riviere, M., and Lattes, A., *Tetrahedron 36*, 1779 (1980); (f) Fargues-Sakellariou, R., Oliveros, E., Maurette, M. T., Riviere, M.,

and Lattes, A., *J. Photochem.* *18*, 101 (1982); (g) Fargues-Sakellariou, R., Maurette, M. T., Oliveros, E., Riviere, M., and A. Lattes, *Tetrahedron 40*, 2381 (1984); (h) Amarouche, H., de Bourayne, C., Riviere, M., and Lattes, A., *C. R. Acad. Sci. 298*, 121 (1984).

3. (a) Rico, I., and Lattes, A., *Nouveau J. Chim. 8*, 429 (1984); (b) Rico, I., and Lattes, A., *J. Colloid and Interface Sci. 102*, 285 (1984).

4. Berson, J. A., Hamlet, Z., and Mueller, W. A., *J. Am. Chem. Soc. 84*, 297 (1962).

5. Cayias, J. L., Schechter, R. S., and Wade, W. H., *J. Colloid and Interface Sci. 59*, 31 (1977).

6. Friberg, S. E., *Colloids and Surfaces 4*, 201 (1982).

7. Woodward, R. B., and Baer, H., *J. Am. Chem. Soc. 70*, 1161 (1948).

8. Hopff, H., and Rautenstrauch, C. W., U.S. Patent 2,262,002 [*Chem. Abstracts 36*, 1046 (1942)].

9. Sternbach, D. D., and Rossana, D. M., *J. Am. Chem. Soc. 104*, 5853 (1982).

10. (a) Rideout, R. C., and Breslow, R., *J. Am. Chem. Soc. 102*, 7816 (1980); (b) Breslow, R 1 Maitra, U., and Rideout, D., *Tetrahedron Lett. 24*(18), 1901 (1983); (c) Breslow, R., and Maitra, U., *Tetrahedron Lett. 25*, 1239 (1984).

11. (a) Grieco, P. A., Garner, P., Yoshida, K., and Hoffman, M. C., *Tetrahedron Lett. 24*, 3807 (1983); (b) Grieco, P. A., Yoshida, K., and Garner, P., *J. Org. Chem. 48*, 3139 (1983); (c) Grieco, P. A., Garner, P., and He, A. M., *Tetrahedron Lett. 24*(18), 1897 (1983).

12. Laszlo, P., and Luccheti, J., *Tetrahedron Lett. 25*, 1567 (1984).

13. Escoula, B., Hajjaji, N., Rico, I., and Lattes, A., *J.C.S. Chem. Commun. 18*, 1233 (1984).

14. Gonzalez, A., and Holt, S. L., *J. Org. Chem. 47*, 3186 (1982).

15. (a) Lagues, M., Ober, R., and Taupin, C., *J. Phys. Lett. 39*, 487 (1978); (b) Lagues, M., and Sauterey, C., *J. Phys. Chem. 84*, 3503 (1980).

25

Water/Sodium Dodecylsulfate/1-Pentanol/N-Dodecane Microemulsions. Realm-of-Existence and Transport Properties

M. CLAUSSE   *U.A. CNRS No. 858, Département de Génie Biologique, Université de Technologie de Compiègne, France*

A. ZRADBA* and L. NICOLAS-MORGANTINI†   *Laboratoire de Thermodynamique et Energétique, Institut Universitaire de Recherche Scientifique, Université de Pau et des Pays de l'Adour, Pau, France*

I. INTRODUCTION

A pseudoternary phase diagram study of the system water/sodium dodecylsulfate/1-pentanol/n-dodecane was performed, at T = 25°C, for different values of the surfactant/alcohol ratio. It was thus possible to determine the features of the microemulsion domain V. While being an all-in-one block volume, built on the realm-of-existence of the water/surfactant/alcohol monophasic solutions, V presents two dissymmetric lobes, that both extend to phase tetrahedron confines corresponding to n-dodecane-rich compositions. In an attempt to probe into microemulsion structure, studies of microemulsion electrical conductivity and apparent viscosity were carried out. It appears possible to define, within the microemulsion domain, regions corresponding to different structural states.

As demonstrated in a previous paper [1], two categories of water/sodium dodecylsulfate/alkanol/hydrocarbon systems may be defined. The type S systems [2] are characterized by the fact

---

Present affiliations:
*Ecole Normale Supérieure de Casablanca, Morocco.
†Département de Chimie Macromoléculaire et Colloidale, L'Oreal, Aulnay-sous-Bois, France.

that, in a ternary phase diagram, the realm-of-existence of the
water/sodium dodecylsulfate/alkanol monophasic solutions consist of
two disjoined areas, L_1 which corresponds to "direct" solubilization
(aqueous solutions of alkanol) and L_2 which corresponds to "inverse"
solubilization (alkanolic solutions of water); consequently, the microemulsion three-dimensional domain consists of two disjoined volumes;
V_1, the extension of the L_1 area, corresponds to "direct" microemulsions (hydrocarbon in water), and V_2, the extension of the
L_2 area, corresponds to "inverse" microemulsions (water in hydrocarbon). The type U systems [2] are characterized by the fact that
the realm-of-existence of the water/sodium dodecylsulfate/alkanol
monophasic solutions is a large area L, which, in the ternary phase
diagram, is stretched continuously from the W apex (100% water) to
the C apex (100% alkanol). Consequently, the microemulsion three-
dimensional domain is a vast all-in-one block volume V that generally
spans the greater portion of the phase tetrahedron. The existence
of two distinct categories of systems is confirmed by the existence
of two distinct types of microemulsion electroconductive and viscous
behavior. In the case of systems whose ionic surfactant is sodium
dodecylsulfate (SDS) and cosurfactant a normal alkanol, the transition from type S to type U systems occurs, whatever the nature of
the hydrophobic hydrocarbon, when 1-pentanol is substituted for
1-hexanol (the threshold value of n_a, the alkanol number of carbon
atoms, is therefore $n_a = 6$).

Since the microemulsion three-dimensional domain of type U systems takes up most of the phase tetrahedron volume, it may be inferred that the structure of type U microemulsions is diverse and
that transitions from one kind of structure to another are smooth.
In an attempt to assess the structural diversity of type U microemulsions, a thorough study of the type U system water/sodium dodecyl-
sulfate/1-pentanol/n-dodecane was carried out. On the grounds of
the data obtained as to microemulsion domain configuration and microemulsion electroconductive and viscous behavior, it appears possible
to define several adjacent composition ranges corresponding to different microemulsion structural states.

II. EXPERIMENTAL

A. Compounds

Freshly bidistilled water was used (electrical conductivity σ around
10^{-4} Sm^{-1}).

The surfactant was sodium dodecylsulfate (99% purity grade)
purchased from Touzart and Matignon (France).

The cosurfactant was 1-pentanol (99% purity grade) and the hydrocarbon n-dodecane (98% purity grade), both purchased from Fluka A.G. (Switzerland).

B. Methods

The microemulsion pseudo-ternary domains were delimited inside WHS-type triangular diagrams [3,4], according to the experimental procedure explained in Chap. 2 [1]. Some refinements, however, were brought to the general method. When appropriate, the sections of the microemulsion pseudo-ternary domain boundaries corresponding to microemulsion-liquid crystal transitions were determined by recording (versus composition) the optical appearance of samples placed between crossed nicols and progressively enriched in one of the system constituents (usually water). For instance,

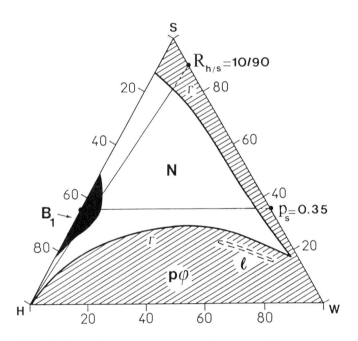

FIG. 1 Pseudo-ternary phase diagram of the water/sodium dodecyl-sulfate/1-pentanol/n-dodecane system. (T = 25°C; k_m = 1/2 or k_x = 1/6.54). N and l are microemulsion regions. B_1 is a region of existence of D-type (lamellar), liquid crystals. $p\phi$ is a region of diphasic media of the Winsor I type (microemulsion in equilibrium with an upper excess organic phase).

this particular procedure was used to demarcate, inside the ternary diagram of Figure 1, the region B_1 corresponding to the existence of mesomorphous media. As illustrated by the conductivity plots of Figures 2 and 3, it was observed [5] that the microemulsion-liquid crystal transition always induces a drastic change in the trend of the electrical conductivity variations versus composition. Therefore, when appropriate, electroconductimetry was also used to determine very accurately the frontiers between microemulsion and liquid crystal regions.

The specific mass ρ, apparent viscosity η, and electrical conductivity σ of microemulsions were measured according to the methods reported in a previous article [1].

All experiments were performed at T = 25°C.

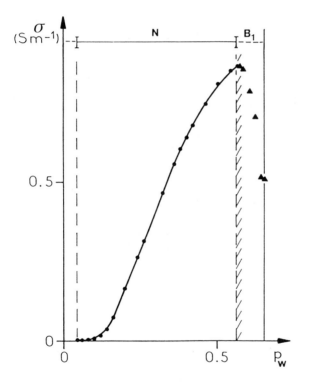

FIG. 2 Variations of the electrical conductivity σ along a p_s-type experimental path running across the regions N (microemulsion) and B_1 (liquid crystal) of the pseudo-ternary phase diagram of Figure 1 (p_s = 0.35, T = 25°C). p_w is the water mass fraction.

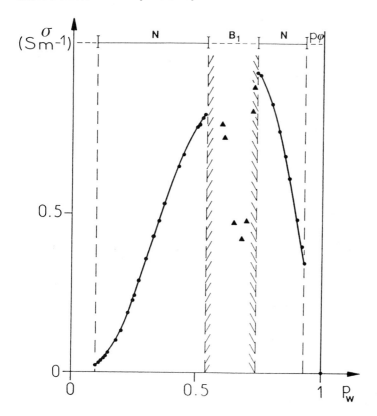

FIG. 3 Variations of the electrical conductivity σ along a $R_{h/s}$-type experimental path running across the regions N (microemulsion) and B_1 (liquid crystal) of the pseudo-ternary phase diagram of Figure 1 ($R_{h/s}$ = 10/90, T = 25°C). p_W is the water mass fraction.

III. PHASE DIAGRAMS

A. Water/Sodium Dodecylsulfate/1-Pentanol System

Figure 4 gives a partial view of the internal organization of the ternary diagram of the water/sodium dodecylsulfate/1-pentanol system, at T = 25°C. The realm-of-existence of the system monophasic solutions is an all-in-one-block domain L in which three distinct areas may be identified. The areas labeled L_1 and L_2 define ranges of compositions corresponding to aqueous solutions of 1-pentanol (L_1), and to pentanolic solutions of water (L_2). L_1 and L_2 are connected by a narrow channel C_L that delimits a range of compositions that

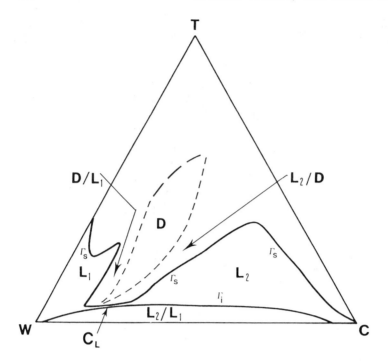

FIG. 4 Ternary phase diagram of the water/sodium dodecylsulfate/
1-pentanol system, (T = 25°C). L_1 and L_2 are, respectively, the
region of "direct" solubilization and the region of "inverse" solubilization. C_L is the channel connecting L_1 and L_2. D is the region of "monophasic" lamellar liquid crystals.

cannot be a priori associated to a specific kind of solubilization.
Taking into account, on the one hand, the rather poor mutual affinity of water and 1-pentanol [6] and, on the other hand, the fact
that the L_1 and L_2 areas are connected, it may be suggested that
the monophasic solution microstructure varies progressively with
composition, according to the following rough pattern, (water content decreasing): water-rich molecular solutions → direct micellar
solutions → solution-type media with ill-defined structure → inverse
micellar solutions → 1-pentanol-rich molecular solutions.

The composition range delimited by the frontier Γ_i of the monophasic solution domain and the CW edge of the triangular diagram
(see Fig. 4) corresponds to the existence of diphasic media of the
L_2/L_1 kind.

The region labeled D on the phase diagram of Figure 4 is the
realm-of-existence of monophasic mesophases of the classical lamellar

type [7]. This region is flanked by two composition ranges corresponding to diphasic media of the L_2/D and D/L_1 kinds (see Fig. 4). X-ray diffraction experiments [7] showed that the lamellar liquid crystal repeat distance δ increases as the overall content of water increases. This is exemplified by the following figures that concern samples characterized by $k_m = 1/2$ (or $k_x = 1/6.54$)*:

Overall mass fraction of water, p_W	0.55	0.70	0.77
Type of medium	L_2/D	D	D/L_1
δ (Å)	38.4	50.9	62.6

These data are in agreement with results reported by other authors for similar ternary systems [8–10].

B. Water/Sodium Dodecylsulfate/1-Pentanol/ n-Dodecane System

The set of diagrams of Figure 5-2 shows the evolution of the microemulsion pseudoternary domain configuration, as a function of the surfactant-to-cosurfactant ratio. The values of k_m are listed in Figure 5-1 and the corresponding values of k_x are reported in Figure 5-2.

From the set of diagrams of Figure 5-2, it is evident that in the phase tetrahedron, the microemulsion three-dimensional domain forms an all-in-one block volume V built on the area of mutual solubilization existing in the ternary diagram of the water/sodium dodecylsulfate/1-pentanol system. A good image of V is given by the diagram (g) of Figure 5-2. V is an irregular body presenting two dissimilar protusions that both aim at the H apex (100% hydrocarbon) of the phase tetrahedron. Because of the highly irregular shape of V, the configuration of the microemulsion pseudo-ternary domain varies greatly with the value of the surfactant-to-cosurfactant ratio. When k_m is either small ($k_m < 1/3$ or so; diagrams (a) to (c) of Figure 5-2), or large ($k_m > 1$ or so; diagrams (h) and (i) of Figure 5-2), the microemulsion pseudo-ternary domain consists of two disjoined areas, N_1, that a priori may be considered as a realm-of-existence of "direct" microemulsion-type media S_1, and N_2, that a priori may be considered as a realm-of-existence of "inverse" microemulsion-type media S_2. It is readily seen from the diagrams (a), (b), (c),

---

*k_m is the surfactant/cosurfactant mass ratio; k_x is the molar ratio.

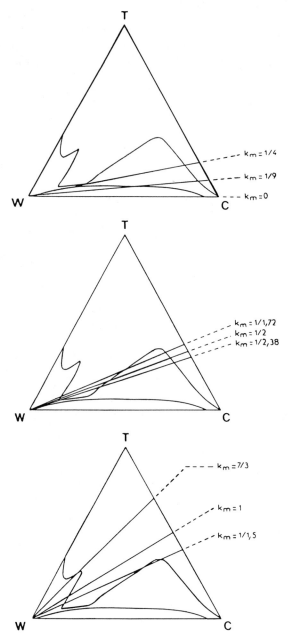

FIG. 5-1 Water/sodium dodecylsulfate/1-pentanol system (T = 25°C). Sections of the phase diagram corresponding to different values of the surfactant/cosurfactant mass ratio k_m.

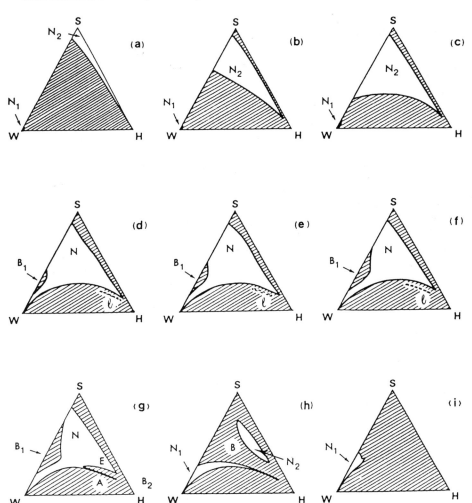

FIG. 5-2 Water/sodium dodecylsulfate/1-pentanol/n-dodecane system
(T = 25°C). Influence of the surfactant-to cosurfactant ratio upon
the configuration of the microemulsion pseudo-ternary domain (blank
areas).

(a) $k_m = 0$; $k_X = 0$ (f) $k_m = 1/1.72$; $k_X = 1/5.64$
(b) $k_m = 1/9$; $k_X = 1/29.45$ (g) $k_m = 1/1.5$; $k_X = 1/4.91$
(c) $k_m = 1/4$; $k_X = 1/13.09$ (h) $k_m = 1/1$; $k_X = 1/3.27$
(d) $k_m = 1/2.38$; $k_X = 1/7.78$ (i) $k_m = 1/0.43$; $k_X = 1/1.40$
(e) $k_m = 1/2$; $k_X = 1/6.54$

(h), and (i) of Figure 5-2 that low values of k_m are more favorable to the formation of "inverse" microemulsions and high values of k_m more favorable to the formation of "direct" microemulsions. It must be noted that for $k_m = 1$ (diagram (h) of Fig. 5-2) the N_1 region is needle-shaped and reaches confines of the phase diagram corresponding to hydrocarbon-rich compositions. This raises a problem as to the structure of the media belonging by their composition to the tip of the N_1 region, since owing to their high hydrocarbon content, they cannot be considered as classical direct microemulsions. At intermediary values of k_m, ($1/2.5 < k_m < 1/1.5$, diagrams (d), (e), and (f) of Fig. 5-2), the microemulsion pseudo-ternary domain consists of a large region N that spans a great part of the diagram, continuously from its W apex to the close vicinity of its HS edge, and of a small lenticular region l that lies close to the main region N but is fully inserted in the part of the diagram corresponding to the existence of polyphasic media. The region N appears to be the result of the merging of a S_1-type microemulsion region with an S_2-type microemulsion region, which evidently raises the question of the structure of the microemulsions whose representing composition points belong to the central part of N. As k_m increases, the lenticular region l becomes connected to the main region N, the microemulsion pseudo-ternary domain then being an all-in-one block region that presents an appendicular singularity A. This peculiar configuration, observed at $k_m = 1/1.5$ (diagram (g) of Fig. 5-2), appears as a prelude to the splitting of the microemulsion pseudo-ternary domain into two distinct regions N_1 and N_2, which occurs when k_m is high enough ($k_m = 1$; diagram (h) of Fig. 5-2). The logic that emanates from the diagrams of Fig. 5-2 proves that the lenticular regions l, the appendicular singularity A, and the needle-shaped region N_1 are different cross-sections of the microemulsion three-dimensional domain lobe that springs from the L_1 region of the ternary diagram of the water/sodium dodecylsulfate/1-pentanol system. This implies that the media whose representing points belong to these peculiar regions, in particular to lenticular regions l, are of the direct kind, even if they are hydrocarbon-rich. These peculiar media however cannot be classical direct microemulsions and therefore possess an "exotic" structure of some sort, compatible with the fact that their hydrocarbon content exceeds the threshold value corresponding to the random close packing of identical hard spheres [11], or even that corresponding to the face-centered cubic close packing [12]. This question will be reconsidered later, in light of data concerning microemulsion electrical conductivity and apparent viscosity.

Owing to the continuity existing between the water-rich and water-poor composition regions of the microemulsion domain, it may be predicted that microemulsion structure varies greatly but progressively as the composition varies. The question of the recognition of microemulsion structure in different regions of the microemulsion domain will be considered and discussed later, in the light of data concerning microemulsion electroconductive and viscous behavior.

The three diagrams of Figure 6 show the influence of k_m upon the configuration of the liquid crystal region. At $k_m = 1/2$ (diagram (a) of Fig. 6) the pseudo-ternary phase diagram presents two distinct ranges of compositions corresponding to the existence of monophasic mesmorphous media or of polyphasic media one phase of which is a mesophase.

The first range (labeled B_1 on the diagram) is simply the extension of the region of lamellar liquid crystals existing in the ternary diagram of the system water/sodium dodecylsulfate/1-pentanol (see Fig. 4). X-ray diffraction experiments have proved that as

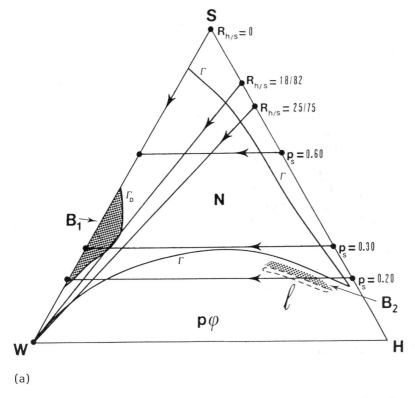

(a)

FIG. 6 Pseudo-ternary phase diagrams (T = 25°C) showing the respective locations of the microemulsion regions (blank areas) and lamellar liquid crystal regions (shaded areas). (a) $k_m = 1/2$; $k_x = 1/6.54$ (b) $k_m = 1/1.5$; $k_x = 1/4.91$ (c) $k_m = 1/1$; $k_x = 1/3.27$.

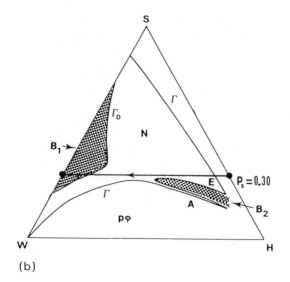

(b)

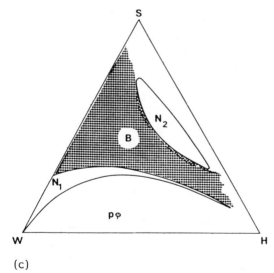

(c)

FIG. 6   (Continued)

the ternary mesophases, the quaternary mesophases, (n-dodecane added), indeed are classical D-type lamellar liquid crystals whose repeat distance δ increases as the overall content of water increases [7]. This is exemplified by the following data, obtained on samples characterized by a value of 0.02 of p_h, the n-dodecane mass ratio, (k_m = 1/2, or k_x = 1/6.54):

Overall mass fraction of water, p_W	0.60	0.68	0.73
Type of medium	S_2/D	D	D/S_1
δ (Å)	46.3	54.8	61.9

Figure 7 shows details of the internal organization of the B_1 region. Its central part corresponds to the existence of monophasic D-type mesophases. This D region is flanked by two regions corresponding to the existence of either S_2/D or D/S_1 diphasic media. It thus clearly appears that the B_1 region defines a range of compositions over which the $S_2 \leftrightarrow S_1$ transition, (microemulsion phase inversion), occurs according to the following pattern,

$$S_2 \longrightarrow S_2/D \longrightarrow D \longrightarrow D/S_1 \longrightarrow S_1$$

This is consistent with one of the possible phase-inversion processes listed in Winsor's scheme [13]. It is worth mentioning that the internal boundaries between the S_2/D, D, and D/S_1 subregions converge on a certain point P that also belongs to the curve Γ_D, that is, the borderline separating the B_1 region from the main microemulsion region N. In samples whose composition is defined by point P, the interface between the liquid crystal and microemulsion phases is highly wrinkled and spots of liquid crystal remain indefinitely suspended in the microemulsion phase. This suggests that both of the coexisting phases have the same specific mass and that the interfacial tension between them is very low. According to Roux [14], the singular point P could be a point of azeotropy.

The region that separates the microemulsion main area N and small lenticular area l defines a narrow composition range over which the system phase behavior is highly complex [5,15]. A certain subregion B_2 of this narrow region corresponds to the existence of mesophases. The internal organization of this B_2 subregion is similar to that of the B_1 region, with a central part corresponding to the existence of monophasic mesophases and two flanking areas corresponding to the existence of diphasic media whose lower and upper phases are mesophases. This is consistent with certain observations reported by other authors [16,17]. X-ray diffraction experiments [7] tend to

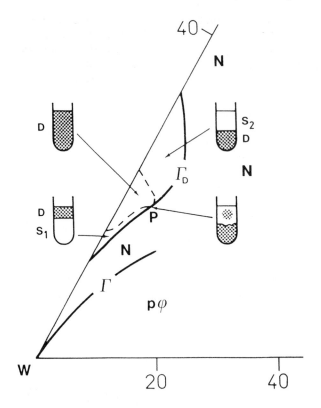

FIG. 7   Detailed view of the internal organization of the B_1 region of the $k_m = 1/2$ pseudo-ternary phase diagram [see diagram (a) of Fig. 6].

indicate that the mesophases of the media belonging to the B_2 region have a lamellar structure whose repeat distance is ∿150 Å. This result suggests that the B_2 mesophases could be lamellar liquid crystals of the "mucous woven type" labeled type B mesophases by Ek-wall et al. [18]. The existence of mesophases of the same kind was observed also by Dvolaitzky et al. [19,20] in the case of the water/sodium dodecylsulfate/1-pentanol/cyclohexane system. Based on data obtained by using the spin-labeling technique, Di Meglio et al. [20] argued that these peculiar lamellar mesophases are microemulsion precursors.

Quite similar situations were observed for $k_m = 1/2.38$ and $k_m = 1/1.72$, (diagrams (d) and (f) of Fig. 5-2). At $k_m = 1/1.5$, (diagram

(b) of Fig. 6), the B_1 region is larger than that corresponding to $k_m = 1/2$ or so and the B_2 region entirely covers the fingerlike composition range that protrudes in between the appendicular region A and terminal region E of the microemulsion pseudo-ternary domain. As proved by x-ray diffraction experiments [7], the mesophases of samples belonging to this B_2 region are type B liquid crystals. When k_m increases, the B_1 and B_2 regions become larger and closer to each other until they eventually merge to form a continuous strip (see diagram (c) of Fig. 6), which proves that there is in the phase tetrahedron a continuous range of compositions corresponding to the existence of lamellar mesophases (either monophasic or belonging to polyphasic media). This implies that lamellar structures may incorporate higher and higher amounts of hydrocarbon without collapsing; they simply swell in a continuous way and their structure changes accordingly, passing from the classical type D to the type B.

IV. MICROEMULSION ELECTROCONDUCTIVE AND VISCOUS BEHAVIOR AS MICROEMULSION STRUCTURE PROBE

It was demonstrated in a previous chapter that constant correlations do exist between system type and microemulsion electroconductive and viscous behavior [1]. This experimental evidence vindicates the use of electroconductimetry and viscosimetry as methods of investigation into microemulsion structure [21]. It is worth recalling here that, by studying transport phenomena in microemulsions, several groups of scientists [22-41] have obtained, in the case of water/ionic surfactant/alcohol/hydrocarbon systems, quite useful information on the structure of monophasic microemulsions of or microemulsion phases belonging to polyphasic media of the Winsor's classification [42].

As a follow-up of previous studies devoted to microemulsions of type U systems [43-49], a thorough investigation into the electroconductive and viscous behavior of water/sodium dodecylsulfate/1-pentanol/n-dodecane microemulsions was undertaken, in an attempt to probe microemulsion structure in different regions of microemulsion pseudo-ternary domains. Most of the data reported thereafter concern microemulsions characterized by $k_m = 1/2$. The pseudo-ternary diagram corresponding to this value of the surfactant-to-cosurfactant mass ratio is actually quite interesting because its microemulsion domain consists of a large region N that spans a very wide range of compositions and of a small lenticular region l that, as previously suggested, probably corresponds to microemulsions of some "exotic" kind, (see diagram (a) of Fig. 6).

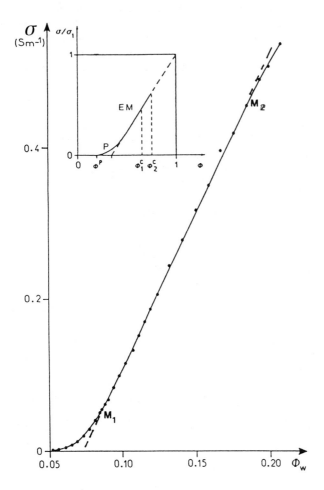

FIG. 8 Variations of microemulsion electrical conductivity σ versus ϕ_W along the [$k_m = 1/1.5$, $p_S = 0.30$] experimental path, at $T = 25°C$ [see diagram (b) of Fig. 6]. The insert shows the variations of the reduced electrical conductivity σ/σ_1, as ϕ, the conductor volume fraction, increases in a randomly heterogeneous binary conductor-insulator composite medium [see Eqs. (1), (2)].

A. Main Microemulsion Region N

Figure 8 shows the variations of σ, the microemulsion electrical conductivity, versus ϕ_w, the water volume fraction, along a p_s-type experimental path. The sodium dodecylsulfate titer being constant, the conductivity increase mainly results from and plainly reveals the modification that microemulsion structure undergoes as the water content (or water–oil ratio WOR), increases. It is readily seen from Fig. 8 that the experimental conductivity plot exhibits features quite similar to those of the theoretical conductivity curve reported in the insert. This theoretical curve shows how, according to the percolation [50–55], and effective medium [56–58], theories, the electrical conductivity σ of a randomly heterogeneous binary conductor-insulator composite medium varies as ϕ, the volume fraction of the conducting constituent, increases (σ_1 is the conductor electrical conductivity). As long as ϕ is smaller than a certain ϕ^P, called the "percolation threshold," the sample is nonconducting because there is no connection between the disperse conducting particles. ϕ^P is the value of ϕ at which the medium suddenly becomes conducting, owing to the formation of an "infinite cluster" of conducting particles. In the upper close vicinity of ϕ^P, the electrical conductivity of the medium increases with ϕ according to a power law [52–55]:

$$\sigma \propto (\phi - \phi^P)^t \tag{1}$$

where t is an exponent whose value depends only on the system dimensionality and not, contrary to ϕ^P, on the physical details of the composite medium. For a three-dimensional system, the value of t is 1.5–1.6 [53]. As ϕ further increases, the power law is no longer valid and the electrical conductivity increases according to the following linear law:

$$\sigma = 3\sigma_1 (\phi - 1/3)/2 \tag{2}$$

The above equation is a simplified form of the effective medium theory (EMT) formula successively derived, in different ways, by Bruggeman [56], Bottcher [57], and Landauer [58]. Equation (2) remains valid as long as ϕ is smaller than the close-packing volume fraction ϕ^C corresponding to the geometric shape of the conducting particles. In the case of uniformly sized hard spheres, the value of ϕ^C corresponding to the random close packing is $\phi_1{}^C = 0.637$ [11], and that corresponding to the face-centered cubic close-packing is $\phi_2{}^C = 0.741$ [12]. When ϕ exceeds ϕ^C, the electrical conductivity variation departs from that predicted by Eq. (2).

It was shown in detail elsewhere [21,43,59] that the electroconductive behavior of type U microemulsions containing 50% or less of water may indeed be depicted through the percolation and effective

medium theories*. That the initial electrical conductivity increase is
very drastic and conforms to a power law is here clearly illustrated
by the $\log \sigma$ versus ϕ_w curve of Figure 9. The insert in this figure
shows how the exponent t and percolation threshold may be determined. In the present case, the value of the parameter x   that
corresponds to the best linearization of the $\sigma^{1/x}$ versus ϕ_w plot is
1.5, which is in very good agreement with Kirkpatrick's theoretical
prediction for three-dimensional systems [53]. The corresponding
percolation threshold value is, in terms of water volume fraction,
$\phi_w{}^p$ = 0.06. This value is comparable to those found by Lagües
et al. [28–30], in the case of water/sodium dodecylsulfate/1-pentanol/
cyclohexane microemulsions.

Zradba [5] found that the general features of microemulsion
electrical conductivity variation along p_S-type experimental paths
are the same whatever the [sodium dodecylsulfate-pentanol] content.
This experimental fact is illustrated here by Figures 10 to 12. When
p_S is sufficiently low (Fig. 10), the structuring of the conductivity
plot clearly indicates that, as the water content increases, σ, the
microemulsion electrical conductivity varies according to three successive modes. The initial drastic increase (see Fig. 11) reveals
the existence of a percolation phenomenon that can be ascribed to
inverse microdroplet aggregation. The next linear increase, which
is depictable through the effective medium theory, can be interpreted as the consequence and expression of the formation of aqueous microdomains resulting from the partial fusion of clustered inverse microdroplets. The final nonlinear increase of σ probably
reveals that the medium undergoes further structural modifications
and becomes bicontinuous [61–63], owing to progressive growth and
interconnection of the aqueous microdomains. At higher values of
p_S (see Fig. 12) the final nonlinear increase does not exist, which
suggests that microemulsions rich in surfactant and alcohol cannot
be in a strict bicontinuous structural state. Figure 12 also shows
that when p_S is high, the onset of the microemulsion electrical conductivity linear increase is ill defined, which tends to indicate that
the transition between the inverse microdroplet clustering and fusion
processes is very progressive.

Figures 13 to 15 show the variations of microemulsion electrical
conductivity σ as ϕ_w, the water volume fraction, increases along

---

*It may be wondered whether the percolation model developed to re-
present electrical conduction in heterogeneous solids is still applicable
to mixtures of liquids, in which brownian motion takes place. From
the analysis carried out by Lagües thereupon [60], it emerges that
there is no significant difference between the values of the exponent
t corresponding to the two cases.

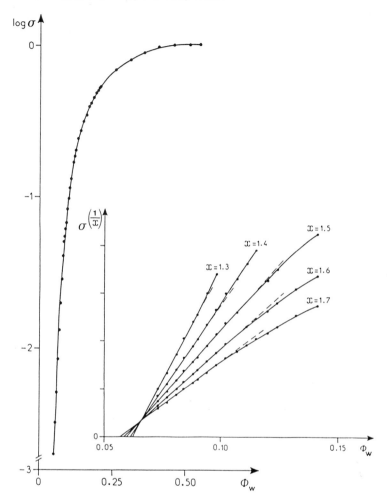

FIG. 9 Variations of $\log \sigma$ versus ϕ_w, the water volume fraction, (see Fig. 8). The insert shows how the critical exponent t and percolation threshold ϕ_w^p may be determined.

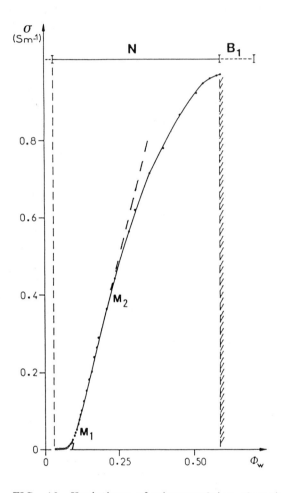

FIG. 10 Variations of microemulsion electrical conductivity σ versus ϕ_W, the water volume fraction, along the [$k_m = 1/2$, $p_S = 0.30$] experimental path, at T = 25°C [see diagram (a) of Fig. 6].

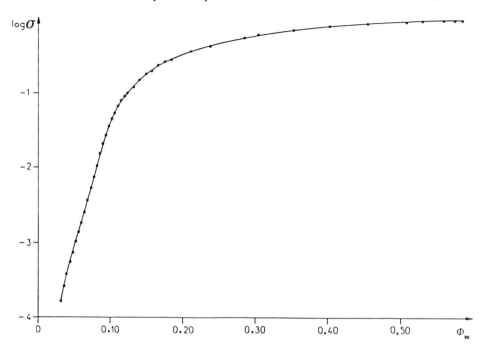

FIG. 11 Variations of logσ versus ϕ_W, the water volume fraction (see Fig. 10).

$R_{h/s}$-type experimental paths. The same analysis as above may be made concerning the ascending branches of the σ versus ϕ_W curves. The structuring of these ascending branches is quite similar to that of the p_s-type electrical conductivity plots (see Fig. 10), and consequently may be used in the same way, to mark out inside the microemulsion pseudo-ternary domain, the composition ranges corresponding to the existence of clustered inverse microdroplets (initial nonlinear increase of σ), aqueous microdomains (linear increase of σ), and bicontinuous structures (final nonlinear increase of σ). It should be remarked that, on the conductivity plots corresponding to low values of $R_{h/s}$ (see Figs. 14 and 15), the transition from the initial nonlinear section and the linear section is not sharp, which is consistent with the observation reported previously as concerns p_s-type conductivity plots characterized by a high value of p_s. Since the B_1 region of the pseudo-ternary diagram (a) of Figure 6 defines a range of compositions over which microemulsion phase inversion takes place, it is quite evident that the descending branches of the

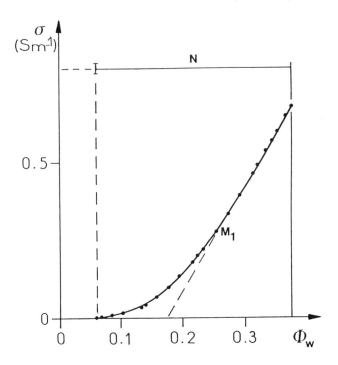

FIG. 12 Variations of microemulsion electrical conductivity σ versus ϕ_W, along the [k_m = 1/2, p_S = 0.60] experimental path, at T = 25°C [see diagram (a) of Fig. 6].

conductivity plots of Figures 14 and 15 correspond to the existence of water-continuous microemulsion-type media. The decrease of σ merely results from the fact that the continuous aqueous phase is progressively diluted with water*. It follows then that, when the conductivity plot is a continuous archlike curve (see Fig. 13), the descending branch corresponds to the existence of water-continuous microemulsion-type media. This implies that, on R_h/s-type experimental paths that do not intersect the B_1 region, the composition points corresponding to electrical conductivity maxima mark the transition from preferentially hydrocarbon-continuous to preferentially water-continuous microemulsion-type media.

---

*It will be shown further on that water-rich microemulsions could have in fact a rather complex electroconductive behavior.

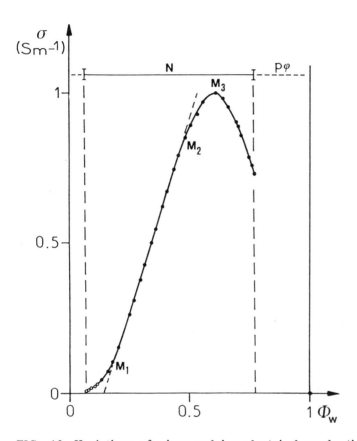

FIG. 13 Variations of microemulsion electrical conductivity σ versus ϕ_w, along the $[k_m = 1/2, R_{h/s} = 25/75]$ experimental path, at T = 25°C [see diagram (a) of Fig. 6].

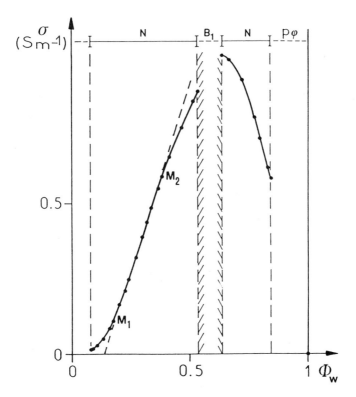

FIG. 14 Variations of microemulsion electrical conductivity σ versus ϕ_W, along the $[k_m = 1/2,\ R_{h/s} = 18/82]$ experimental path, at T = 25°C [see diagram (a) of Fig. 6].

On the basis of the preceding considerations, it appears possible to use microemulsion electroconductive behavior as a probe into microemulsion structure. In particular, the boundaries of the composition ranges corresponding to different microemulsion structural states may be determined by plotting, within the main microemulsion region N, the composition points at which the variation mode of microemulsion electrical conductivity changes (see Fig. 16). The basic division of the main microemulsion region N into two broad areas is obtained with the line C_3. Since it is defined by the points M_3 that on $R_{h/s}$-type conductivity curves correspond to maxima of microemulsion electrical conductivity (see Fig. 13), C_3 represents the dividing line between the realm-of-existence N_d of preferentially water-continuous microemulsion-type media (lower part of N), and the realm-of-existence N_i of preferentially hydrocarbon-continuous microemulsion-type media. In this respect, it is quite gratifying to observe that the point P on which the B_1 region internal boundaries and the boundary Γ_D between the B_1 and N regions converge (see Fig. 7) belongs to the line C_3. N_i itself is partitioned by the lines C_1 and C_2 into three adjacent subareas, labeled CIM, MIM, and BIP in Figure 16. C_1 and C_2 are defined by points M_1 and M_2, respectively, that on conductivity curves of the p_s-type (open and filled squares) and $R_{h/s}$-type (open and filled circles), correspond to the onset and end of microemulsion electrical conductivity linear increase, (see Figs. 10 and 13). The subarea CIM which is the part of N_i contained by C_1 and the upper branch of the boundary Γ of the main microemulsion region N, defines the composition range over which the inverse microdroplets form clusters whose existence is revealed by the percolation phenomenon. The subarea MIM, bounded by the lines C_1 and C_2, represents the range of compositions corresponding to the existence of aqueous microdomains that result from the partial merging of clustered inverse microdroplets. In between

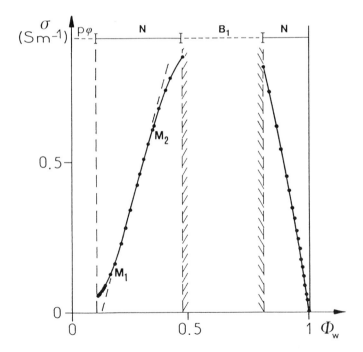

FIG. 15 Variations of microemulsion electrical conductivity σ versus ϕ_w, along the [$k_m = 1/2$, $R_{h/s} = 0$] experimental path, at T = 25°C [see diagram (a) of Fig. 6].

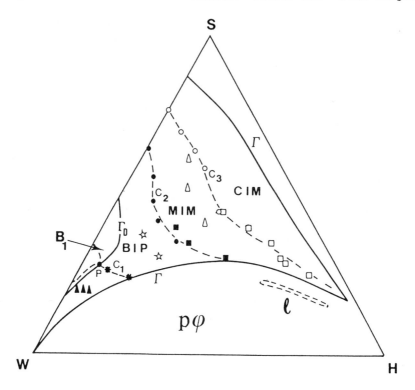

FIG. 16 Pseudo-ternary phase diagram of the system water/sodium dodecylsulfate/1-pentanol/n-dodecane ($k_m = 1/2$, T = 25°C). Partition of the microemulsion main region N into different adjacent areas corresponding to various microemulsion structural states.

the lines C_2 and C_3, lies the subarea BIP over which, as the result of the interconnecting of the aqueous microdomains, the microemulsions are in a bicontinuous structural state that is inherent to microemulsion progressive-phase inversion. As concerns the area N_d, located below the line C_3, the delimiting of composition ranges corresponding to different microemulsion structural states is not so easy as in the case of the N_i area, first because N_d is relatively small, as compared to N_i, and secondly because the descending branch of the $R_{h/s}$-type conductivity curves do not exhibit any obvious singularity (see Figs. 13 to 15). As illustrated by Figure 17, it however appears possible to get a certain idea of the internal organization of N_d by considering the variation of Λ, Λ is the microemulsion equivalent conductance, versus $C^{1/2}$ (C, expressed in moles per cubic meter, being the concentration of sodium dodecyl-sulfate in water). On the one hand, the fact that different sections of the Λ versus $C^{1/2}$ curves are straight segments confirms that the N_d area defines a range of compositions corresponding to water-continuous media. On the other hand, the structuring of the Λ versus $C^{1/2}$ curves reveals the structural diversity of the water-rich microemulsion-type media. The sections of the Λ versus $C^{1/2}$

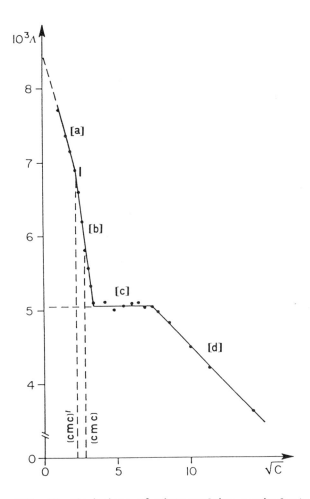

FIG. 17 Variations of microemulsion equivalent conductance Λ, expressed in Sm^2 $mole^{-1}$, along the part of the [$k_m = 1/2$, $R_{h/s} = 0$] experimental path corresponding to water-rich compositions, at $T = 25°C$ [see diagram (a) of Fig. 6].

plot labeled [a] and [b] (see Fig. 17) are easily identifiable. The sudden change in Λ decrease mode that occurs at a certain value of C (point I) is characteristic of a molecular solution $\leftrightarrow$ micellar solution transition. It consequently appears that section [a] defines a range of compositions corresponding to the existence of molecular solutions and section [b] a range of compositions corresponding to the existence of "direct" micellar solutions. The value of C corresponding to the breakpoint I defines (cmc)', that is the critical micelle concentration of sodium dodecylsulfate in 1-pentanol aqueous solutions characterized by the value 1/6.54 of k_x, the molar ratio of sodium dodecylsulfate to 1-pentanol. The value found for (cmc)' (expressed in terms of mass fraction) is 0.145%, which value of course is lower than 0.235%, the value of the sodium dodecylsulfate (cmc) in pure water at the same temperature (T = 25°C) [64-66]. Over a certain range of compositions, defined by the segment [c] in Figure 17, Λ keeps to within less than 1%, a constant value of 5.04 Sm^2 $mole^{-1}$, which is very close to the value of the sodium cation limit equivalent conductance, namely $\lambda_c = 5.02$ Sm^2 $mole^{-1}$ [67]. This phenomenon may be interpreted as the expression of the existence of bulky "direct" spherical micelles ("direct" spherical microdroplets) over the said composition range. The electrical conduction would be almost entirely due to sodium counterions belonging to the Gouy–Chapman layers of the "direct" microdroplets, the contribution of the bulky macroions formed by the inner parts of the direct microdroplets being negligible. The final section of the Λ versus $C^{1/2}$ plot ([d] on Fig. 17) corresponds to a linear decrease of the microemulsion equivalent conductance as the concentration of sodium dodecylsulfate increases. This suggests that over the composition range defined by [d] the "direct" microemulsions are in a new structural state which, as a prelude to microemulsion phase inversion, could correspond to the existence of highly oblong "direct" particles or even of bicontinuous structures of some kind. From the above considerations, it comes out that the area N_d, that is, the part of the main microemulsion region N that is the realm-of-existence of preferentially water-continuous microemulsion-type media, may be partitioned into four adjacent subareas corresponding to aqueous molecular solutions of sodium dodecylsulfate, 1-pentanol and n-dodecane, direct micellar solutions, water-continuous microemulsions containing "direct" spherical microdroplets and water-continuous microemulsions with some sort of structure precluding microemulsion phase inversion.

Some additional information as to the structural diversity of microemulsions belonging by their compositions to the main microemulsion region N can be obtained by analyzing microemulsion viscous behavior. As illustrated by the plots reported in Figures 18 and 19, η, the microemulsion apparent viscosity, varies in a

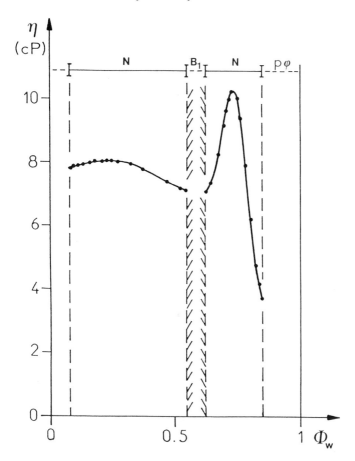

FIG. 18 Variations of microemulsion apparent viscosity η versus ϕ_w, along the $[k_m = 1/2, R_{h/s} = 18/82]$ experimental path, at T = 25°C [see diagram (a) of Fig. 6].

nonmonotonic way when the water content increases along $R_{h/s}$-type experimental paths. The viscosity plot of Figure 18 tells that η reaches a local maximum at two composition points that are located on either side of the B_1 region and, as the boundary Γ_D of the B_1 region is approached, tends toward a virtual minimum corresponding to a certain composition whose representing point would belong to the B_1 region, namely, the composition range over which microemulsion phase inversion occurs. It thus appears possible to relate the occurrence of microemulsion phase inversion to the existence of local

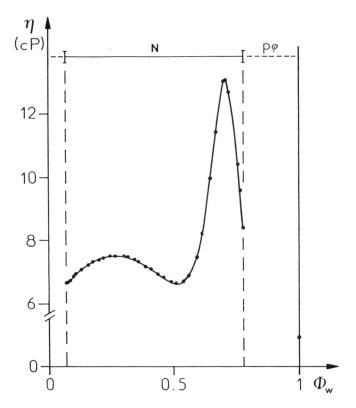

FIG. 19 Variations of microemulsion apparent viscosity η versus ϕ_w, the water volume fraction, along the [$k_m = 1/2$, $R_{h/s} = 25/75$] experimental path, at $T = 25°C$ [see diagram (a) of Fig. 6].

minima of microemulsion viscosity. Therefore, it is quite gratifying to observe that the composition points, (open stars in Fig. 16), at which the microemulsion-apparent viscosity η reaches a local mini-mum (see Fig. 19) belong to the region BIP which can be considered, on the grounds of the results concerning microemulsion electrocon-ductive behavior as the range of compositions corresponding to micro-emulsion bicontinuity and progressive phase inversion. Figure 19 shows that the composition points at which η, the microemulsion-apparent viscosity, reaches its first local maximum (water content increasing), belong to the subarea MIM delimited by the lines C_1 and C_2, (open triangles in Fig. 16). This is consistent with the alleged existence of water microdomains that result from the merging of clustered inverse microdroplets and preludes to the formation of bicontinuous structures. As for the viscosity second local maxima, whose corresponding composition points belong to the N_d area of the main microemulsion region N (filled triangles in Fig. 16), their interpretation is more complex, because electroviscous effects ought to be taken into account [68,69]. However, it is not unreasonable to consider that the existence of these viscosity maxima reveal that, close to the line C_3, the water-continuous microemulsions are in a structural state of some exotic kind, compatible with a progressive phase-inversion process. To add a final touch to the preceding considerations on microemulsion viscosity, it is worth mentioning that two-peaked plots such as that of Figure 19 have been reported by other scientists [34-36], as concerns the viscous behavior of microemulsion phases belonging to WI, WII, and WIII media. Local minima of microemulsion viscosity were observed at composition points belonging to the central part of the realm-of-existence of WIII triphasic media, that is when the microemulsions (middle phases) most probably are bicontinuous [34-36,70-71]. This finding is consistent with the results reported above as to the location of the composition points corresponding to viscosity local minima (open stars in Fig. 16), and thus indirectly confirms that, within the main microemulsion reion N, the subarea labeled BIP defines a range of compositions corresponding to microemulsions in bicontinuous state. As for the local maxima of microemulsion viscosity, these were observed at composition points belonging to the realms-of-existence of either WI diphasic media, whose lower phases are microemulsions of the "direct" kind, or WII diphasic media, whose upper phases are microemulsions of the "inverse" kind. This again is consistent with the results reported above concerning the location, within the main microemulsion region N, of the composition points corresponding to local maxima of microemulsion viscosity (see Fig. 16).

B. Small Lenticular Region *l*

From the study of the influence of the sodium dodecylsulfate-to-1-pentanol ratio upon microemulsion pseudo-ternary domain configuration, (see Fig. 5-2), it is found that small lenticular regions *l*, such as that existing in the phase diagram of Figure 6, belong to the three-dimensional microemulsion domain highly elongated lobe springing from the "direct" solubilization region L_1 of the diagram of the water/sodium dodecylsulfate/1-pentanol system, (see Fig. 4). This implies that the microemulsions belonging by their compositions to lenticular regions *l* are water continuous, even if their hydrocarbon contents are very high. That these peculiar microemulsions certainly possess a continuous aqueous phase is strongly supported by certain results obtained as to their electroconductive and viscous properties.

Figures 20 and 21 show the variations, versus the water volume fraction ϕ_w, of microemulsion electrical conductivity σ and viscosity η,

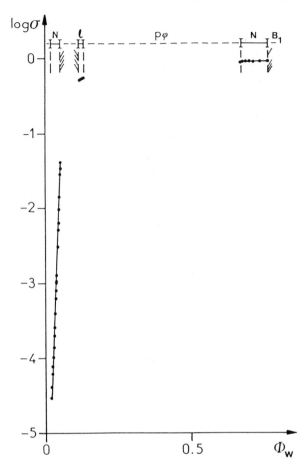

FIG. 20 Variations of microemulsion electrical conductivity σ versus
ϕ_w, along the [$k_m = 1/2$, $p_s = 0.20$] experimental path, at T = 25°C
(see diagram (a) of Fig. 6).

along a p_s-tpye experimental path that cuts across the main microemulsion region N and small lenticular region l of the $k_m = 1/2$
phase diagram (see diagram (a) of Fig. 6). It is readily seen from
the electrical conductivity plot of Figure 20 that the surfactant concentration being fixed ($p_s = 0.20$) the microemulsions belonging by
their compositions to the region l (water content ∿10%) are almost
as conducting as the water-rich microemulsions (water content above
70%) whose composition points belong to the upper part of the area

N_d of the main microemulsion region N (see Fig. 16). In contrast, they are ten times more conducting than microemulsions, quite comparable to them by composition, whose representing points belong to the main microemulsion region N. As concerns microemulsion viscous behavior, it appears from Figure 21 that the viscosity of the l microemulsions, first, decreases when the water content increases, as does the viscosity of the water-rich microemulsions, and, second is of the same order as that of the most water-rich microemulsions. The same peculiar electroconductive and viscous behavior of l microemulsions was observed at other values of p_s [5].

Thus, the results obtained as to the electroconductive and viscous properties of microemulsions belonging by their composition to a lenticular region l converge with data concerning the configuration

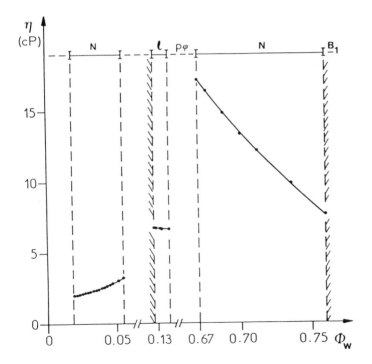

FIG. 21 Variations of microemulsion apparent viscosity η versus ϕ_w, along the [$k_m = 1/2$, $p_s = 0.20$] experimental path, at T = 25°C [see diagram (a) of Fig. 6].

of the three-dimensional microemulsion domain in indicating that these peculiar microemulsions are water-continuous media. They however cannot be classical "direct" microemulsions, (assemblies of identical globules), because their hydrocarbon contents generally exceed that corresponding to the random close packing [11] and even the face-centered cubic close packing [12] of identical hard spheres. This implies that they possess an "exotic" structure of the bicontinuous sort. Since their compositions are highly unbalanced in favor of hydrocarbon, it may be wondered whether l microemulsions are bi-continuous in the usual sense, which is more suited to microemulsions containing water and hydrocarbon in comparable amounts [63]. A possible alternative model for the structure of l microemulsions would be that of a dynamic random arrangement of organic polyhedral mi-crodomains separated by thin aqueous films [66]. This model, en-visaged by Lissant [72] to depict emulsions whose disperse phase content is very high, also has been used by Chen et al. [73] to explain the apparently anomalous electroconductive and viscous be-havior of water-poor three-component microemulsions incorporating didodecyldimethylammonium bromide as ionic surfactant and no cosurfactant.

V. GENERAL CONCLUSION

The results reported and commented upon in the preceding sections lend strong support to certain of the broad conclusions that were previously derived as concerns the relations existing between micro-emulsion domain configuration, transport properties, and microstruc-ture [1], and throw some light upon the structural transformations that type U microemulsions undergo as the composition varies.

It is experimentally demonstrated that, in accordance with some general principles governing the influence of the nature of the al-kanol cosurfactant upon certain microemulsion physicochemical prop-erties [1–4], the water/sodium dodecylsulfate/1-pentanol/n-dodecane system belongs to the type U category [1,2], which means that its microemulsion three-dimensional domain is an all-in-one block volume V that spans a wide range of compositions. That the said system is type U is confirmed by the fact that electrical conduction in water-poor microemulsions is of the percolation kind.

The microemulsion three-dimensional domain V presents two high-ly dissymmetric lobes that both extend to phase tetrahedron confines corresponding to n-dodecane-rich compositions. One of the lobes is preferentially connected to the region of "direct" solubilization L_1 of the ternary diagram of the water/sodium dodecylsulfate/1-pentanol system, and the other to the region of "inverse" solubilization L_2. The region of the phase tetrahedron where the two lobes merge cor-respond to n-dodecane-poor compositions.

At certain values of the surfactant-to-cosurfactant mass ratio k_m, ($k_m = 1/2$), the cross-section of the phase tetrahedron is such

that the microemulsion pseudo-ternary domain consists of two dissimilar regions, N and *l*. The main microemulsion region N spans a
great part of the pseudo-ternary diagram, continuously from the W
apex (100% water) to the close vicinity of the HS edge, (part of the
diagram corresponding to water-poor compositions). This implies
that, over this N region, the microemulsion local structure is highly
diverse, but varies smoothly with composition. The small lenticular
region *l* belongs to the lobe of the microemulsion three-dimensional
domain that is connected preferentially to the region L_1 of "direct"
solubilization of the system water/sodium dodecylsulfate/1-pentanol;
l consequently appears to be a region of water-continuous microemulsion-type media although it corresponds to n-dodecane-rich compositions. This assertion is strongly supported by the fact that the
electroconductive and viscous properties of *l* microemulsions are
similar to those of water-rich microemulsions.

By carefully analyzing microemulsion electroconductive and viscous behavior, it appears possible to divide the main microemulsion
region N into seven composition ranges that correspond to different
microemulsion structural states. As the water content decreases,
microemulsion structure varies progressively according to the following sequence: molecular solutions of sodium dodecylsulfate, 1-pentan-
ol and n-dodecane in water → "direct" micellar solutions → classical
"direct" microemulsions → preferentially water-external microemulsions
of ill-defined intermediary structure (bicontinuous ?) → preferentially
hydrocarbon-external bicontinuous microemulsions → "inverse" microemulsions containing aqueous microdomains → classical "inverse"
microemulsions in which the microdroplets tend to cluster. This
pattern of structural evolution in type U monophasic *nonsaturated*
microemulsion-type media is consistent with Winsor's scheme [13] and
with propositions made by other groups as to the structural evolution in monophasic *saturated* microemulsions whose cosurfactant is a
"short" alkanol or in microemulsion phases belonging to polyphasic
media of the Winsor's classification [42]. It is thus fully demonstrated that, as has been repeatedly suggested for 10 years or more
[2,15,21,44-49,74,75], electroconductimetry and viscosimetry are
reliable methods that may be used to probe into microemulsion structure (at least in a qualitative way), and to delimit, within the vast
microemulsion domains of Type U systems, composition ranges corresponding to different microemulsion structural states. That could
be quite useful to practitioners dealing with the formulation of
specialty fluids of industrial interest, for instance of complex fluids
intended to be used as multifunctional solvents or as vehicles of
chemically or biologically active elements [76].

ACKNOWLEDGMENTS

A. Zradba wishes to thank the Ministère des Relations Extérieures
(France), for the financial help provided to him from October 1980

to July 1983. L. Nicolas-Morgantini is grateful to the Ministère de la Recherche et de l'Industrie (France) for the research scholarship granted to him from October 1981 to October 1983. Professor M. Clausse expresses to NATO Scientific Affairs Division (Belgium), his deep appreciation for the support granted by this institution (Contract No. 020.82).

The authors wish to express to Mrs. A. M. Schmitt their sincere appreciation for her meticulous typing of the present article.

REFERENCES

1. Clausse, M., Nicolas-Morgantini, L., Zradba, A., and Tourand D., *this volume*, pp. 15–62.
2. Clausse, M., Peyrelasse, J., Boned, C., Heil, J., Nicolas-Morgantini, L., and Zradba, A., in *Surfactants in Solution*, Mittal, K. L., and Lindman, B. (eds.), Plenum Press, New York, 1984, vol. III, pp. 1583–1626.
3. Clausse, M., Nicolas-Morgantini, L., and Zradba, A., in *Proc. World Surfactants Congress*, (Munich, May 6–10, 1984), Kürle Druck and Verlag, Gelnhausen, West Germany, 1984, vol. III, pp. 209–219.
4. Clausse, M., Heil, J., Zradba, A., and Nicolas-Morgantini, L., *Jorn. Com. Esp. Deterg. 16*, 497 (1985).
5. Zradba, A., Thèse de Doctorat de 3ème cycle, Université de Pau, France, 1983.
6. Ginnings, P. M., and Baum, R., *J. Am. Chem. Soc. 59*, 1111 (1937).
7. Friberg, S. E., private communication, 1983.
8. Mandell, L., Fontell, K., and Ekwall, P., *Adv. Chem. Ser. 63*, 89 (1967).
9. Ekwall, P., Mandell, L., and Fontell, K., *J. Colloid Interface Sci. 31*, 508 (1969).
10. Ekwall, P., Mandell, L., and Fontell, K., *J. Colloid Interface Sci. 31*, 530 (1969).
11. Finney, J. L., *Nature 266*, 309 (1977).
12. Günther, K., and Heinrich, D., *Z. Physik. 185*, 345 (1965).
13. Winsor, P. A., *Chem. Rev. 68*, 1 (1968).
14. Roux, D., private communication, 1984.
15. Clausse, M., Zradba, A., and Nicolas-Morgantini, L., *C. R. Acad. Sci. Paris 296*, Sér. II, 237 (1983).
16. Roux, D., Bellocq, A. M., and Leblanc, M. S., *Chem. Phys. Lett. 94*, 156 (1983).
17. Roux, D., and Bellocq, A. M., in *Macro- and Microemulsions*, Shah, D. O. (ed.), ACS Symposium Series 272, American Chemical Society, Washington DC, 1985, pp. 105–118.

18. Ekwall, P., Mandell, L., and Fontell, K., *Mol. Crystals Liquid Crystals 8*, 157 (1968).
19. Dvolaitzky, M., Ober, R., Billard, J., Taupin, C., Charvolin, J., and Hendricks, Y., *C. R. Acad. Sci. Paris 292*, Sér. II, 45 (1981).
20. Di Meglio, J. M., Dvolaitzky, M., Ober, R., and Taupin C., *C. R. Acad. Sci. Paris 296*, Sér. II, 405 (1983).
21. Clausse, M., in *Encyclopedia of Emulsion Technology*, Becher, P. (ed.), Marcel Dekker, New York, 1983, vol. I, pp. 481-715.
22. Shah, D. O., and Hamlin, R. M., *Science 171*, 483 (1971).
23. Shah, D. O., Tamjeedi, A., Falco, J. W., and Walker, R. D., *AIChE J. 18*, 1116 (1972).
24. Falco, J. W., Walker, R. D., and Shah, D. O., *AIChE J. 20*, 510 (1974).
25. Shah, D. O., Bansal, V. K., Chan, K., and Hsieh, W. C., in *Improved Oil Recovery by Surfactant and Polymer Flooding*, Shah, D. O., and Schechter, R. S. (eds.), Academic Press, New York, 1977, pp. 293-337.
26. Bansal, V. K., Chinnaswamy, K., and Shah, D. O., *J. Colloid Interface Sci. 72*, 524 (1979).
27. Bansal, V. K., Shah, D. O., and O'Connell, J. P., *J. Colloid Interface Sci. 75*, 462 (1980).
28. Lagües, M., Ober, R., and Taupin, C., *J. Physique (Paris) Lett. 39*, L-487 (1978).
29. Dvolaitzky, M., Lagües, M., Le Pesant, J. P., Ober, R., Sauterey, C., and Taupin, C., *J. Phys. Chem. 84*, 1532 (1982).
30. Lagües, M., and Sauterey, C., *J. Phys. Chem. 84*, 3503 (1980).
31. Larche, F., Rouvière, J., Delord, P., Brun, B., and Dussossoy, J. L., *J. Physique (Paris) Lett. 41*, L-437 (1980).
32. Cazabat, A. M., Chatenay, D., Langevin, D., and Pouchelon, A., *J. Physique (Paris) Lett. 41*, L-441 (1980).
33. Cazabat, A. M., Langevin, D., Meunier, J., and Pouchelon, A., *J. Physique (Paris) Lett. 43*, L-89 (1982).
34. Bennett, K. E., Hatfield, J. C., Davis, H. T., Macosko, C. W., and Scriven, L. E., in *Microemulsions*, Robb, I. D. (ed.), Plenum Press, New York, 1982, pp. 65-84.
35. Kaler, E. W., Bennett, K. E., Davis, H. T., and Scriven, L. E., *J. Chem. Phys. 79*, 5673 (1983).
36. Kaler, E. W., Davis, H. T., and Scriven, L. E., *J. Chem. Phys. 79*, 5685 (1983).
37. Sjöblom, E., Edberg, T., and Stenius, P., *J. Dispersion Sci. Technol. 4*, 123 (1983).
38. Van Nieuwkoop, J., and Snoei, G., *J. Colloid Interface Sci. 103*, 400 (1985).
39. Van Nieuwkoop, J., and Snoei, G., *J. Colloid Interface Sci. 103*, 417 (1983).

40. Lindman, B., Stilbs, P., and Moseley, M. E., *J. Colloid Interface Sci. 83*, 569 (1981).
41. Lindman, B., and Stilbs, P., in *Proc. World Surfactants Congress*, (Munich, May 6–10, 1984), Kürle Druck und Verlag, Gelnhausen, West Germany, 1984, vol. III, pp. 159–167.
42. Winsor, P. A., *Trans. Faraday Soc. 44*, 376 (1948).
43. Lagourette, B., Peyrelasse, J., Boned, C., and Clausse, M., *Nature 281*, 60 (1979).
44. Clausse, M., Peyrelasse, J., Heil, J., Boned, C., and Lagourette, B., *Nature 293*, 636 (1981).
45. Heil, J., Thèse de Doctorat de 3ème Cycle, Université de Pau, France, 1981.
46. Heil, J., Clausse, M., Peyrelasse, J., and Boned, C., *Colloid Polymer Sci. 260*, 93 (1982).
47. Clausse, M., Heil, J., Peyrelasse, J., and Boned, C., *J. Colloid Interface Sci. 87*, 584 (1982).
48. Peyrelasse, J., Boned, C., Heil, J., and Clausse, M., *J. Phys. C.: Solid State Phys. 15*, 7099 (1982).
49. Clausse, M., and Heil, J., *Nuovo Cimento Lett. 36*, 369 (1983).
50. Frisch, H. L., and Hammersley, J. M., *J. Soc. Ind. Appl. Math. 11*, 894 (1963).
51. Shante, V. K. S., and Kirkpatrick, S., *Adv. Phys. 20*, 325 (1971).
52. Kirkpatrick, S., *Phys. Rev. Lett. 27*, 1722 (1971).
53. Kirkpatrick, S., *Rev. Mod. Phys. 45*, 574 (1973).
54. Cohen, M. H., and Jortner, J., *Phys. Rev. Lett. 30*, 696 (1973).
55. Webman, I., Jortner, J., and Cohen, M. H., *Phys. Rev. B. 11*, 2885 (1975).
56. Bruggeman, D. A. G., *Ann. Phys. Leipzig 24*, 636 (1935).
57. Bottcher, C. J. F., *Rec. Trav. Chim. Pays-Bas 64*, 47 (1945).
58. Landauer, R., *J. Appl. Phys. 23*, 779 (1952).
59. Clausse, M., Boned, C., Peyrelasse, J., Lagourette, B., McClean, V. E. R., and Sheppard, R. J., in *Surface Phenomena in Enhanced Oil Recovery*, Shah, D. O. (ed.), Plenum Press, New York, 1981, pp. 199–228.
60. Lagües, M., *J. Physique (Paris) Lett. 40*, L-331 (1979).
61. Scriven, L. E., *Nature 263*, 123 (1976).
62. Scriven, L. E., in *Micellization, Solubilization, and Microemulsions*, Mittal, K. L. (ed.), Plenum Press, New York, 1977, vol. II, pp. 877–893.
63. Kaler, E. W., and Prager, S., *J. Colloid Interface Sci. 86*, 359 (1982).
64. Flockhart, B. D., *J. Colloid Sci. 16*, 484 (1954).
65. Moroi, Y., Nishikido, N., Uehara, H., and Matuura, R., *J. Colloid Interface Sci. 50*, 254 (1975).

66. Nicolas-Morgantini, L., Thèse de Doctorat de 3ème Cycle, Université de Pau, France, 1984.
67. Crow, D. R., *Principles and Applications of Electrochemistry*, Chapman and Hall, London, 1979.
68. Ghosh, B. N., *J. Indian Chem. Soc. 56*, 700 (1979).
69. Deynega, Y. F., Popko, K. K., and Kovganich, N. Y., *Heat Transfer (Soviet Res.) 10*, 50 (1978).
70. Auvray, L., Cotton, J. P., Ober, R., and Taupin, C., *J. Physique (Paris) 45*, 913 (1984).
71. Auvray, L., Cotton, J. P., Ober, R., and Taupin, C., *J. Phys. Chem. 88*, 4586 (1984).
72. Lissant, K. J., in *Emulsions and Emulsion Technology*, Lissant, K. J. (ed.), Surfactant Science Series, Vol. VI, Schick, M. J., and Fowkes, F. M. (eds.), Marcel Dekker, New York, 1974, Part 1, Chap. 1, pp. 1–69.
73. Chen, S. J., Fennell Evans, D., and Ninham, B. W., *J. Phys. Chem. 88*, 1631 (1984).

Index